Ahab's
ROLLING
SEA

A Natural History of *Moby-Dick*

RICHARD J. KING

The University of Chicago Press

Chicago and London

The University of Chicago Press, Chicago 60637
The University of Chicago Press, Ltd., London

Published 2019
Printed in the United States of America

28 27 26 25 24 23 22 21 20 19   1 2 3 4 5

ISBN-13: 978-0-226-51496-3 (cloth)
ISBN-13: 978-0-226-51501-4 (e-book)
DOI: https://doi.org/10.7208/chicago/9780226515014.001.0001

Library of Congress Cataloging-in-Publication Data

Names: King, Richard J., author.
Title: Ahab's rolling sea : a natural history of
Moby-Dick / Richard King.
Description: Chicago : University of Chicago Press,
2019. | Includes bibliographical references and index.
Identifiers: LCCN 2019018442 | ISBN 9780226514963
(cloth) | ISBN 9780226515014 (ebook)
Subjects: LCSH: Melville, Herman, 1819–1891.
Moby Dick. | Sea in literature.
Classification: LCC PS2384.M63 K56 2019 |
DDC 813/.3—dc23
LC record available at https://lccn.loc.gov/2019018442

♾ This paper meets the requirements of ANSI/NISO Z39.48-1992
(Permanence of Paper).

For Lisa

I fully believe in both, in the poetry and in the dissection.

RALPH WALDO EMERSON,
"The Naturalist," 1834

CONTENTS

INTRODUCTION

Each age, one may predict, will find its own symbols in *Moby-Dick*. Over that ocean the clouds will pass and change, and the ocean itself will mirror back those changes from its own depths.
  Lewis Mumford, *Herman Melville*, 1929[1]

On the morning that Captain Ahab is going to die, he stands aloft at the masthead for one last time. In a few hours the line attached to the harpoon that he's going to hurl at the White Whale will snatch around his own neck and pull him overboard to drown. Suspended some ten stories above the waves, Ahab says to himself: "But let me have one more good round look aloft here at the sea; there's time for that. An old, old sight, and yet somehow so young; aye, and not changed a wink since I first saw it, a boy, from the sand-hills of Nantucket! The same!—the same!—the same to Noah as to me."[2]

Was Ahab's ocean really the same as Noah's? Was it the same as ours? Herman Melville (1819–1891) completed *Moby-Dick; or, The Whale* in 1851. He set the story about a decade earlier. His mid-nineteenth century was a period of tremendous upheaval and revelation about humanity's place in the natural world, and his novel was by far the most profound American literary work about the ocean at the time. It would remain so for at least another century, and perhaps it still is. This natural history in front of you is the story of how *Moby-Dick* serves as a gauge to capture that American knowledge and perception of the ocean and its inhabitants—and how that view of the sea has changed up through today. *Moby-Dick* was the first novel, for example, to feature ocean animals to suggest such significant metaphorical and spiritual implications for our own behavior. Reading *Moby-Dick* in the twenty-first century, now well in the Anthropocene, we can read this novel as a proto-Darwinian, proto-environmentalist masterpiece of ocean nature writing that still has much to say, even when applied to our current global crises.[3]

Ishmael explains multiple times in *Moby-Dick* that two-thirds of the earth is covered by water. Geographers today estimate it at almost seventy-one percent. Melville understood or intuited, as remains true, that the sea drives our climate, our biodiversity, our economy, our international politics, and our imaginations. The ocean remains the most expansive, fascinating, complex, and sublime ecosystem on our planet. The ocean still supports—to us—some of the strangest and least known life-forms on Earth.[4]

When I first went to sea in 1993 out of Vancouver, British Columbia, I was twenty-two years old, a few months older than Melville when he first sailed on a whaleship from New Bedford in 1841. When I boarded my ship, a three-masted barquentine named *Concordia*, I too was bound for the South Pacific. A freshly minted teacher of English, I sailed out of Vancouver and spent eleven months with North American high school students, tracing an enormous figure-eight around the Pacific with the Hawaiian Islands at the center. We sailed to, among other places, Hilo, Majuro, Darwin, Papua New Guinea, Bali, Oahu, Fiji, Sydney, and Pitcairn Island, finishing in San Francisco. I read *Moby-Dick* for the first time sitting in a thatched chair in Moorea, French Polynesia. I read for four straight days, unaware that Melville himself had ambled around that same island, maybe even that exact beach.

One afternoon, a few weeks into teaching *Moby-Dick* and some nine months into that first voyage, I was rereading the novel as we sailed for Easter Island. I needed some time and space, so I climbed up the rigging to see the curvature of the horizon. I climbed aloft all the way to the royal yard, over one hundred feet above the surface. Leaning out over the starboard royal yardarm, the absolute highest and farthest place I could get from the deck, I saw far off the bow what I believed to be the forward, single puff of a sperm whale. I had been so nose deep in the novel—studying Christian symbolism, Fascist parallels, and connections to Milton's *Paradise Lost*—that the sight of a living sperm whale bordered on miraculous. I didn't shout down to the deck. The puff of mist seemed an offering to me alone. I watched the whale's flukes as it dove.

What I realized in the days that followed was that for me *Moby-Dick* is first and foremost a novel about the living, breathing, awe-inspiring global ocean and its inhabitants. Most explorations of this great American novel breeze too quickly past the marine life at sea that Melville so treasured and illuminated.

This natural history aims to provide that background by moving roughly chronologically through the voyage of the *Pequod*, exploring topics in marine biology, oceanography, and the science of navigation as Ishmael takes them up in *Moby-Dick*.

Now more than twenty-five years from my first voyage, I, like so many other readers of *Moby-Dick*, perceive the world's oceans to be vulnerable, fragile, and in need of our stewardship. We have overfished the water, overdeveloped coastal habitats, and rocketed the rate of the introduction of marine invasive species. We have polluted the sea with oil spills, chemical runoff, and pervasive plastics. The amount of carbon dioxide in the atmosphere has increased by over seventy percent since Melville's years at sea, and is still rising. Since the first records in 1895, this carbon dioxide has not only increased the average US air temperature by between 1.3°F to 1.9°F, it has also altered the entire chemistry and temperature of the ocean itself—a seemingly impossible conceit. We hurl harpoons at a faceless ocean, slowly killing ourselves as we drag down entire ecosystems of life along with us. Because of sea level rise and the ungraspable phantom of melting polar and glacial ice, Melville sailed over a Pacific Ocean in the 1840s that was likely at least eight inches lower than it is today.[5]

Yet as we watch the ice melt on television documentaries, as we try to mitigate the erosion of our coasts due to sea level rise and the effect climate change has had on island, coastal, and Arctic communities, as we prepare for the next hurricane, and as we passively, Ishmael-like, forgive ourselves for buying just one more plastic bottle of water or another bite of tuna sushi, we somehow simultaneously perceive the ocean as more than simply vulnerable and in need of our protection. Because we still, somehow simultaneously, perhaps even because of novels such as *Moby-Dick*, continue to revere our twenty-first-century ocean in a similar way as did Noah, Jonah, and Ahab. We still envision the sea, even with all our technological advancements and scientific knowledge, as relentless and indifferent and immortal and sublime and eager to lure us in with a trace of sympathy and kindness then kick our ass and not even look back.

For example: many years after I sailed aboard the *Concordia*, on the afternoon of February 17, 2010, about 300 nautical miles off the coast of Brazil, a powerful squall caught the ship. The helmsman adjusted course to run before it, but too slowly. The wind heeled the *Concordia* so far on her side, so quickly, that hatches, doorways, and vents—which in retrospect should've been closed—began to downflood with ocean as the ship was knocked down on its side. The yard on which I'd leaned so many years earlier on my first voyage now crashed and stabbed the surface of chaotic seas. Sails filled with sea water. From belowdecks students and crew scrambled up along bulkheads, now sideways, desperate to get out. Somehow every single student, teacher, and member of the professional crew made it into four inflatable life rafts. The bosun swam to recover the rescue beacon. The wind blew the rafts crammed with people to leeward. Sometime after they floated away in terror, their *Concordia*, my *Concordia*, sank to the bottom. The sixty-four castaways floated for thirty-six hours without any steerage or any knowledge whether anyone anywhere knew what had happened. Following their radio and GPS signals, and then flares, the Brazilian Navy and two merchant ships, the *Crystal Pioneer* and the *Hokuetsu Delight*, rescued all hands.[6]

Tragedy at sea in the twenty-first century is less of an aberration than you might think. The sea still takes our largest steel-hulled ships. Eighty-five large ships were lost around the world in 2015. That was the fewest in a decade. One of those lost that year was the nearly 800' long US merchant vessel *El Faro*, which was overwhelmed in a hurricane and drowned. The ship lost power, lost steerage, and sank in waters that were about three miles deep. Mitchell Kuflik, aged twenty-six, was one of the thirty-three men who died. Mitchell grew up in my town of Mystic, Connecticut. His fiancé was my daughter's first babysitter.[7]

In *Moby-Dick*, Herman Melville wrote of both the beauty and the cruelty of the ocean. He summed up the nineteenth-century view of the sea and foreshadowed Ahab and his crew's death in a single spiked club of a sentence that he tucked into a brief chapter about, of all things, zooplankton. I think this sentence is the most profound summary in the English language about the human relationship with the ocean—pre-Darwin, pre-Carson, and before any introduction of the concept of the Anthropocene. Melville slipped this single sentence into the single best novel ever written about life at sea—and about sea-life:

> But though, to landsmen in general, the native inhabitants of the seas have ever been regarded with emotions unspeakably unsocial and repelling; though we know the sea to be an everlasting terra incognita, so that Columbus sailed over numberless unknown worlds to discover his one superficial western one; though, by vast odds, the most terrific of all mortal disasters have immemorially and indiscriminately befallen tens and hundreds of thousands of those who have gone upon the waters; though but a moment's consideration will teach, that however baby man may brag of his science and skill, and however much, in a flattering future, that science and skill may augment; yet for ever and for ever, to the crack of doom, the sea will insult and murder him, and pulverize the stateliest, stiffest frigate he can make; nevertheless, by the continual repetition of these very impressions, man has lost that sense of the full awfulness of the sea which aboriginally belongs to it.[8]

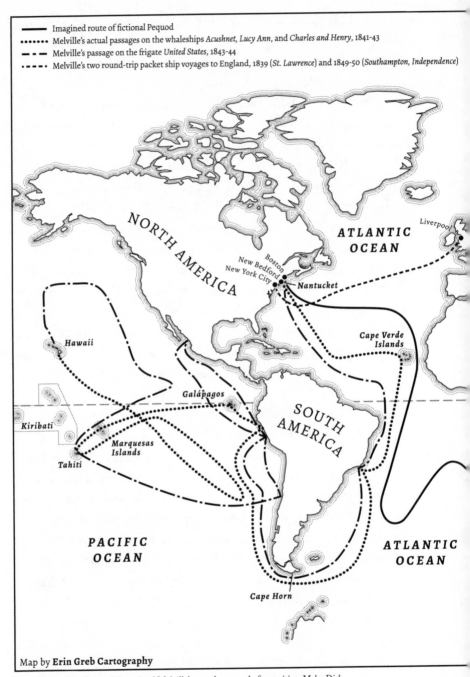

——— Imagined route of fictional Pequod
••••••• Melville's actual passages on the whaleships *Acushnet, Lucy Ann,* and *Charles and Henry,* 1841-43
— ·· — Melville's passage on the frigate *United States,* 1843-44
— ··· — Melville's two round-trip packet ship voyages to England, 1839 (*St. Lawrence*) and 1849-50 (*Southampton, Independence*)

NORTH AMERICA

ATLANTIC OCEAN

Liverpool

Boston
New Bedford
New York City
Nantucket

Cape Verde
Islands

Hawaii

Galápagos

SOUTH
AMERICA

Kiribati

Marquesas
Islands

Tahiti

PACIFIC
OCEAN

ATLANTIC
OCEAN

Cape Horn

Map by **Erin Greb Cartography**

FIG. 1. Track of the fictional *Pequod* and Melville's actual voyages before writing *Moby-Dick*.

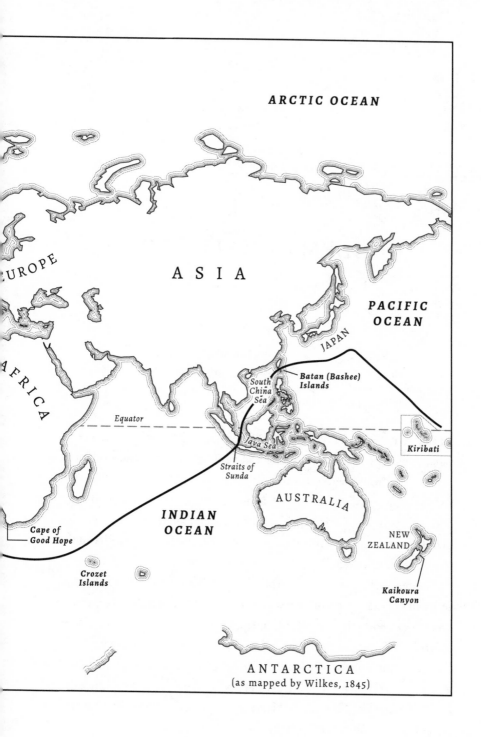

ARCTIC OCEAN

ASIA

EUROPE

PACIFIC
OCEAN

AFRICA

JAPAN

South
China
Sea

Batan (Bashee)
Islands

Equator

Java Sea

Kiribati

Straits of
Sunda

AUSTRALIA

INDIAN
OCEAN

Cape of
Good Hope

NEW
ZEALAND

Crozet
Islands

Kaikoura
Canyon

ANTARCTICA
(as mapped by Wilkes, 1845)

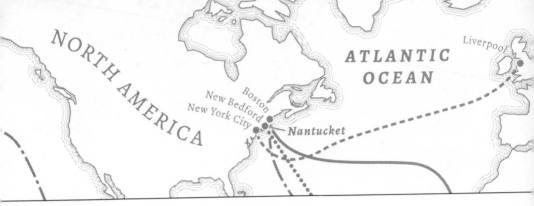

Ch. 1

HERMAN MELVILLE

Whaleman, Author, Natural Philosopher

> In one of those southern whalemen, on a long three or four years' voyage, as often happens, the sum of the various hours you spend at the mast-head would amount to several months.
> Ishmael, "The Mast-Head"

In the first chapter of *Moby-Dick*, "Loomings," Ishmael is on the verge of suicide, eager to quit urban life and get back out onto the wild, open ocean. He strolls the New York City waterfront in the mid-1800s before the raising of the Statue of Liberty, before the Brooklyn Bridge, and before the population of Manhattan had ballooned to a half million people as immigrants continued to form and build the largest city in the Western Hemisphere.[1]

Ishmael travels from New York City to New Bedford, Massachusetts, where the sea is a place to make a living, but also a place to die, and one where God reigns, with his agent the whale. In New Bedford, he meets his Polynesian soulmate and the story's human hero, Queequeg, a Pacific Islander. In Nantucket, he chooses their ship, the *Pequod*, named after what he believes to be an extinct Native American tribe. Once out at sea, Ishmael in "The Mast-Head" comes to grips with the immediate potential for his death, not by the whale, but by the searching itself.

Historians estimate that on average at least one man died at sea per mid-nineteenth-century whaling voyage, figuring a thirty-man crew and a three-and-a-half-year average trip to sea. Half of the dead perished from disease, and the other half died from some type of accident, which included deaths while engaged with hunting the whale, but also fatalities by falling from aloft when working the ship or looking for these animals. For example, not long after Melville a young man named William Allen sailed out of New Bedford aboard another whaleship. He wrote in his journal in 1842 of a moment when he was aloft looking for whales. His shipmate, George Stevens, whizzed by him from above with "inconceivable velocity." Allen wrote: "He struck the water face downwards with a terrible crash; the water flew as high as the foot of the foresail!" The captain ordered a whaleboat to be lowered with a few men to look for the body, but then he ceased the search far too quickly for the crew's comfort. As they sailed away, the sailors went aft and asked the captain to have the royal yards rigged back in place "as all sperm whaler's do" in order to give them more to hold on to. The captain denied their request, consenting only to an additional rope strung across the shrouds.[2]

Toward the end of *Moby-Dick* in "The Life-Buoy," Ishmael describes a shipmate falling from aloft. After the sun rises and the watch changes over, a man rolls out of his hammock and climbs straight up for his shift. "He had not been long at his perch," Ishmael explains, "when a cry was heard—a cry and a rushing—and looking up, they saw a falling phantom in the air; and looking down, a little tossed heap of white bubbles in the blue of the sea."[3]

Well before the death of this sailor in the Pacific, it becomes clear that Ishmael's sense of height and depth is essential to understanding *Moby-Dick* and the nineteenth-century whaleman's relationship with his watery world. After the *Pequod* first leaves Nantucket, the sailors do not visit a port. They barely sight land for the rest of the story. Yet Ishmael spends far fewer words than you'd expect talking about time or distance or vast horizons. His descriptions of life at sea in *Moby-Dick* are primarily vertical. Ishmael ponders the visions aloft at the masthead, the height of the sky, and the metaphors of the heavens and clouds as lofty, philosophical thoughts. Then, by contrast, he dwells on the depths of the sea, the deepest dives under the surface, and the metaphors of Hell and bottomless madness. In the chapter "The Castaway," it is the one imaginary God-given glimpse of the ocean beneath the surface, deep down, that transforms Pip, the boy with the least power on board, into a raving lunatic. Pip had jumped a second time from a whaleboat, and his shipmates left him behind to float alone on the ocean. His insanity comes not from the unbroken horizon or from his distance from the ship, but from when his soul, his previous stand in reality, is now lost, sinking beneath the surface.[4]

Melville drove the plot of his novel with Ahab's vendetta against one individual sperm whale and with Ishmael's intellectual quest to understand the largest predator on Earth. Ishmael says this species stays below "for an hour or more, a thousand fathoms in the sea." Throughout the novel Ishmael and Ahab regularly reference the whale's ability to dive into darkness. Today we know that sperm whales, along with some beaked whales, dive deeper and longer than any other mammals on the planet. Melville and his contemporaries suspected this of the sperm whale, but they could not definitively confirm it. They did not have sonar or radio transmitter technology to tell the exact depths of the sea or to follow a whale, but they clocked the sperm whale diving for at least eighty minutes, and individuals that they had harpooned could take out 4,800 feet of line beneath the surface. Biologists have since tracked sperm whales foraging as deep as 6,500 feet, staying under for as long as 138 minutes.[5]

Ishmael's ocean is "unsounded" and "bottomless." The year before the publication of *Moby-Dick*, the American government for the first time sent out a ship for the sole purpose of searching for deep sea soundings. With a steel wire they recorded a spellbinding depth in the middle of the North Atlantic of 34,200 feet, deeper than any peak is high on land. This would turn out to be incorrect at that location, but an equivalent depth has since been measured in a few other ocean trenches, in both the Atlantic and the Pacific. Oceanographers estimate today that the *average* depth of the sea is about 12,450 feet, while the average elevation on land is barely 2,755 feet. The deepest place on Earth, the Mariana Trench, is not far from where we might imagine the *Pequod* sails on its way to meet its final end in the equatorial Pacific. The Mariana Trench is 36,200 feet deep, providing enough water to drown the height of Mount Everest—with Mount Washington scooped on top.[6]

Melville found inspiration for his deep diving when he attended a lecture in Boston by Ralph Waldo Emerson in the winter of 1849. Although he found Emerson a bit too optimistic and self-satisfied, spouting ideas too far out even for him, Melville was still impressed. He wrote a letter to a friend after the lecture:

> And, frankly, for the sake of the argument, let us call him a fool;—then had I rather be a fool than a wise man. —I love all men who *dive*. Any fish can swim near the surface, but it takes a great whale to go down stairs five miles or more; & if he dont attain the bottom, why, all the lead in Galena can't fashion the plummet that will. I'm not talking of Mr Emerson now—but of the whole corps of thought-divers, that have been diving & coming up again with blood-shot eyes since the world began.[7]

The following year in *The Literary World*, praising Nathaniel Hawthorne—to whom he would dedicate *Moby-Dick*—Melville wrote that a man of lofty genius who can "soar to such a rapt height" must also have "deep and weighty" meanings. Hawthorne, Melville declared, was a genius of "great, deep intellect, which drops down into the universe like a plummet."[8]

For *Moby-Dick* Melville found ideal metaphors in the heaven-bound masthead and the deep-diving sperm whale.

THE WHALESHIP *CHARLES W. MORGAN*

Although it's a warm spring morning, Mary K. Bercaw Edwards wears a thick fisherman's sweater under her climbing harness. She stands on the deck of the *Charles W. Morgan* beside a massive stretch of chain that leads to a monstrous rusted hook, which was once used for peeling the blubber off whales. The whaleship is docked in the estuary at Mystic Seaport, a maritime museum in Mystic, Connecticut.

"It can get cold up there," she says, looking aloft toward the hoops.[9]

Bercaw Edwards is the foreman of the group of museum staff members who show visitors the traditional arts and jobs of the sailor. For the visitors' experience, for example, she has over the years often climbed up to the hoops to shout "Whale ho!" She is also, not coincidentally, a professor and a Melville scholar.

"This first step is the hardest," she says, "because of the distance."

She hoists herself off the rail and onto the ratlines, which are the rope rungs tied between vertical cables wrapped in tar and twine. These cables, called shrouds, taper up to the mast under the first platform, which is called the lower top.

Melville would not have sailed with a harness, of course, nor were his ship's masts supported by wire and steel standing rigging—shipwrights began using these materials on working ships a few decades later—but otherwise the climb aloft on the *Charles W. Morgan* is nearly identical to the path Melville would've climbed when aboard his whaleships.

Bercaw Edwards climbs up and then clips in. To get around and onto the lower top, Bercaw Edwards has to clamber nearly horizontally to the water and then around the edge to stand up on the small platform. We're about four stories above the water now. The deck of the whaleship from aloft looks like a fish lying on its side. But it's more similar to a mahi-mahi, since unlike nearly all other types of ship designs, the bow of the American whaleship was not sharply pointed, nor is the widest part of the hull toward the middle. The whaleship hull is widest forward. This shape is in part to provide more storage for casks of oil, water, and food. Keep this in mind when we consider the final scene of the novel when Moby Dick smashes his head into the bow of the *Pequod*. The hull of the American whaleship, in a sort of convergent evolution, had the shape of the sperm whale's head itself, which also, squarish up front, tapers down underneath to a narrow keel-like lower jaw.

Here's the thing about the *Charles W. Morgan*: this whaleship is practically Herman's boat. Shipwrights launched this ship in July 1841 in New Bedford, Massachusetts. In the fall of the previous year, a shipyard in Mattapoisett, just five miles to the east, launched the *Acushnet*. This *Acushnet* was the whaleship aboard which twenty-one-year-old Melville sailed on the third of January, 1841, out of the Acushnet River. The *Charles W. Morgan* is nearly identical to the *Acushnet* in tonnage, rig, and all functional parts (see plate 1). The *Charles W. Morgan* is also similar to the *Pequod*, which is Melville's imaginary creation of a more fantastical, older model—a "cannibal of a craft." So when you walk the decks of the *Charles W. Morgan*, you're experiencing something as close as possible to that which inspired and transported Melville to the Pacific and then later to the writing of *Moby-Dick*. In this way, the *Charles W. Morgan* is one of the most significant artifacts in American literature. Wrapping your hands around the shrouds of the *Morgan* aloft is equivalent to drumming your fingers on the railing in Harper Lee's courthouse in Monroeville, Alabama. Or, if it still existed, to placing your palms on the frosty window in the cabin that Henry David Thoreau built beside Walden Pond.[10]

"All right. Let's keep climbing," Bercaw Edwards says.

MELVILLE'S EXPERIENCE AT SEA

Melville was born in New York City, the third of eight children. His father was an upper-middle-class merchant who went bankrupt when Herman was a child and then died a few years later "in delirium," as Bercaw Edwards puts it, when Herman was twelve. His mother somehow took care of the family, leaning on Herman's oldest brother and the reluctant generosity of their extended family. Melville was far from an intellectual prodigy, although he showed some early aptitude for applied math and spent about two years at one of the best academies for science in Albany, New York. But he had to leave because of family circumstances. As a teenager he read and learned through other community institutions when possible, and he worked as a bank messenger and on his uncle's farm. At nineteen he trained to be a surveyor and civil engineer but wasn't able to get a job. In early June 1839, he signed on as a foremast hand aboard a merchant ship named the *St. Lawrence*. He sailed trans-Atlantic to Liverpool to help deliver cotton, and then returned a couple months later to New York harbor with the ship now loaded with metal bars, spools of rope, sewing supplies, and thirty-two passengers.[11]

He tried teaching school for a while, then visited an uncle in Galena, Illinois. He next moved to New York City to try to work in an office. This didn't go well. According to his older brother, "Herman has had his hair sheared & whiskers shaved & looks more like a Christian than usual," but his illegible handwriting and inconsistent spelling did not place him in a favorable position to find work. Meanwhile, Melville had read the sea novels of James Fenimore Cooper and was recently reading Richard Henry Dana Jr.'s bestseller *Two Years before the Mast* (1840), one of the first realistic narratives of a working sailor's life on a merchant ship. Melville had a couple relatives who had been to sea and were at the time working on naval vessels and whaleships. So with few other options and surely a mixture of the call for adventure and a semi-suicidal ambivalence about his safety and future such as Ishmael would later describe in "Loomings," young Melville decided to go back out on the ocean. This time he found a whaleship.[12]

Aboard the *Acushnet*, Melville stood his lookout every day or two with his feet on a mere set of spreaders. These "t'gallant crosstrees" amount to a couple wood boards bolted across the mast. Shipwrights did not begin to install the iron hoops until decades later. (See fig. 2.)

About two years before sitting down to write *Moby-Dick*, Melville read *Etchings of a Whaling Cruise* (1846), written and illustrated by J. Ross Browne. He reviewed the narrative for the *Literary World*. Browne, an aspiring journalist, advocated for the rights of working whalemen by showing the brutality and injustice on board. Yet Browne wrote romantically of shifts on lookout for whales: "There was much around me to inspire vague and visionary fancies: the ocean, a trackless waste of waters; the arched sky spread over it like a variegated curtain; the seabirds wheeling in the air; and the myriads of albacore [tuna] cleaving

FIG. 2. Illustration from J. Ross Browne's *Etchings of a Whaling Cruise* (1846).

their way through the clear, blue waves, were all calculated to create novel emotions in the mind of a landsman."[13]

A few years later in *Moby-Dick*, Melville wrote through Ishmael in "The Mast-Head," sharing Browne's sentiments: "In the serene weather of the tropics it is exceedingly pleasant—the mast-head; nay, to a dreamy meditative man it is delightful. There you stand, a hundred feet above the silent decks, striding along the deep, as if the masts were gigantic stilts, while beneath you and between your legs, as it were, swim the hugest monsters of the sea."[14]

On the way to the central Pacific and the Marquesas, a passage of about a year and a half, Herman Melville and his shipmates aboard the *Acushnet* likely stopped in only three ports: Rio de Janeiro in Brazil and Santa and Tumbez in Peru (see fig. 1). After leaving Tumbez, the *Acushnet* sailed for more than six months without entering any port. The ship cruised for whales around the Galápagos, anchored, and then sailed west along the eastern equatorial Pacific, known then as the Offshore Ground. The captain of the *Acushnet*, deliciously named Valentine Pease Jr., could have reasonably made the leg from the Galápagos to the Marquesas Group in less than three weeks. Instead Captain Pease dawdled for 141 days, zigging and zagging back and forth across the equator, searching for whales with little success as the *Acushnet* meandered over some of the most open ocean on Earth.[15]

After that passage, perhaps because of it, in July 1842 Melville and a shipmate deserted the *Acushnet* on the island of Nuku Hiva in the Marquesas. Meanwhile, the first major expedition of exploration funded by the United States, led by Charles Wilkes, had just returned from its four-year circumnavigation. John James Audubon, aware of this and toward the end of his career, had written to Secretary of State Daniel Webster to request a position to illustrate and supervise the specimens, "but would be better pleased if our government would establish a natural history institution to advance our knowledge of natural science, and place me at the head of it."[16]

On Nuku Hiva Melville and his shipmate spent a month living among the islanders. Hopping from ship to ship was not uncommon

among whalemen. After he claimed to have escaped from the Polynesians, who he said were cannibals, Melville secured himself a berth
on a small Australian whaleship named the *Lucy Ann*. This ship had a
sick captain and was undermanned. The crew mutinied by refusing duty
when they were ordered to sail aimlessly off Papeete Harbor. With the
others, Melville was placed into a French-run Tahitian jail.[17]

After some leisurely confinement, he and another one of his shipmates escaped.

"Which wasn't hard," Bercaw Edwards says. "He just walked away
one night and sailed off to an island nearby."

Melville tried potato farming on the island of Moorea. When that
lost its appeal, Melville hopped aboard a Nantucket whaleship named
the *Charles and Henry*. He worked on this ship for some five months up
to Hawaii.[18]

Ashore in Honolulu, Melville worked odd jobs, including resetting
pins in a bowling alley. After the *Acushnet* and Captain Pease arrived
in the harbor and then took off whaling again, Melville, perhaps partly
homesick and partly fearing prosecution for desertion, signed aboard an
outbound ship of the Navy, the *United States*. Aboard the *United States*,
Melville sailed for fourteen months back around Cape Horn and up to
Boston. In early October 1844, he walked off with his sea bag and an
entirely different worldview.

By that fall of 1844, less than twenty miles inland of Boston Harbor,
Emerson had purchased the land beside Walden Pond. He'd soon allow
his disciple, Thoreau, to build a cabin there. Across the Atlantic, Darwin had finished a first draft of an essay on natural selection, which he
would tinker with and expand and sit on for another fifteen years.

Melville made his way back to New York City and began to write
down his stories. His first book, *Typee* (1846), was an embellished account of his time among the "cannibals" of Nuku Hiva. He followed
this with *Omoo* (1847), about the mutiny on the *Lucy Ann* and his exploration of Moorea. "Then he wrote three more books in two years,"
Bercaw Edwards says, "each connected to his different voyages at sea."
He wrote *Mardi* (1849), another narrative of the South Pacific that had

some whaleship material and in which Melville began to explore in depth his blossoming interest in natural philosophy. Then he quickly wrote *Redburn* (1849), derived from that first trans-Atlantic voyage, and *White-Jacket* (1850), about life on a naval vessel derived from his time on the *United States*.

In the fall of 1849, Melville sailed on one more voyage before beginning *Moby-Dick*. For his first ocean trip as a passenger, Melville traveled from New York to London aboard the *Southampton*. Melville was carrying his manuscript of *White-Jacket* to sell directly to a British publisher.

On the first morning out of sight of land on this trans-Atlantic, he climbed up the ratlines. Within an absolutely extraordinary pair of journal entries, Melville described what he'd later echo in *Moby-Dick* with Ahab: "[I climbed aloft] to recall the old emotions of being at the masthead. Found that the ocean looked the same as ever." The very next day, Melville was the first to spot and alert all hands that a man had fallen overboard. Melville threw over a block and tackle for the man to grab. The man did so. But then he let go with a "merry" expression on his face. Melville wrote: "Saw a few bubbles, & never saw him again." The captain told Melville he'd seen four or five similar suicides.[19]

Back in New York, Melville began working on *Moby-Dick*. By May 1850, Melville wrote to Richard Henry Dana Jr.: "About the 'whaling voyage'—I am half way in the work, & am very glad that your suggestion so jumps with mine [about choosing to write a whaling book]. It will be a strange sort of a book, tho', I fear; blubber is blubber you know; tho' you may get oil out of it, the poetry runs as hard as sap from a frozen maple tree; —& to cook the thing up, one must needs throw in a little fancy, which from the nature of the thing, must be ungainly as the gambols of the whales themselves. Yet I mean to give the truth of the thing, spite of this."[20]

This was exactly what Melville tried to do: to present the natural history of ocean life as accurately as he could, learned from his own experience and from his deep readings of ocean-going naturalists and sailors, then occasionally throwing in a little fancy for the sake of driving his story and exploring larger truths about human life.

Over one year later, after writing that passage to Dana, Melville was thirty-one years old, overwrought with debt, and now a father to young children. He assessed himself a failure as a writer. He'd moved out of the city to try to raise corn and potatoes on a small farm as he tried to complete *Moby-Dick*. He wrote to Hawthorne: "What I feel most moved to write, that is banned,—it will not pay. Yet, altogether, write the *other* way I cannot. So the product is a final hash, and all my books are botches."[21]

MELVILLE AS NATURAL PHILOSOPHER

With the mast hoops at her waist, Bercaw Edwards stands aloft on the starboard side of the topmast. She can see the roofs of all the buildings at the museum and all the way down the Mystic River to the railroad bridge—the track laid down in the 1850s that would forever alter the ecology of the coastline by cutting so much of the marshes off from Long Island Sound. She sees an osprey soar over to the other side of the estuary, toward the opposite side of the river and the suburban forest that fills up along the hills. Those hills in Melville's time were completely bare, entirely clear-cut for firewood and shipbuilding.

In a 2011 issue of the *Biological Bulletin*, published by the Marine Biological Laboratory in Woods Hole, Massachusetts, the physicist Harold Morowitz published an article titled "Herman Melville, Marine Biologist," in which he argued in fun that if Melville had the chance to attend a university, he would've majored in English and minored in biology. Ishmael declares famously in "The Advocate" that "a whale-ship was my Yale College and my Harvard." Morowitz pointed out that 17 of the 135 chapters of *Moby-Dick* "deal primarily with the anatomy, physiology, ecology, metabolism, and ethology of the sperm whale, *Physeter macrocephalus*, and various assorted cetaceans as well as seals, squid, sharks, albatrosses, and other marine birds."[22]

We'll be visiting each of these topics in the chapters to follow, but how much of this overall interest in natural history is Ishmael's character, crafted for the purpose of the novel? With the recognition that the bifurcation between the humanities and the sciences was just starting to split in the mid-nineteenth century and that our current partitioning of right- and left-brained people is a construction of the late twentieth century, it is still fair to ask whether Melville was a sailor-naturalist himself as a young man at sea. Did he, say, curl up in his bunk to sketch *Sargassum* in his journal, drying and pressing the fronds into the pages?

"I've not heard anything like that before," Bercaw Edwards says. "He had a famous explorer uncle who had an interest in natural history, and he spent a summer with him when he was young. But biographers haven't found any particular scientific interest before his voyages. And none of Melville's journals or letters from those years in the Pacific have survived."[23]

It's tempting to compare young Melville's time at sea to that of a slow, transformative voyage of scientific discovery like that of Charles Darwin. But as Robert Madison, an emeritus professor of the Naval Academy, once told me, it's probably more likely Melville was curled in his bunk reading British poetry or art history, rather than scanning the horizon for a new species of seabird. For example, in the earliest sea journal of Melville's that remains, from that trip to England aboard the *Southampton* in 1849 as a passenger, he wrote of weather and seamanship a couple times, but he barely mentioned marine life at all aside from a couple sightings of land birds.[24]

So we need to be careful to not give Melville and all our early mariners the title of "biologist" or "field naturalist" just because they went to sea for a long stretch—just as some fishermen today are true students of the marine environment and some very much are not, no matter how much experience at sea they've had.

Between 1830 and 1850, an average of roughly, perhaps, eight thousand men a year stood at the masthead on American whaleships around the world, all individuals with different backgrounds and interests. Particularly for mariners going to sea for first voyages under sail—there was much else to occupy the sailor's attention as he learned how to work and how to cope with all the social aspects and technical demands of life on a ship. This was an age, too, before the modern field guide and before photography without a tripod. That said, scholars have discovered an extraordinarily high literacy rate among sailors: from seventy-five to ninety percent. Reading cultures thrived aboard ships. Maritime narratives since at least William Dampier's *A New Voyage Around the World* (1697) regularly included natural history descriptions into every kind of adventure at sea, which reflected and primed sailor-readers toward marine observation.[25]

For example, consider the second mate of the maiden voyage of the *Charles W. Morgan* in 1841, a young man from Martha's Vineyard named James Osborn, who recorded the over seventy-five books that he read during his three-and-a-half years at sea. The first on Osborn's list was John Mason Good's *The Book of Nature* (1826), which has sections on geology, taxonomy, animal senses, and human sleep. Melville referenced Good's popular science book by name, also first, in his chapter "A Man-of-war Library" in *White-Jacket*, saying it was very good but "not precisely adapted to tarry tastes." Whalemen gathered and traded natural history objects, and made folk art out of marine animal bones and teeth and out of the rostra, fins, feathers, wings, and feet of other animals. Many men, like Osborn, painted whaling scenes in their journals. They often passed their journals and illustrations between each other, comparing and even drawing in each other's books. Several logbooks and journals reveal that individuals were interested in natural observation and even shell-collecting during liberty ashore (see fig. 3). The early natural history collections first developed and curated in American port cities are full of contributions from mariners of all stripes. Of the mariners' wives who went to sea, of which there were hundreds, several kept journals, too, which often have observations and paintings concerned with natural history.[26]

Certainly, by the very business and processes of whaling, successful captains, mates, and harpooners who returned for multiple voyages became adept at reading the surface of the water. They needed to differentiate between species. They learned each type of whale's diving and migration habits. They learned the anatomy of a few species of whales and other marine mammals by dissecting them down through the blubber

FIG 3. Journal drawing by whaleman Dean C. Wright (c. 1841–45). Clockwise from the top: Blackfish (e.g., short-finned pilot whale, *Globicephala macrorhynchus*), Shark (e.g., silky shark, *Carcharhinus falciformis*), Sun Fish (*Mola* spp.), Albaco[re?] (tuna, e.g., *Thunnus alalunga*), Shovel Nose Shark (hammerhead, *Sphyrna* spp.), Right Whale (*Eubalaena* spp.), Blunt Nose Porpoise (e.g., harbor porpoise, *Phocoena phocoena*, Chilean dolphin, *Cephalorhynchus eutropia*), Sword Fish (*Xiphius gladius*), Bill Fish (smaller swordfish, sailfish, or marlin), Sharp Nose porpoise (e.g., common bottlenose dolphin, *Tursiops truncatus*).

layers. The men extracted teeth, baleen, and they occasionally probed or cut into the internal organs for meat, for ambergris, or simply for curiosity. The whalemen experienced a level of contact with marine mammals in the wild unmatched by even the most accomplished and devoted marine biologists of today. This hunter's knowledge extended to the entire ocean ecosystem. In the centuries of ocean sailing before GPS, radar, and accurate charts, mariners learned to navigate in part by recognizing coastal bird species and changes in currents, clouds, air pressure, water temperature, and water color.

In the South Pacific, Melville traveled entirely under sail. He rarely sailed more than ten miles per hour. This isolation and slow speed at sea is unprecedented in any commercial or even nearly all recreational enterprises in the twenty-first century. Containerships and oil tankers now cross the entire Pacific in as little as two weeks.

Surrounded by all this at sea, then so clearly interested in the implications of scientific developments as he wrote *Mardi* and then *Moby-Dick*, Melville can safely be given the broad title of natural philosopher. Melville's years as a working sailor on the Pacific Ocean were the most profound and expansive time of his life. The opening lines of his first book, *Typee*, declare his pride: "Six months at sea! Yes, reader, as I live, six months out of sight of land; cruising after the sperm-whale beneath the scorching sun of the Line, and tossed on the billows of the wide-rolling Pacific—the sky above, the sea around, and nothing else!"[27]

To write the scenes and details of ocean life in *Moby-Dick*, the author first drew from his own experience. He privileged the sailor's perspective over the "old naturalists." Melville cared to get his marine biology, weather, and seamanship correct. He created Ishmael as his know-it-all pedagogue, a narrator who very much fancied himself a doctor of natural philosophy in cetology.

Ishmael's playful rage against the inaccuracy and unreliability of depictions of whales and the whaleman's experience emerged even in Melville's 1847 review of *Etchings of a Whaling Cruise*. Melville was frustrated with Browne because he depicted a sperm whale "roaring" in pain from a harpoon. Melville wrote: "We can imagine the veteran Coffins and Colemans and Maceys of old Nantucket elevating their brows at the bare announcement of such a thing. Now the creature in question is as dumb as a shad, or any other of the finny tribes. And no doubt, if Jonah himself could be summoned to the stand, he would cheerfully testify to his not having heard a single syllable, growl, grunt, or bellow engendered in the ventricle cells of the leviathan." (It turns out this is only *mostly* true. We'll get to this later.)[28]

Back aloft aboard the *Charles W. Morgan*, Bercaw Edwards looks down past her feet. A school group, ant-like, files up the gangway. A kayaker pauses his paddling to look up at the rigging. He doesn't look high enough to see us.

Bercaw Edwards tells me of a voyage of the *Charles W. Morgan* in 1864, when the son of the captain drowned during a gale to the north of Japan, likely from a fall from aloft. Later in that same voyage another man from the Mariana Islands died from injuries sustained from a fall from the rigging.[29]

Ishmael ends "The Mast-Head" by converting the danger of standing aloft to an existential one, suddenly placing the reader all the way up there alone: "There is no life in thee, now, except that rocking life imparted by a gently rolling ship; by her, borrowed from the sea; by the sea, from the inscrutable tides of God . . . And perhaps, at mid-day, in the fairest weather, with one half-throttled shriek you drop through that transparent air into the summer sea, no more to rise for ever."[30]

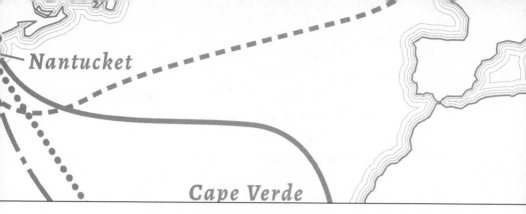

Ch. **2**

NUMEROUS FISH DOCUMENTS

Though of real knowledge there be little, yet of books there are a plenty; and so in some small degree, with cetology, or the science of whales. Many are the men, small and great, old and new, landsmen and seamen, who have at large or in little, written of the whale.

  Ishmael, "Cetology"

In the spring of 1851, having left downtown Manhattan for a farmhouse in Pittsfield, Massachusetts, Melville was reenergized about the potential of his novel about the White Whale. Shut upstairs in his study for long hours with his window looking out toward Mount Greylock, Melville read, scribbled notes in the books he owned, and scrawled page after page of his manuscript. He was surrounded by dozens if not hundreds of volumes, some of which he owned, and most others he borrowed from libraries, friends, and family in New York and Boston.

In "Cetology," Ishmael declares, "I have swam through libraries and sailed through oceans; I have had to do with whales with these visible hands." So in addition to recalling his own experiences, he turned to his books for more inspiration, factual reference, and ideas on the novel's style to help boil the sap. Melville knew his novel was not the first about a whaling voyage. Melville mashed and whirred all of these influences into a slow-burn-story-stew of immense depth of flavor and force. Melville's messy madness did have a method. It was intricately connected to his sea narratives and what he was reading of natural history.[1]

Part of the method derives from the reality that whaling voyages in the nineteenth century regularly lasted between two to five years. They were tedious and slow. Men stood hours each day at the masthead. The novel's digressive girth and meditative, explorative meandering matches the pace of life on a whaleship. Notice how Melville punctuated his long-winded expository chapters with the chaotic catching of whales and the nine evenly spaced sightings of other vessels. Melville's masterpiece is large and long to match not just the whaling voyage, but also his subjects of the whale itself and the long, meditative life at the masthead. As the *Pequod* sails farther and farther out to sea to confront the White Whale, the whalemen aboard conduct their business and kill whales: approximately ten sperm whales and one right whale over the course of the story.

After showing how to find and capture whales, Melville physically and metaphorically dissected the sperm whale, both over the course of his novel and the course of the *Pequod*'s voyage. Ishmael moves from Ch. 67, "Cutting In," on down to Ch. 103, "Measurement of the Whale's Skeleton." Melville often guided his reader through this process by creating tidy chapter pairs. In the first chapter of a pair, he wrote in a narrative style, describing events with his characters in order to explain how things worked on a whaleship or about animal life. Then, in the very next chapter, he zoomed out from the adventure and used an essay style to further explore the philosophy, history, or science of the same topic.

With books piled at his desk and on shelves, Melville stated his references outright through Ishmael, especially in "Extracts" and "Cetology." The novel's narrator regularly refers to a half dozen or so authors and their recent books, which Ishmael dubs his "numerous fish documents." Three nonfiction accounts about American whaling voyages were published within the decade before Melville finished his: *Inci-*

dents of a Whaling Cruise (1841) by Yale graduate Francis Allyn Olmsted, *Etchings of a Whaling Cruise* by Browne, and *The Whale and His Captors* (1850) by a reverend named Henry T. Cheever. Even before these American works, three narratives of life on English whaleships were published by three authors who placed more emphasis on describing whales and other marine life. These were: *An Account of the Arctic Regions, with a History and Description of the Northern Whale Fishery* (1820) by the whaleman-naturalist Williams Scoresby Jr., *The Natural History of the Sperm Whale* (1839) by the ship's surgeon Thomas Beale, and *Narrative of a Whaling Voyage Round the Globe* (1840) by another whaleship surgeon named Frederick Bennett. All six of these nonfiction accounts provided factual information and stylistic influences for Melville when he sat down to write his fictional voyage. Ishmael completes his list of whale authors—and there were still many more out there—by declaring that even Beale and Bennett only touched on the life of the sperm whale. "The sperm whale, scientific or poetic, lives not complete in any literature," Ishmael says. He steps forward as the man for the job. Sometimes Melville copied material outright from one of these authors or he cribbed their structures or arguments. Sometimes, however, what we read now in these other sources seems just like what Melville wrote not because he copied it, but because he and the other author experienced similar events.[2]

For reference material on the biology of sperm whales and other pelagic sea life, Melville seems to have turned most often to the narratives by the surgeons Beale and Bennett and to an entry on whales in *The Penny Cyclopædia* (1843), which in turn took much of its material from Beale and from Scoresby. Most of Beale and Bennett's observations on sperm whale biology and behavior have held true for modern whale ecologists.[3]

In February 1834, the surgeon Thomas Beale had recently returned from a voyage on two whaleships, upon which he'd served as the ship's doctor. He had been home for a year, was about twenty-seven years old, and was working at a post as assistant surgeon at St. John's British Hospital in London. Beale worked on his monograph of the sperm whale, which he would publish the following year, earning an award among his peers. During that same February 1834, Frederick Bennett was still out to sea sailing aboard the whaleship *Tuscan* as their ship's doctor. Bennett's ship had rounded Cape Horn while Darwin's HMS *Beagle* rested nearby at anchor off Tierra del Fuego.

In 1839 Beale published his revised and expanded book-length version, *The Natural History of the Sperm Whale*. A young man trying to earn his place in London's scientific community, toiling away at a hospital for the poor, Beale was obsequious in his bowing to the expertise of other naturalists of his day. He quoted large swaths of their work directly into his book. Beale loved to slip in his Latin anatomical terms, and he was regularly the hero of his own story. For example, he writes of how he healed native people around the South Pacific rim and even how at one point his keen memory for geographic details rescued a boatload of his shipmates. Bennett, meanwhile, had returned home in 1836 and presented material on sperm whales in public scholarly lectures, which Beale incorporated verbatim into his natural history. Bennett then published his own book the following year.

Historians don't know much else about these two surgeons.[4] Bennett strikes me as less self-righteous, and far more curious and empirical about absolutely everything: whales, birds, fish, invertebrates, seaweeds, and human cultures. Bennett captured fish and examined their stomachs to expand his inquiries into the nature of bioluminescence. Bennett described getting a tattoo in French Polynesia, which he seemed to have endured purely just to see what it was all about: "I gratified a wish to observe the process and effects of the tatoo [*sic*] by having a figure thus impressed upon myself." He chose a circular pattern that he saw on the body of his Tahitian tattoo artist and asked for it on his upper arm. He described the process clinically, with a stiff upper lip.[5]

Melville's method of tidy chapter pairs, along with the entire format of *Moby-Dick*, is in many ways a direct combination of the well-established structure employed by Beale and Bennett, who both wrote sections and chapters devoted to matters of natural history while also writing companion material devoted more broadly to the voyage and the adventure. Beale's first part is devoted to anatomy, including chapters titled "Of the Brain," "Of the Ear," and "Of the Sexual Organs," while his second part is a "Sketch of a South-Sea Whaling Voyage," which includes chapter headings such as "Storm ensues" and "we kill a female Whale." Bennett flipped this structure for his book. His first part is the chronology of his adventure. His second has an appendix with sections on "Cetaceans," "Birds," "Fishes," "Mollusca," and "Marine Phosphorescence."

Melville slurried these two forms together—the scientific descriptions and the voyage narrative. Harvard's library has Melville's own copy of Beale's *The Natural History of the Sperm Whale*. Several of Ishmael's comments and ideas in *Moby-Dick* can be traced directly to this volume. For example, in "Less Erroneous Pictures of Whales," Ishmael declares, "All Beale's drawings of this whale are good, excepting the middle figure in the picture of three whales in various attitudes, capping his second chapter." In Melville's actual copy of the book he put an "x" under the picture and scrawled in pencil at the bottom: "There is some sort of mistake in the drawing of Fig: 2. The tail part is wretchedly crippled + dwarfed, & looks altogether unnatural. The head is good."[6] (See fig. 4.)

Beale and Bennett are hard to keep straight—I'll refer to them throughout as Surgeon Beale and Doctor Bennett—and really all of Melville's fish documents and their authors tend to run together for the modern reader. See Figure 5 if you'd like some help. But getting all these guys and their books confused is part of the point. Melville created *Moby-Dick* within a crowded market of popular sea voyage narratives in which copying pages of others' writing was common and even scholarly.

Moby-Dick is often a laborious, digressive mess. But it is a stew of ingredients and styles from various well-known and reliable chefs of his day, and there's a lot more to the stew and its progression than just throwing it all into the pot. There is a method to his madness, which was centered on the imaginative, careful, and exhaustive exploration of ocean life, especially whales. In 1851 most of his readers weren't ready

CHAPTER II.

HABITS OF THE SPERM WHALE.

Fig. 1.   *Fig.* 2. X   *Fig.* 3.

IT is a matter of great astonishment that the consideration of the habits of so interesting, and in a commercial point of view of so important an animal, should have been so entirely neglected, or should have excited so little curiosity among the numerous, and many of them competent observers, that of late years must have possessed the most abundant and the most convenient opportunities of witnessing their habitudes. I am not vain enough to pretend that the few following pages include a perfect sketch of this subject, as regards the sperm whale; but I flatter myself that somewhat of novelty and originality will be found justly ascribable to the observations I have put together; they are at all events the fruit of long and attentive consideration.—For convenience of description, the habits of this animal are given under the heads of feeding, swimming, breathing, etc.

X There is some sort of mistake in
the drawing of Fig: 2. The tail part
is undoubtedly cropped & dwarfed, &
looks altogether unnatural. The head is good.

c 2

FIG. 4. Melville's notes in a page of his copy of Thomas Beale's *The Natural History of the Sperm Whale* (1839).

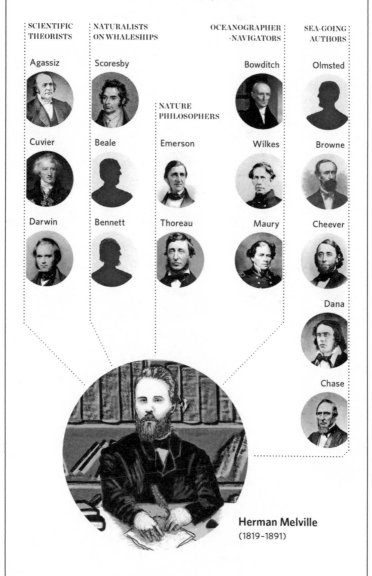

Ishmael's "Fish Documents"

Herman Melville's natural history
sources for *Moby-Dick*

SCIENTIFIC
THEORISTS

NATURALISTS
ON WHALESHIPS

OCEANOGRAPHER
-NAVIGATORS

SEA-GOING
AUTHORS

Agassiz

Scoresby

Bowditch

Olmsted

NATURE
PHILOSOPHERS

Cuvier

Beale

Emerson

Wilkes

Browne

Darwin

Bennett

Thoreau

Maury

Cheever

Dana

Chase

Herman Melville
(1819–1891)

FIG. 5. A guide to Ishmael's "Fish Documents" in *Moby-Dick*.

for a work of fiction like this. A critic for the *Southern Quarterly Review* wrote: "In all the scenes where the whale is the performer or the sufferer, the delineation and action are highly vivid and exciting. In all other respects, the book is sad stuff, dull and dreary, or ridiculous."[7]

The novel did not sell out of the first printing. Today, it is often the chapter "Cetology" that first puts a bone in the reader's throat.

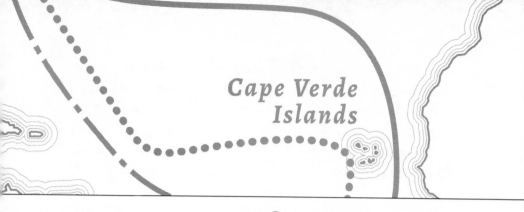

Ch. 3

CETOLOGY AND EVOLUTION

> Be it known that, waiving all argument, I take the good old fashioned ground that the whale is a fish, and call upon holy Jonah to back me.
>
> I give the popular fishermen's names for all these fish, for generally they are the best.
> Ishmael, "Cetology"

Ishmael begins "Cetology" apologizing for a break in the action since the *Pequod* is finally at sea and rolling along under sail. He has established how dangerous and honorable it is to be a whaleman, but before he continues to the high drama of revealing Ahab and his true mission of lusty revenge against the White Whale, he pauses here on the North Atlantic for what he deems "a matter almost indispensable": to define whether a whale is a mammal or a fish and to classify how sperm whales fit into the broader taxonomy of cetaceans. "Cetology" is his longest direct exposition on science in the novel.

IS THE WHALE A FISH OR A MAMMAL?

Twenty-first-century American lobstermen will often among themselves refer to a catch of the American lobster (*Homarus americanus*) as a crate of "bugs." These fishermen are well aware that lobster are not insects. Most of them could tell you that lobsters are related to crabs, that they are decapods, and that they had some connection to the Crustacea (a subphylum within the phylum Arthropoda, which includes insects). Yet my guess is that most Americans, fishermen or not, would be hard-pressed to explain the difference between a lobster and an insect, other than insects can't live underwater and perhaps have a different number of legs.

In "Cetology," Ishmael sides with the fishermen, the Nantucket whalemen, who use the colloquial name of "fish" to describe whales. He is joking—mostly. It's a debate with some history.

On January 1, 1819, the year of Melville's birth and the year the ill-fated whaleship *Essex* sailed away from Nantucket, the newspapers in New York City reported the final verdict of an entire court case that hinged around this very question of whether a whale should be considered a fish or a mammal. The case was about money: a merchant refused to pay a penalty for not having his "fish oil" inspected. The merchant had purchased three barrels of *whale* oil. The leading ichthyologist in New York City, Samuel Mitchell, was a lead witness, as was a whale-ship captain named (true story) Preserved Fish. Although Captain Fish believed, like Professor Mitchell, that the whale was a mammal, he had a difficult time defending the idea under a lawyer's quick-witted cross examination. Another working whaleman, James Reeves, the only other person brought to the stand who had seen whales at sea, spoke as a witness with a different opinion than Captain Fish, deciding from *his* three voyages that the whale was a fish. Seaman Reeves was not confident of the nature of the spout, for example. Perhaps whales did breathe water? In the end, the jury was not convinced either. Thus, in 1819 in New York City, by the court of law and in the dockside realm of oil inspection, the whale was still a fish.[1]

Melville read of this case decades later, yet he somehow resisted including this in *Moby-Dick*. The case was mentioned in his numerous fish documents. For example, Dr. Frederick Bennett wrote directly of the debate in 1841, disappointed that the American jury did not listen to "the learned distinctions of science."[2]

By the time Melville went to sea in the 1840s, most whalemen and the general public knew that whales breathe air, are warm-blooded, nurse live young, and so on—as Ishmael delineates in "Cetology," quoting Linnaeus (likely from his encyclopedia), who had written of these traits nearly a century earlier. Later in "The Blanket," Ishmael writes that "like man, the whale has lungs and warm blood." But in the America in which Melville grew up, *fish* was a broader term than we use it today. Fish were simply animals that live in the water all the time, derived straight from the Bible's grouping of the birds, beasts, and fishes. Consider the names, for example, crayfish, or starfish. Out at sea the whalemen referred to female whales as "cows," the males as "bulls," and infant whales as "cubs." Yet collectively the whalemen still called the whales fish: because the animals lived in water all the time, never hauling out on the beach as did "amphibious" seals.[3]

Today, for lobstermen, the word *bug* rolls off the tongue better, their catch does look like giant insects, and the nickname diminishes the creatures they capture, rendering their endeavor an easier, if a seemingly lowlier task. Perhaps the word *fish* did the same for the American whalemen, too. Whether a whale was a mammal or a fish was simply a different term with little practical value. If anything, calling it a mammal at the time was a bit haughty. The name mammal in Melville's day even carried a somewhat salacious connotation as it brought up the image of a woman breastfeeding.[4]

Beginning with "Cetology" and then throughout the novel, Ishmael positions the practical hunter's knowledge of the whalemen above that of the "learned naturalists ashore," those pale closet naturalists of the world who sat in preservative-choked laboratories receiving specimens to analyze, men who never had any direct experience with the animals alive. Although genuinely interested in their findings and endeavors, Melville seems to have had a career-long desire to deride, or at least cynically question, what he saw as at times a soulless mainstream scientific community. So when in doubt, Ishmael sides with the whalemen.[5]

Ishmael reflects accurately that even in the 1840s the terminology question remained an active one in the forecastles of the American whaling fleet. In *Etchings of a Whaling Cruise*, for example, it was such common knowledge that a whale was a mammal that Browne used this to make fun of the ignorance of a fellow greenhand. Browne's sailor says, when looking over the rail at the first sperm whale caught of the voyage: "Why, some folks says whales isn't fish at all. I rayther calculate they are, myself. Whales has fins, so has fish; whales has slick skins, so has fish; whales has tails, so has fish; whales ain't got scales on 'em, neither has catfish, nor eels, nor tadpoles, nor frogs, nor horse-leeches. I conclude, then, whales *is* fish. Every body had oughter call 'em so. Nine out of ten *doos* call 'em fish."[6]

Within the scientific community by the 1850s, when Melville sat with his fish documents in his study in his Pittsfield farmhouse, the matter was firmly settled. Surgeon Beale wrote of whales as mammals without deigning to address the issue. Dr. Bennett began his general comments on whales explaining there was no reason they could *not* be mammals. In his *Book of Nature*, Good explained that whales were in the seventh order of the mammals, as put forth by Linnaeus. He also wrote that Baron Cuvier had a newer system of three mammalian orders, divided by types of feet: hooves, clawed, or fin-like. Melville's *Penny Cyclopædia* also mentions this system by Cuvier, the "great zoologist."[7]

No dictionary, encyclopedia, or any book of natural history at the time left out Baron Georges Cuvier. Ishmael calls him, sarcastically, "the great Cuvier," probably because of that encyclopedia entry, but also because Beale points out so many of Cuvier's errors when it came to whales. (Georges' younger brother Frederic was also responsible for that "squash" of a whale illustration that Ishmael mocks later in the novel.) Baron Georges Cuvier, a French paleontologist, was the Western world's most influential naturalist in the early 1800s. He largely invented the concept of comparative anatomy, a focus on skeletal systems, and the proof that certain species on Earth had actually lived and then gone extinct—which required some new explanations to account for Noah's flood and the Biblical age of the planet. In addition to leaning on Cuvier, the authors of books of natural history for the general public in the midcentury all wrote of whales as mammals, but they still all seemed to feel the need to discuss the decision. Good, for example, wrote: "there is some force in introducing these sea-monsters into the same class with quadrupeds," due to the heart, lungs, backbone, and teats. On the other hand, Good did agree, whales do not have feet, hair, or proper nostrils, and they live in water and mostly act and look like fish.[8]

In January 1851, one of Melville's brothers gave him a translated edition of Baron Cuvier's book on fish, one of the fifteen volumes of *The Animal Kingdom*. We still have Melville's annotated copy, which includes his underlines and checks in the section in which Cuvier explains—on the same page with a lengthy, opinionated footnote about the New York City court case—that confusion still existed regarding the terminology of whales as fish. Cuvier chastised: "The definition of fish, such as we find it in the writings of modern naturalists, is perfectly clear and precise. They are *vertebrated animals with red blood, breathing through the medium of water by means of branchiæ.*" In "Extracts," Ishmael cites Cuvier stating that "the whale is a mammiferous animal without hind feet." Ishmael's definition of a whale in "Cetology" seems a direct parry, a fingers-flick under the chin, sent back across the Atlantic to the grave of Baron Cuvier in Paris: a whale, Ishmael states, also in italics, is "*a spouting fish with a horizontal tail.*"[9]

"Cetology" is Ishmael's first of many scenes in *Moby-Dick* in which he deliberately defies the scientists of his day. The distrust of the scientific community, or at least the sense that scientists are too sequestered in their ivory towers or computer-lined labs, remains prevalent among large portions of American fishing communities today, including the American lobster fleet. Many fishermen still feel that marine biologists do not have enough direct experience with animals and ecosystems, and, just like in that court case in New York City, this distrust can have economic implications, since fisheries biologists, beginning in the late nineteenth century, have had significant influence and authority over the regulatory framework and the laws that govern the fishermen and the food that arrives in our kitchens and restaurants. It would be nearly a century after *Moby-Dick* that scientists began to have a significant voice in regulating international whaling.

WHALE TAXONOMY BEFORE *ON THE ORIGIN OF SPECIES*

Rob Nawojchik stands on the balcony of Great Mammal Hall at Harvard's Museum of Natural History. He is looking at whale skeletons. Nawojchik teaches the subject of marine mammal classification better than anybody I've ever met. I once saw him squeeze through the door of a lecture hall an enormous branch of a tree to illustrate the idea of evolution. "We're looking at just the leaves up top here," he told the students, facing the foliage in their direction. "Each leaf is a species. Evolutionary systematics is trying to figure out the pattern of branches connecting the leaves, without being able to see the branches directly. It's a detective story."[10]

Great Mammal Hall was built in the 1870s as an addition to the original building that opened its doors in 1859, thanks to the star-power and fund-raising of Louis Agassiz, whom America had quickly adopted as its most famous naturalist. Agassiz had arrived on a lecture tour only a decade earlier from Switzerland, after which he was convinced to stay at Harvard during the same years that Melville had begun his writing career and met and married into the prominent Boston family of Elizabeth Shaw.

Ishmael actually drops Agassiz's name in *Moby-Dick* when referring to how the scratches on a sperm whale's skin look like the "violent scraping" on coastal rocks from ice, referring to Agassiz's most lasting scientific contribution: the expansion and popularization of the theory of how ice ages crafted geological features. Agassiz, a direct mentee of Cuvier, was a passionate scientist and man of God. He was all over the newspapers and just the type of naturalist Melville would seem to have applauded. Agassiz was an intrepid, ambitious man who scaled mountains and was quick to propose far-reaching theories about God's design. Agassiz advocated for putting down your books and relying on ob-

servations out in the field. He passionately believed in teaching natural philosophy to the general public.[11]

Construction on Great Mammal Hall, where Nawojchik now stands, finished just before Agassiz died. The space has been restored to its original Victorian character. Curators have squeezed into glass cases along the walls the taxidermy and skeletal specimens of a vast range of animals, such as penguins, koalas, and foxes. Hanging from the tall ceiling at various levels, above glass cases of taxidermy buffalo, zebras, and white-maned mountain goats, are skeletons of a range of marine mammals, including three full-sized whale specimens: a sperm whale, a North Atlantic right whale (*Eubalaena glacialis*), and a fin whale (*Balaenoptera physalus*). Rigged below them are the other hanging marine mammal skeletons, including a narwhal (*Monodon monoceros*), a harbor porpoise (*Phocoena phocoena*), a pygmy sperm whale (*Kogia breviceps*), and the extinct Steller's sea cow (*Hydrodamalis gigas*), which curators knew had been hunted to eradication as early as 1768.[12]

On the upper balcony, Nawojchik stands between the right whale and sperm whale. He can reach out and touch them. The right whale's long black plates of baleen extend down from the upper jaw. Known often as "whalebone" in Melville's time, baleen is indeed hard, but it's made instead of the protein *keratin*, the stuff of mammalian fingernails, horns, hair, and hooves. (Keratin is easily confused with *chitin*, which constitutes, for example, the exoskeleton of lobsters and the beaks of squid.) As Ishmael describes later in "The Right Whale's Head," the inner edges of baleen separate into hairy fibers within the mouth in order to sieve small organisms. The vertical plates are indeed like the slats window blinds. The sperm whale's skull, by contrast, has no baleen the upper jaw. The sperm whale's skull has instead two long narrows of thick, white, conical teeth on its lower jaw—the better to grab 's leg with.

Stand ere on the balcony beside these skeletons, Nawojchik says: "I reread Dick and the 'Cetology' chapter, and I can see how Melville, thro hmael, is being satirical, making fun of the scientists. But I thin el gets a little *too* dismissive—almost like a guy on a

barstool who just throws up his hands. You have these same kinds of conversations today: *Nah, it doesn't matter. Who cares? It's this, or it's that. What's the big deal? How could you possibly classify them? Yeah, he's got teeth, but he's big. Or he's got this or he's got that.* Ishmael in that way is not representing the scientific thinking of his time, for sure."[13]

In other words, just because Beale, Bennett, and Scoresby were all vexed by how to classify whales, does not mean they did not want to be able to or did not approach the questions with a plan. Ishmael, it seems, is more annoyed with the likes of Cuvier and an English naturalist named John E. Gray for their condescending confidence on the matter. In a section that Melville marked and underlined and scribbled beside, Surgeon Beale ridiculed French naturalist Bernard Germaine de Lacépède for claiming there to be no less than eight separate species of sperm whales.[14]

Nawojchik appreciates Melville's frustration in 1851 regarding what might seem the arbitrary divisions of Linnaean classification, or really any historical classification of the natural world going back to Aristotle. Taxonomy was continually in flux. It was hard, as it is today, for a layman to understand how the systems were and are derived. Cuvier, and then Agassiz, saw no branches underneath the leaves. Agassiz taught that after various catastrophes, such as Noah's flood or a few creeping Ice Ages, God had created whole new worlds of life, ever improving until He settled on humans as the highest in the great chain of being. Agassiz did not believe in this burbling new idea about the transmutation of species. Cuvier and Agassiz both knew, for example, what Stubb, the second mate of the *Pequod*, jokes about in *Moby-Dick*: the bones inside a whale's fin correspond to the fingers of the human hand but the naturalists spoke of these as "affinities" in God's design, tool box. Agassiz, with Harvard naturalist Augustus Gould, wrot their 1851 college textbook that the tail and fin of the whale corresd to the limbs of mammals and that their muscles work to run a m in a similar way. What Agassiz would not fathom at the time ould never accept even decades after *On the Origin of Species*, is t fore-paws on one species could over millions of years slowly, ntally, *alter into* a fin on an eventually new swimming animal.[15]

Without a clear reason for any classification that convincingly rationalized the choice of skeletal features, visible similarities, behaviors, and/or habitats, Ishmael from his barstool, or more appropriately his capstan above the forecastle, admits befuddlement. He argues, *why not simply use size?*

With a hand on the rail of the balcony Nawojchik says, "If we threw a bunch of objects in a room—tennis balls, automobile mufflers, plastic pens, and so on—or a bunch of different leaves to remain with that analogy—different people would classify them in many different ways, by size, shape, color, and so forth."

Nawojchik volunteers that Charles Darwin is his personal hero. Darwin published *On the Origin of Species* in 1859, ironically the same year as the opening of Agassiz's Museum of Comparative Zoology (which is now one of the research arms of the Harvard Museum of Natural History). In *Origin of Species* Darwin explained the *mechanism* of this transmutation—evolution by natural selection—by which animals over millions of years had changed. As most scientists began to quickly accept and understand Darwin's explanation of the process, taxonomy changed forever: physical and behavioral traits were now understood to connect species together by common ancestry, by shared branches. Victorian taxonomists began to organize the animal kingdom on paper and in their museums with a methodology that focused on this ancestry, drawing these lineages of descent, not just organized by atemporal choices of shared characteristics or habitats. Nawojchik likes to point out that the only illustration in *Origin of Species* is a simple tree diagram that models speciation over time, in the same way that you could show the descent of the Galápagos finches that evolved different traits over time due to the isolation of different islands and different ecological and environmental conditions. Post-*Origin* taxonomists searched for meaningful characteristics that unified groups from shared progenitors. Today Great Mammal Hall has a large sign "The Evolution of Hoofed Mammals" with a treelike diagram about the ungulates, in which cows, pigs, whales, camels, and rhinos are all evolved from an ancient common ancestor. Even as early as the late 1600s, English anatomists had noticed the similarities between the stomach, reproductive organs, and other parts in whales and in hoofed mammals, but they did not imagine any shared ancestry.[16]

In the first edition of *Origin of Species*, Darwin wrote of the potential of a land mammal evolving into a marine one: "In North America the black bear was seen by Hearne swimming for hours with widely open mouth, thus catching, like a whale, insects in the water. Even in so extreme a case as this, if the supply of insects were constant, and if better adapted competitors did not already exist in the country, I can see no difficulty in a race of bears being rendered, by natural selection, more and more aquatic in their structure and habits, with larger and larger mouths, till a creature was produced as monstrous as a whale." Although the veracity of this initial account from a Canadian fur-trapper was later questioned, which is why Darwin removed it from subsequent editions, scientists now believe that whales evolved from a long-line of transitional forms which can be traced back to a sort of amphibious, wolflike pakicetids that began foraging in streams some fifty million years ago, adjusting its form to more plentiful food found in the water. There's strong evidence, too, to suggest that pinnipeds (seals, sea lions, and walruses) also evolved from land, likely a bearlike ancestor, and bears are their closest living relatives. Nawojchik explains that it would take another full century—the discovery of various fossils, isotope dating of these fossils, continued research with embryology, the development of phylogenetic systematics, and then DNA analysis techniques—to really narrow down the traits that revealed that this supposition about whales evolving from land mammals was actually true. The hippopotamus (*Hippopotamus amphibius*) is now considered the closest living relative to current whales.[17]

In light of evolution then, Ishmael's use of size as a way to classify the whales is simplistic, but in Melville's time it wasn't really that much more arbitrary than other organizational schemes. In his discussion of the fin whale in "Cetology," Ishmael recognizes, but deems nonsense, that many naturalists had divided the toothed whales and the "whalebone" whales. The toothed vs. baleen whales split has proven to be accurate. Baleen is a meaningful trait, and it's one of the most significant in whale evolution. We know today that some thirty-five to forty million years ago, in the Eocene, environmental conditions began to favor some of the archaic toothed whales that began to evolve baleen plates in their mouths (separate from their teeth, which would atrophy to nothing over time). By about thirty to thirty-four million years ago, archaic whales had diverged into the two clear lineages, two branches, that we recognize today as the mysticetes, the baleen whales, and the odontocetes, the toothed whales.[18] (See fig. 6.)

After Ishmael rejects any classification based on visible traits such as teeth, baleen, humps, or fins, he then rejects Cuvier's methodology of comparing skeletal structures. Ishmael says:

> But it may possibly be conceived that, in the internal parts of the whale, in his anatomy—there, at least, we shall be able to hit the right classification. Nay; what thing, for example, is there in the Greenland whale's anatomy more striking than his baleen? Yet we have seen that by his baleen it is impossible correctly to classify the Greenland whale. And if you descend into the bowels of the various leviathans, why there you will not find distinctions a fiftieth part as available to the systematizer as those external ones already enumerated. What then remains? nothing but to take hold of the whales bodily, in their entire liberal volume, and boldly sort them that way.[19]

Nineteenth-century scientists on both sides of the Atlantic had been learning that there were clear and numerous distinctions between skeletal systems of baleen and toothed whales. In Melville's *Penny Cyclopædia*, there were illustrations of the baleen whale's skeleton and skull a few pages away from that of the sperm whale's skull. The most obvious difference is the concave bowl of the sperm whale's skull in comparison to the convex arched-pole skull of the baleen whale. So in this case, Ishmael does not represent the scientific thought of his time. Toothed whale skeletons were noticeably different for the "systematizer" or the lay person. (See fig. 7.)

Nawojchik shows me three other characteristics of the sperm whale's

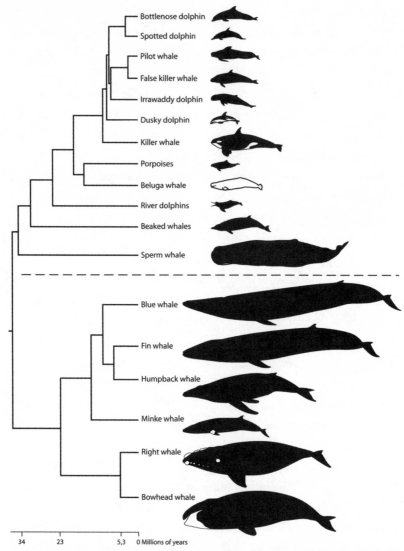

Bottlenose dolphin
Spotted dolphin
Pilot whale
False killer whale
Irrawaddy dolphin
Dusky dolphin
Killer whale
Porpoises
Beluga whale
River dolphins
Beaked whales
Sperm whale
Blue whale
Fin whale
Humpback whale
Minke whale
Right whale
Bowhead whale

34   23   5,3 0 Millions of years

FIG. 6. A modern cetology by Emese Kazár (2013) from McGowen, Spaulding, and Gatesy (2009). No one drew trees of descent for any groups of animals before Darwin's *On the Origin of Species* in 1859. Ishmael names 14 whale species, with an opening for more; modern taxonomies name more than 80 cetaceans.

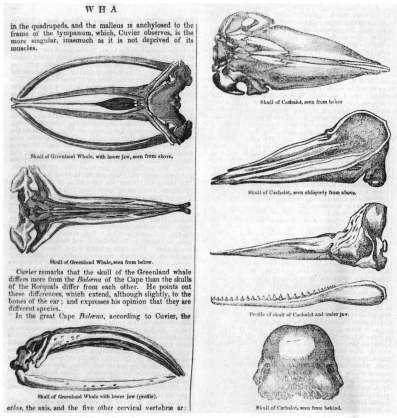

W H A

in the quadrupeds, and the malleus is anchylosed to the frame of the tympanum, which, Cuvier observes, is the more singular, inasmuch as it is not deprived of its muscles.

Skull of Greenland Whale, with lower jaw, seen from above.

Skull of Greenland Whale, seen from below.

Cuvier remarks that the skull of the Greenland whale differs more from the *Balæna* of the Cape than the skulls of the Rorquals differ from each other. He points out these differences, which extend, although slightly, to the bones of the ear; and expresses his opinion that they are different species.

In the great Cape *Balæna*, according to Cuvier, the

Skull of Greenland Whale with lower jaw (profile).

atlas, the axis, and the five other cervical vertebræ are

Skull of Cachalot, seen from below.

Skull of Cachalot, seen obliquely from above.

Profile of skull of Cachalot and under jaw.

Skull of Cachalot, seen from behind.

FIG. 7. The images of baleen whale skulls (left) and sperm whale skulls (right) that Melville saw, separated by a few pages, in the "Whale" entry in his copy of his *Penny Cyclopædia* (1843).

skull that show it to be a sperm whale, as distinct from other toothed whales, and even from the dwarf and pygmy sperm whales (*Kogia* spp.). But to tell you the truth, I can't follow everything that he says, despite his lively ability to teach and even with the entire skeletons right there in front of us. I appreciate Melville's frustration with the taxonomy business, especially in an age before *Origin of Species*, when species were considered static and designed perfectly by God. My point here is that you have to cut an educated Christian sailor like Ishmael some slack here for his skepticism of the classification systems for whales.

Today scientists continue to refine and alter our systems based on new information. Species are always evolving in blurry lines—from bacteria to belugas—which was one of Darwin's major points. A debate over subspecies or species is not simply pompous scientists bickering or turf-defending—although that can be part of it, certainly—but it's more deciding how far and fast change and evolution is happening. Delineating species is crucial today for management purposes—so we can decide about regulations regarding hunting or fishing, and to decide when and how to act on endangered species or invasive ones. Yet nearly any sort of regulatory framework and the necessity of these lines for management purposes were practically unknown in Melville's time.

Fundamentally for Melville, Cuvier, Linnaeus, and Aristotle, naming and organizing animals was foremost for learning about the diversity of life, which remains true today if you ask a biologist like Nawojchik. The scientific nomenclature is useful so that we can be sure that we're all talking about the same organism, since common names vary so greatly even within countries and across regions. That's why I include scientific names parenthetically throughout this natural history after I mention an animal for the first time. I also insert the binomial nomenclature in the hopes it makes me appear a little smarter and more earnest. In "Cetology," Ishmael pokes fun of this tendency, suggesting the Linnaean scientific name of the sperm whale is "Macrocephalus of the Long Words." Macrocephalus means big head in Greek. The description is useful since the sperm whale's head can be up to a full third the length of its body, especially in males, but scientific names in zoology still remain almost exclusively the language of specialists.[20]

ISHMAEL'S "BIBLIOGRAPHICAL SYSTEM" OF WHALES

Just as Rob Nawojchik recognizes classification isn't the sexiest of topics, albeit a necessary one, Ishmael knows he needs to add flair and humor to his own lecture. After choosing magnitude as the simplest, "practicable" system of whales, Ishmael then organizes within a "Bibliographical system" of the paper sizes used in book publishing, a joke more recognizable to the nineteenth-century reader who saw far more range in sizes of books at libraries and book stalls. Folio whales are the largest. (Think of the oversized section in your library.) Ishmael explains in a footnote, that quarto are the next largest—but this is a more square-shaped book, so that doesn't work. Ishmael then has octavos and duodecimos for the medium-sized and smaller rectangular books of whales.

Reading the whale, the wild animal in a variety of forms, as a text, adds all sorts of layers, too: about perception, about interpretation, and about theological exploration, which was often referred to as seeking to read God's "Book of Nature." Thus Dr. John M. Good, the author, surgeon, and devout son of a minister, named his popular scientific work *The Book of Nature*. Louis Agassiz, too, had his own way of conflating classification with Christian and literary endeavor. Agassiz and Gould taught that all animals were an expression of "divine thought, as carried out in one department of that grand whole which we call Nature." They wrote that the student of natural history, given only to the highest form of man, should approach the study in the same way the student would look at a work of literature, by first endeavoring "to make ourselves acquainted with the genius of the author." This meant understanding God's previous works and paths, the now fossilized worlds between floods, ice ages, and volcanic eruptions, in order to understand the final, current, highest form of humanity.[21]

Despite all the forecastle, populist cynicism about classification and laboratory naturalists, Ishmael actually provides in "Cetology" a reasonably accurate and representative synopsis of the whales as known to mid-nineteenth-century mariners, with names and descriptions that align truthfully to those written in mariner's logbooks and narratives published by both author-sailors and professional naturalists. In fact, Ishmael's architecture is indeed more complete, earnest, and organized in the study of the varieties of whales than anything included, quite genuinely, in most of the sources published at the time. Ishmael's taxonomy is not useful in terms of understanding the branches of evolution, of course, but it is a reasonably accurate record of common names and how working mariners knew these animals.[22]

See Figure 8 at the end of this chapter for a quick study about which whales Ishmael is likely referring to in "Cetology," in comparison to how we know them today. Below are more of the subtleties and further explanation, recognizing that Melville wanted to get it correct, but he didn't mind Ishmael injecting some anthropomorphic humor, a little flourish, and some poetry along the way, especially in order to keep the reader interested and to make some loftier points.

FOLIO I. SPERM WHALE

Ishmael makes it clear that no one thought that the spermaceti oil in the head of sperm whales was sperm of the sexual emission variety: that was "absurd."[23]

Ishmael explains that the sperm whale is "without doubt, the largest inhabitant of the globe." Moby Dick as the Earth's apex giant serves his story, but Surgeon Beale wrote the exact same. This suggests that not everyone under sail ever got close enough to get a sense of the full size of a blue whale (*Balaenoptera musculus*). Dr. Bennett, however, knew of baleen whales that grew to over one hundred feet, which is true.[24]

FOLIO II. RIGHT WHALE

As Melville was composing *Moby-Dick*, he knew of rumblings about a difference between the right whales of the Northern Hemisphere and those of the Southern Hemisphere, and that the right whales of the Arctic, often called the Greenland whale, were also their own species. Whalemen and naturalists had been settling the latter, that the bowhead whale (*Balaena mysticetus*) was in fact separate from right whales. Bowheads have, true to their name, a steeper bump and slope to their foreheads. Their baleen is longer. In contrast to the heads of right whales, bowhead skin is also smooth and free of any callosity or invertebrate hitchhikers. Bowheads live only in the Arctic and down to the far northern Canadian Maritimes. After dividing the bowheads from the right whales, the splitting of the right whales into three different species would take over a century after *Moby-Dick*, although the idea was emerging in the 1850s. Today, thanks in large part to DNA analysis, biologists nearly all agree that three species of right whales swim in the global ocean: the North Atlantic right whale, the North Pacific right whale (*Eubalaena japonica*), and the southern right whale (*Eubalaena australis*), none of which currently mix geographically or genetically.[25] (See fig. 9.)

While Melville worked on a whaleship in the 1840s, whaleman called the bowheads and the right whales collectively, as Ishmael says, the right whale, the black whale, the true whale, or even simply *the* whale. Indeed, this whale was the first hunted commercially, offshore, by the Basques on both sides of the North Atlantic by the 1500s, if not earlier. Coastal, localized hunting of these whales had been conducted for centuries before in Europe, and perhaps North America, likely by a variety of other peoples, along with the Basques. Since it was slow, coastal, and plump with oil and baleen, hunters seem to have called it the "right whale," as it was the best one to chase.[26]

Ishmael summarily dismisses all of the proposals of splitting the right whales into multiple species: "Some pretend to see a difference between the Greenland whale of the English and the right whale of the Americans. But they precisely agree in all of their grand features." Here Ishmael jabs his contemporary scientists, perhaps in particular the English naturalist John E. Gray, the Keeper of Zoology at the British Museum, who was a notorious splitter, especially of the baleen whales. Ishmael says: "It is by endless subdivisions based upon the most inconclusive differences, that some departments of natural history become so repellingly intricate."[27]

FOLIO III-VI. FIN-BACK, HUMP BACK, RAZORBACK, AND SULPHUR BOTTOM

Ishmael's classification and the naming in "Cetology" show the challenges of observing and naming baleen whales. Even today it's hard for people standing on the deck of a ship to identify a particular species of whale at a distance. Ishmael's Fin-Back, Hump Back, Razorback, and Sulphur Bottom are collectively referred to today as the rorquals, which comes from the Norwegian word for furrow, since they all have pleated skin under the jaws to expand, pelican-like, for feeding. Rorquals make up the baleen whale family Balaenopteridae. Ishmael referred to them

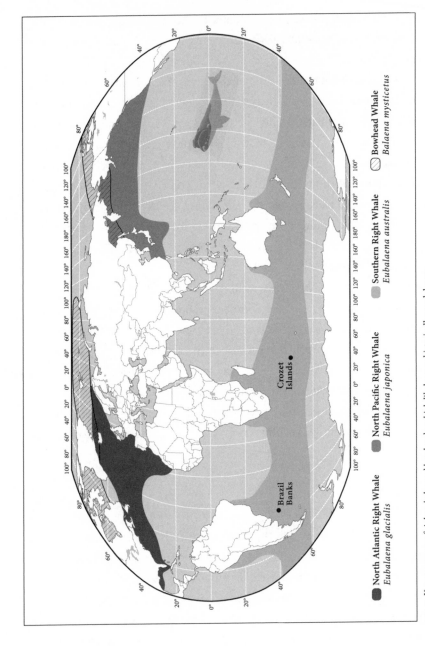

FIG. 9. Known ranges of right whales and bowheads, which likely were historically much larger.

North Atlantic Right Whale
Eubalaena glacialis

North Pacific Right Whale
Eubalaena japonica

Southern Right Whale
Eubalaena australis

Bowhead Whale
Balaena mysticetus

Crozet
Islands

Brazil
Banks

as the "uncapturable whales." Their spouts can all look similar, tall and columnar. They generally swam too quickly and were too elusive, and even if the whalemen were able to harpoon these animals, they could almost never haul them in. To chase these whales was a waste of time. Ishmael uses this as a metaphor later in the novel when a luckless captain named Derick chased after a whale that he would never catch: "Oh! many are the Fin-Backs, and many are the Dericks, my friend."[28]

Authors and mariners in the mid-nineteenth century used the same common names for Ishmael's folio whales: finbacks, humpbacks, sulphur bottoms, and less often razorbacks. Most of what Ishmael describes in "Cetology" about these rorquals is accurate to how we describe and name them today, even if he lumped them into only four species, compared to the eight different rorquals now delineated. Fin whales do have notably sharp, dorsal fins—but these do not grow nearly as tall as Ishmael claims—and fin whales do indeed have a tall, straight spout. Humpbacks do have more of a hump, and they do more actively breach and flop their enormous fins or show their full flukes when diving. (Humpbacks are also easier to identify and hunt because they tend to be less evasive than the other rorquals; their spouts tend to be shorter and bushier, and they have exceptionally large and long flippers.) Ishmael's "Sulphur Bottoms," which we know today as blue whales, do indeed often have a yellow "brimstone belly," which scientists learned in the 1920s is actually a coating of diatoms, a type of microscopic algae.[29]

Ishmael's "Razor Backs" are harder to pin down. Ishmael says that they only show their back, rising "in a long sharp ridge." Sei whales, diagnostically, do not arch their back as much as other whale species, and they seem to show their flukes less often when going down.[30]

Different whalemen in Melville's time, however, surely used, perhaps interchangeably based on what they could see in a brief moment, all of these names, including when they saw Bryde's whales (*Balaenoptera edeni*), the much smaller minkes (*Balaenoptera acutorostrata*), or maybe even gray whales (*Eschrichtius robustus*), which have more of a dorsal ridge. All of these species of baleen whales have various close similarities when spotted in the wild, particularly at a distance. Other than the humpbacks, the whalemen rarely if ever had the rorquals alongside. Ishmael concedes to "several varieties" of these "whalebone whales."[31]

THE OCTAVOES: MEDIUM-SIZED TOOTHED WHALES

Ishmael's "Black Fish" is fairly straightforward. These are pilot whales (*Globicephala* spp.), which are still called blackfish—although to keep things confusing, blackfish today is sometimes used to mean killer whales, false killer whales (*Pseudorca crassidens*), melon-headed whales, (*Peponocephala electra*), and others. The common name of blackfish for pilot whales was well known on both sides of the Atlantic and through-out the mariners' voyages at sea. As Ishmael explains accurately, whale-men often caught these whales for harpooning practice and for smaller amounts of oil. James Osborn, for example, described them often dur-ing his voyage aboard the *Charles W. Morgan*. He drew one in his jour-nal. Dean C. Wright drew one prominently on his page of sea animals (see earlier fig. 3).[32]

Ishmael describes the upturned mouth of the pilot whale as devilish, a "Mephistophelean grin." He seems to have made up the name "Hyena Whale" to match. The grin is emphasized in an illustration Melville saw in Dr. Bennett's narrative. (See fig. 10.) Today, we usually anthropomor-phize this facial feature of the medium and smaller toothed whales as a friendly smile.

Ishmael's nineteenth-century octavo common names for grampus, killer, and thrasher seem to be as fluid and regional as the international variants for lobster, crawfish, and scampi. The name grampus seems at the time to have been a fairly fluid term for a medium-sized toothed whale that was *not* easily identified as a pilot whale. Grampus was also used for killer whales (*Orcinus orca*), which was also synonymous with thrashers for some. A grampus was also used for Risso's dolphins, which are smaller and a lighter gray, named *Grampus griseus* by Cuvier in 1812. Historian Michael Dyer has found only one period logbook illustration labeled as a grampus, and this looks much more like a beaked whale (*Mesoplodon* spp. or *Ziphius cavirostris*). When lecturing on the natural

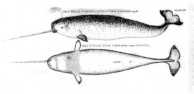

FIG. 10. Illustrations in Frederick Bennett's *Narrative of a Whaling Voyage Round the Globe* (1840), left and right, and William Scoresby Jr.'s *An Account of the Arctic Regions* (1820), middle.

history of whales in New Bedford in the 1830s, the merchant Charles W. Morgan, who'd later have the ship named after him, explained that the killer whale was merely a *type* of grampus, a "small whale." That said, some mariners and authors, such as Reverend Cheever, did indeed differentiate between all three names at the time, but any period universal agreement did not exist.[33]

THE DUODECIMOS: DOLPHINS AND PORPOISES

Although used occasionally for the mammal in the scientific literature of Melville's time in the same way we do today, the term dolphin was more commonly used until recently to mean more often the species of fish that's also known today as the dorado or mahi-mahi (*Coryphaena hippurus*). (See fig. 11.)

When American whalemen said "porpoise," they were referring to a species in either of the groups we differentiate colloquially today: the smaller, coastal, blunt-nosed porpoises *and* the larger, pelagic, long-snouted dolphins. Ishmael uses them interchangeably. It can be confusing. (If only he had used scientific names!) When he is talking about the old bookbinding design of a dolphin wrapped around an anchor in "Monstrous Pictures of Whales," he's describing the marine mammal. When Ishmael writes of the prolific waters of the Indian Ocean in "Stubb kills a Whale," teeming with "porpoises, dolphins, flying-fish, and other vivacious denizens," he means the dolphin fish, known to feed on flying fish and featured regularly in maritime literature because of their brilliant rainbow scales.[34]

Some of the most common large pelagic dolphins that are found

Whale.

Nar Whale.

Shark.

Dolphin.

Cod.

Lamprey.

Salmon.

Sturgeon.

Lobster.

Crab.

Publishd by Vernor & Hood Jan.r 1.st 1803.

FIG. 11. Plate in *Goldsmith's History of the Earth and Animated Nature* (1807). Note how the mahi-mahi is labeled "dolphin." In this plate is also the whale and narwhal illustration that Ishmael makes fun of directly in "Monstrous Pictures of Whales."

globally and that swim under the bows of ships are the bottlenose dolphins (*Tursiops* spp.). Several other species bow-ride, too. These would
all likely fit under Ishmael's "Huzza Porpoises."[35]

Ishmael recognizes that the sailors found the bow-riding of the dolphins irresistible, needing to shout "huzza!" (i.e., hurray! whoopee!) at
what seemed to be raucous play behavior. Certainly today, it never gets
old: leaning over the rail and watching dolphins effortlessly glide and
swerve and leap around the bow of your sailing ship is magical out at
sea. Yet Ishmael shifts quickly from this frivolity to explain that dolphins provide good oil and good eating. We'll pick this up again in regard to "The Whale as a Dish," but it's telling how, within a little over
a century, the idea of eating dolphin will for most of American culture
be akin to murdering and eating a pet dog or cat.

The name of the "Algerine Porpoise" appears in a couple whalemen's
journals. Perhaps these were just a larger dolphin species? Melville
likely remembered this name from his sailing days, because none of his
published fish documents mention it. Fierce pirates from Algeria were
often the subject of stories at the time, so Ishmael has fun at least with
this aspect.[36]

Ishmael's remaining octavos, the "Narwhale" and the "Mealy-
mouthed" or "Right Whale Porpoise," were well-known and distinct
enough that they can be easily identified then and today. Despite the
notorious Nar Whale in *Goldsmith's Animated Nature*, Melville saw
accurate illustrations and read of these two smaller toothed whales (see
figs. 10, 11 above). The narwhal provided plenty of opportunity, too, for
both pit humor and his metaphor of reading the book of whales.

FORECASTLE APPELLATIONS

Ishmael recognizes a further thicket of common names, some of which
the modern reader recognizes. Some are found in Melville's fish documents. A couple I think Melville made up for kicks. I'll leave them up to
future builders to nail down, but to get you started, some environmental historians theorize that the "Scragg Whale," a name used at least in

New England in the 1720s, might have been the Atlantic gray whale, which seems to have gone extinct sometime around then, perhaps due to human hunting.[37]

WHY DID MELVILLE INCLUDE THE CHAPTER "CETOLOGY"?

Why did Melville want to make any of these points at all in a novel about an obsessive, dictatorial captain hunting an enormous, mythical whale? What happens to the story if Melville had an editor who convinced him to just cut "Cetology"?

"Cetology" is Ishmael's first natural history essay in *Moby-Dick*, the first chapter with footnotes, and the first chapter in which Ishmael as the traditional narrator seems to fade and alter. This taxonomic treatise shifts the narrator's authority. Ishmael is no longer the goofy greenhand, but the scholar and survivor who has lived this world of whaling, studied whales, and now has something of his own to say about it. In "Cetology" Ishmael starts to build the factual material to craft an epic story about the human-whale relationship that is only more extraordinary in its believability. He declares that only one who has been out there killing whales himself is equipped to tell this story properly. Not only that, but the teller must be an American, since the British and French whaling is bumbling gentlemen's play. Their scientists sit at home in dusty, dark Europe to argue intricacies of bones and baleen, while not properly listening to their surgeons who have returned from the Pacific. In "Cetology," Ishmael declaims that the true epic story of the sperm whale is to be told only by the American whaleman.

Ishmael's admittedly failed effort to create a draft of a whale taxonomy in "Cetology" also emphasizes the fundamental obscurity of the creatures of the ocean. This unknowability builds the significance and drama of his story. Yet he still confidently establishes in this chapter that the sperm whale was the largest and highest on his chain of all animals on earth, continuing to elevate Moby Dick, the king of kings.

Perhaps the greatest significance of "Cetology" to the twenty-first-century reader is this: what if this sets up Ishmael as a character and a natural philosopher whose entire exploration and survival prefigures,

CETOLOGY AND EVOLUTION 57

anticipates, the theories of Charles Darwin, the person who entirely changed the way we perceive the world? In "Cetology," Ishmael favors the ideas of Agassiz and the static chain of being, with sperm whales above all whales, and man above all. But by the end of the story, humans no longer reign. Does Ishmael shift over the course of the telling of his story to be a proto-Darwinian narrator?

A few scenes after "Cetology," in his lower layer to Starbuck in "The Quarter-Deck," Ahab rages against the reality that God might not have chosen humankind above all, after all. The sperm whale, the animal, might be *equal* in the eyes of God, maybe even favored. Or what if there was no God at all? Not fate, but gradual, predictable, occasionally random and rapid, environmental conditions shaping species who amorally struggle to live and reproduce: none of Agassiz's progressivism or hierarchy, but Darwin's tree of life. Darwin ended *Origin of Species* by decentering the human:

> There is grandeur in this view of life, with its several powers, having been originally breathed into a few forms or into one; and that, whilst this planet has gone cycling on according to the fixed law of gravity, from so simple a beginning endless forms most beautiful and most wonderful have been, and are being, evolved.[38]

If Ahab cannot triumph, cannot avenge himself over a single sperm whale—be that whale agent or be that whale principal—what does that mean for humankind?

In 2000, at the turn of the new millennium, three decades into modern American environmentalism, but before much of a public discussion about anthropogenic climate change, the author and scholar Eric Wilson published a short, largely forgotten essay that declared: "Melville was not merely an amateur cetologist but a powerful, innovative philosopher of biology, intuitively (if not empirically) aware of Darwin's most iconoclastic ideas almost ten years before they found print— a harbinger of the scientist's momentous dissolution of the great chain of being."[39]

HM NAME	19C SAILOR/NATURALIST COMMON NAME	19C SCI. NAME (BENNETT, 1840)	21C COMMON NAME	21C SCI. NAME
Folios				
I. Sperm Whale 	Sperm, Spermaceti, Cachalot	*Physeter macrocephalus*, *Catodon macrocephalus*	Sperm Whale	*Physeter macrocephalus*, Linnaeus, 1758
II. Right Whale  	Right, Black, True, Greenland, et al.	*Balena mysticetus* ["Greenland"], *Balena australis* ["Cape Whale/Southern Right Whale"]	North Atlantic Right Whale, North Pacific Right Whale, Southern Right Whale, Bowhead Whale	*Eubalaena glacialis*, Müller, 1776; *Eubalaena japonica*, Lacépède, 1818; *Eubalaena australis*, Desmoulins, 1822; *Balaena mysticetus*, Linnaeus, 1758
III. Fin-Back 	Fin-back	[Bennett gives common name, but no sci. name]	e.g. Fin Whale	*Balaenoptera physalus* (Linnaeus, 1758)
IV. Hump Back 	Humpback	*Balena gibbosa*	Humpback Whale	*Megaptera novaeangliae* (Borowski, 1781)
V. Razor Back 	Razor-back	*Rorqualis borealis*	e.g. Sei Whale	*Balaenoptera borealis*, Lesson, 1828
VI. Sulphur Bottom 	Sulphur Bottom	[Bennett seems to lump with Sei]	e.g. Blue Whale	*Balaenoptera musculus* (Linnaeus, 1758)

FIG. 8. Whale species in Ishmael's "Cetology." (For the modern scientific names, this follows the convention in which parentheses indicate that the species has been shifted to within another genus since first named. See Figure Credits and Notes on page 415 for more details on this figure.)

HM NAME	19C SAILOR/NATURALIST COMMON NAME	19C SCI. NAME (BENNETT, 1840)	21C COMMON NAME	21C SCI. NAME
Octavoes				
I. Grampus	Grampus, Killer	e.g. *Phocaena orca*	e.g. Risso's Dolphin and/or beaked whale, among others	*Grampus griseus* (G. Cuvier, 1812), *Mesoplodon spp.*, and/or *Ziphius cavirostris*, G. Cuvier, 1823, among others
II. Black Fish	Black Fish	*Phocaena sp.*	Pilot Whale or Blackfish	*Globicephala spp.* (Traill, 1809)
III. Narwhale	Narwhal	*Monodon monocerus*	Narwhal	*Monodon monocerus*, Linnaeus, 1758
IV. Killer	Killer, Grampus	*Phocaena orca*	Killer Whale or Orca	*Orcinus orca* (Linnaeus, 1758)
V. Thrasher ?	Thrasher, Killer	[Bennett doesn't mention, but synonymous with "Killer" in Hamilton, 1843; separate species for Cheever, 1850.]		
Duodecimoes				
I. Huzza Porpoise	Porpoise, Common Dolphin	*Delphinus delphis*	e.g. Short-beaked Common Dolphin, Common Bottlenose Dolphin, Spinner Dolphin, and/or others	*Delphinus delphis*, Linnaeus, 1758, *Tursiops truncatus* (Montagu, 1821), *Stenella longirostris* (Gray, 1828), and/or others
	Spinner Dolphin			
II. Algerine Porpoise ?	Another ocean dolphin?	[Bennett doesn't mention; Murphy, 1912, records it as a beaked whale.]		
III. Mealy-mouthed Porpoise	Right Whale Porpoise	*Delphinus peronii*	Southern/Northern Right Whale Dolphin	*Lissodelphis spp.* (Lacépède, 1804)

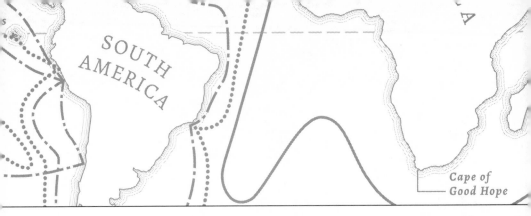

Ch. 4

WHITE WHALES AND
NATURAL THEOLOGY

Whosoever of ye raises me a white-headed whale with a wrinkled brow and a
crooked jaw ... with three holes punctured in his starboard fluke ... he shall
have this gold ounce, my boys!
    Ahab, "The Quarter-Deck-Ahab and all"

On February 19, 1834, while Thomas Beale was at his writing desk
in London working on his sperm whale study and the ornithologist-
botanist Thomas Nuttall (who'd resigned from Harvard and made out
his will) prepared in Philadelphia for a natural history expedition across
the continent to the Pacific Coast, Ralph Waldo Emerson traveled in
a horse-drawn carriage through Boston. Emerson wrote later in his
journal:

> A seaman in the coach told the story of an old sperm whale which he
> called a white whale which was known for many years by the whale-
> men as Old Tom & who rushed upon the boats which attacked him &
> crushed the boats to small chips in his jaws, the men generally escaping
> by jumping overboard & being picked up. A vessel was fitted out at New
> Bedford, he said, to take him. And he was finally taken somewhere off
> Payta [H]ead by the *winslow* or the *essex*. He gave a fine account of a

storm which I heard imperfectly. Only "the whole ocean was all feather white."[1]

Emerson was unaware at the time that he was recording the inspiration for what would be the most famous animal character in American literature.

THE TWO MAIN HISTORICAL YARNS THAT INSPIRED *MOBY-DICK*

As Melville would do later in *Moby-Dick*, the sailor in the coach with Emerson seems to have spliced two yarns together. The whaleship *Essex* was the ship in 1820 that was famously rammed twice by a sperm whale and sunk in the eastern equatorial Pacific. The ship's first mate, Owen Chase, one of the handful of survivors, then became the captain of the whaleship *Winslow*. This ship was his first command, trusted to him directly after the horrors of the *Essex* disaster. Chase sailed the *Winslow* back to the Pacific Ocean. No record suggests, however, that the *Essex* was sunk by a *white* sperm whale; or that Chase aboard the *Winslow* was Ahabically seeking the one individual sperm whale that had sunk the *Essex*; or that Chase was looking for a white whale; or that he ever caught one. Chase's voyage to the Pacific and back aboard the *Winslow* was a profitable and relatively uneventful one, focusing on right whales.

The second strand of the story from Emerson's sailor is about a white whale supposedly named Old Tom. As Ishmael explains in "The Affidavit," he'd heard of a few accounts of named sperm whales that turned and clashed with whalemen in their small boats. Five years after Emerson's journal entry, author and adventurer Jeremiah Reynolds published "Mocha Dick; Or the White Whale of the Pacific," a story about an ornery sperm whale that was an albino or at least all scarred-up to whiteness. Reynolds published this piece, which might have been partly or entirely fictional, in the *Knickerbocker*, a variety magazine. In this story, Mocha Dick is a seventy-foot-long bull sperm whale. Reynolds wrote: "From the effect of age, or more probably from a freak of nature, as exhibited in the case of the Ethiopian Albino, a singular consequence had resulted—*he was white as wool!*" The author of an unverified article in 1892 in the *Chicago Daily Tribune* claimed that among five known "wicked whales," Mocha Dick was real, and that this whale was responsible for the loss of over a dozen boats, the damage of three whaleships, and the death of no less than thirty men during nearly two decades. Although in Reynolds' narrative Mocha Dick is eventually killed off the coast of Chile, the *Daily Tribune* author claims that in 1859 a Swedish whaleship off the coast of Brazil caught Mocha Dick (who was not white, but had a long, white scar), now old, battered, and half-blind.[2]

"Mocha," by the way, is an island off Chile near a popular whaling ground at the time. "Dick" is the now antiquated nickname for Richard. No one knows for certain why Melville chose "Moby." Theories abound.[3]

Before writing *Moby-Dick*, Melville likely read the "Mocha Dick" narrative or at least heard a version in a whaleship's forecastle. After perhaps meeting Chase's son at sea, Melville read Owen Chase's published account of the loss of the *Essex*, and then before completing *Moby-Dick* got his own copy. In *Moby-Dick* Ishmael cites Owen Chase in the "Extracts," then retells the story of the *Essex* among a collection of events "practically or reliably known to me as a whaleman" in "The Affidavit." Ishmael aims in "The Affidavit" to prove that whalemen could indeed identify individual whales by their physical traits and that sailors had seen and killed whales that were white. In a list of four named and "famous whales" of "ocean-wide renown," Ishmael tells of a "New Zealand Tom," who was "the terror of all cruisers," and of "Timor Jack," who was "scarred like an iceberg," both named individual whales that Melville had read about in Dr. Bennett's natural history. New Zealand Tom was, Bennett wrote, "conspicuously distinguished by a white hump."[4]

NATURAL THEOLOGY

You can justly imagine Emerson pretty bundled up in that coach that February day in 1834 when he first heard about Old Tom, because winters were colder and icier in the 1830s in New England than they are today, by some 4°F on average in Boston. A couple months earlier he had given a talk titled "The Uses of Natural History" to the Natural History Society in Boston, in which he explained the life-altering impact of his recent trip to Europe. Emerson had walked through the vast Cabinet of Natural History at Jardin des Plantes in Paris. Emerson explained in his "The Uses of Natural History" lecture that beyond the obvious economic benefit to humankind, the study of God's natural world was a way to improve your mind and body. Learning about natural science, he said, also provided "the delight which springs from the contemplation" of the truths and designs in nature.[5]

Emerson's philosophies, as stated in that lecture and in his transcendentalist manifesto *Nature* (1836), we now broadly define as part of "natural theology," the attempt to weave the teachings of the Bible into developments in biological and geological knowledge. Before the best-selling and scandalous publication of books about the transmutation of species such as *The Vestiges of the Natural History of Creation* (1844) and then later Darwin's *Origin of Species*, there was rarely any public conflict or debate between religious faith and the careful inquiry into the wonders of the natural world. For Emerson, Louis Agassiz, and others, going back at least to Isaac Newton in the early 1700s, exploration into the natural systems designed by God was a way to celebrate the Creator's genius and His benevolence. There was no culture war between science and Christianity in Melville's mid-nineteenth century. Before Darwin shook up everything, Emerson and Agassiz were the leading voices of natural theology in America.[6]

Melville wrote *Moby-Dick* in 1851 from his own version of natural theology, applied to the sea. Ishmael sails right along with inquiries into everything about the sperm whale in order to celebrate the magnificent awe-inspiring creations of God. It is Ahab, the fallen angel, who rages against God and His creations.

Melville scholar Jennifer Baker once wrote that Melville's scientific investigations were a "precondition for his experience of awe, astonishment, and marvel," resulting in Ishmael's aesthetic efforts to move his readers with accuracy. We know that Melville read at least part of

Darwin's *Voyage of the Beagle*. He owned a copy while writing *Moby-Dick*, and he cited the book in the "Extracts." Careful scientific curiosity was in some of Melville's other readings, too, such as the narratives of Frederick Bennett and William Scoresby Jr. As the *Pequod* sails down into the South Atlantic, Ishmael begins to mix his empirical vigor with transcendental observation in order to revere the beauty and terror of God's ocean world, just as Agassiz did for glaciers, combining these approaches to carefully and reverentially describe, as Ishmael dubbed them in "The Affidavit," the ocean's "honest wonders."[7]

In other words, Melville wrote as accurately as possible about marine biology, oceanography, and seamanship, then he whipped it all into fictional life to get that sap flowing with soul and spirit and meaning—but not so far (most of the time) that the facts no longer rang true to the ocean world as he knew it. For example, in one of his chapter pairs, Ishmael yarns with accuracy of the large bull sperm whale's whitish skin color, then in "The Whiteness of the Whale," he explores every facet of the hue's symbolism and broader meanings.

In 1851 you could have your science and your poetry, too. Even your religion. Ishmael certainly does. Ahab's the one that's tied all up in knots about it.

OF WHITE-ISH SKIN AND BLUBBER

Although Ishmael ends "The Whiteness of the Whale," by referring to Moby Dick as the "Albino whale," his Moby Dick is not an albino in the way we define this term today. Ishmael does not describe the animal as completely white, nor does he ever mention his eyes being pink or red. Ishmael explains the coloration in the chapter "Moby Dick":

> For, it was not so much his uncommon bulk that so much distinguished him from other sperm whales, but, as was elsewhere thrown out—a peculiar snow-white wrinkled forehead, and a high, pyramidical white hump ... The rest of his body was so streaked, and spotted, and marbled with the same shrouded hue, that, in the end, he had gained his distinc-

tive appellation of the White Whale; a name, indeed, literally justified by his vivid aspect, when seen gliding at high noon through a dark blue sea, leaving a milky-way wake of creamy foam, all spangled with golden gleamings.[8]

Albinism is a recessive gene, meaning it's a genetic trait mutation that both parents have to pass along. Albinism has variations, too. An albino ape, fur seal, whale, or human might have no pigment at all, revealed by pink eyes. An individual can be "leucistic," with just enough pigment to still have some eye color. Or a single animal can have "piebaldism," with pigment missing in just some areas of its body. The whale character Melville created seems to be only partially white and marbled because of a lack of skin pigment, but also perhaps more from a long life of survival as a male sperm whale.[9]

In "The Blanket" Ishmael begins the biological dissection of the sperm whale. He explains that the skin of the whale is multilayered, with a thick blubber layer underneath a thin, nearly transparent outer layer. He explains that older, bull whales have more scratches on their hides, "probably made by hostile contact with other whales," which accumulate into a "veritable engraving" in the layer beneath the transparent one. The blubber layer, he says, has enormous insulating qualities in both warm and cold weather. Nearly all of this has proved true. But we know far more today about skin and blubber.[10]

Surgeon Beale's first chapter of his *The Natural History of the Sperm Whale* was titled "External form and peculiarities of the Sperm Whale." Melville underlined this section of Beale's: "The skin of the sperm whale, as of all other cetaceous animals, is without scales, smooth, but occasionally, in old whales, wrinkled, and frequently marked on the sides by linear impressions, appearing as if rubbed against some angular body."[11]

I've watched current marine mammal researchers argue over beers as to whether the sperm whale's skin is gray or brown. The published experts, like Hal Whitehead, lean toward gray. Bennett said "dull-black." Beale wrote that sperm whales are dark, even black, then fading into silvery near the belly, adding: "In different individuals there is, however, considerable variety of shade, and some are even piebald. Old 'bulls,' as full-grown males are called by whalers, have generally a portion of grey on the nose immediately above the fore-part of the upper-jaw, and they are then said to be 'grey-headed.'" Larger sperm males might tend to have pale "head-whorls" more often, but smaller males and females can have those, too. In the early 1970s, a Russian biologist named Alfred Berzin, who had studied aboard industrial whaling ships, found that among other color variations, 18 percent of all captured males in the Sea of Japan had bodies that were "slightly whitish." (See fig. 12.) Melville would have seen, too, that sperm whales also often have a white gape around their lips, inside their mouths, and occasional white patches on their bellies. Sexually mature females can have white and callused dorsal fins. And sperm whales of either gender can even appear white when they're sloughing their skin.[12] (See plate 3.)

Certainly not all older sperm whales are partly or mostly white, but Melville's creation of an old, isolated, mostly white, bull sperm whale is more probable than it is fantastical.

It is difficult to calculate the odds of a white or mostly white sperm whale, however, or really any other pigment anomalies in any wild ocean species. Though we can do little at this point to corroborate the facts about the alleged Mocha Dick, we do have more recent records of the occasional white sperm whale, as well as white individuals of about twenty other whale species. Amos Smalley, a Native American whaleman from Martha's Vineyard, killed a white sperm whale in the South Atlantic aboard the New Bedford whaleship *Platina* in 1902. In the 1950s and '60s Russian and Japanese whalemen captured six white sperm whales in the far North Pacific and Antarctic waters, both males and females, some with regular eyes, others with pink eyes. In 1961 a Scottish whalemen off Antarctica photographed a nearly sixty-foot white male sperm whale that they'd killed and winched on deck. In the late 1990s the author and adventurer Tim Severin recorded multiple accounts of a white sperm whale witnessed by Indonesian islanders, although he did not see it for himself. In 1995 off the Azores, Flip Nick-

Copyright 1903
N S Hutchinson + Co.

4

FIG. 12. Cutting into a sperm whale with spades from the cutting stage aboard the whaleship *California*, 1903. Note the white lower jaw and white gape, and the scratches on the sperm whale's head, likely from squid, killer whales, or other sperm whales.

lin photographed an entirely white cub sperm whale, which was perhaps fathered by a large white bull sperm whale filmed in the area the year before. Nicklin thought the calf was a female. A year later Hiroya Minakuchi photographed an adult white sperm whale in the same region, which was almost certainly the same animal full grown. This white sperm whale has since been photographed off the Azores as recently as 2016 (see plates 5 and 6).[13]

In "The Blanket" Ishmael puts forth a lighthearted philosophical quandary as to what exactly is the skin—the thin surface layer alone, or does the thick blubber layer also count as part of the skin, too? Ishmael describes a bookmark he owns that's made from the hardened "isinglass" outer layer of skin of the sperm whale. The curator of maritime history at the New Bedford Whaling Museum, Michael Dyer, a longtime colleague of mine whose hair has turned a brilliant white over the two decades I've known him, once pulled out for me a few pieces of whale skin that had been brought home by early mariners. The pieces of skin varied in shade, from black to light brown. They felt and looked like leaves without the veins, or maybe more like dried seaweed. One was all wrinkled. They had no smell. They were certainly stiff enough for bookmarks and indeed "hard and brittle," as Ishmael says. The skin was translucent when put up to the light, but not "transparent" as Ishmael and other observers described. In 1840 Dr. Bennett explained: "The *epidermis*, or scarf-skin, is exceedingly delicate; being no thicker than the membrane known as 'gold-beaters'-skin.' It is transparent, of a pale-brown colour, and, after the death of the whale, is readily and easily detached from the body." The pieces at the museum are too opaque to be something that you could read through, but, indeed, as Bennett described, fresh skin off a whale can be nearly clear and you can indeed read through it. Melville could not resist the metaphor.[14] (See fig. 13.)

Later in "The Battering Ram," in order to elevate the realism of a sperm whale knocking a hole into the hull of a wooden ship, Ishmael explains how tough the skin of the sperm whale is at its forehead: harpoons simply bounce off this part of the whale. The skin here, the subpapillary layer of the head, is indeed nearly as thick and dense as that

FIG. 13. Sperm whale skin in a petri dish "laid upon the printed page" of "The Blanket."

of the tail. The tough skin on the head might help the whale to reduce water resistance while diving.[15]

Despite regularly shedding their skin, presumably in part for discouraging parasites and fouling organisms, sperm whales do, as Ishmael explains in "The Blanket," retain wounds for their whole life. Melville and the naturalists at the time did not connect the scratches on the skin of sperm whales to the suckers and talons of large squid, but whale biologists today know that these injuries make up a large part of the "hieroglyphics" of scars on sperm whales. Sperm whales also have scratches, *rake marks*, which the whalemen and naturalists did recognize as likely from the teeth of other sperm whales, especially male competitors. Several other species of whales, notably the Risso's dolphins, are known for similar accumulated marks from encounters with their own kind. Biologists today have found that most, if not all, species of cetaceans, from porpoises to blue whales, can also have scars from sharks and killer whales. Just as in humans, as across Ahab's face, scar tissue that heals over a wound on whales remains white.

Beneath the outer layer of the whale is the whale's blubber. (See fig. 14.) Melville understood this to be for thermoregulatory purposes. "It is by reason of this cosy blanketing of his body," Ishmael says, "that the whale is enabled to keep himself comfortable in all weathers, in all

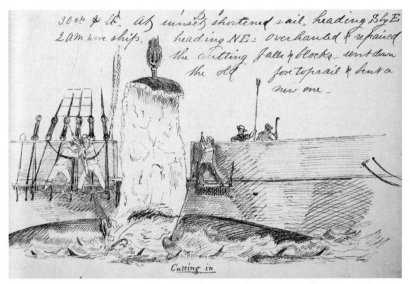

FIG. 14. Journal illustration of cutting in a sperm whale by Robert Weir aboard the *Clara Bell*, 1857. Note the whalemen stabbing at the sharks.

seas, times, and tides." Ishmael's depiction is a bit oversimplified to how we understand blubber today. The baleen whales tend to have a thicker blubber layer than the toothed whales, but blubber thickness is not the only element that aids a whale's thermoregulation. The thickness of the blubber layer also changes within an animal's annual cycle due to migration and ocean temperatures, since whales use their blubber as a reservoir for food and water depending on their energetic needs and prey availability. These ecological details were unknown in Melville's time. Although whalemen knew that, say, a female sperm whale's blubber layer could measure more than four and a half inches thick, they did not know that this whale can survive for three months on this fat. Blubber certainly insulates the animal, but it does not, as Ishmael implies, keep cold air *in* exactly. Whales, in fact, can overheat because of their blubber. Sperm whales pump a great deal of blood across their thin flukes and pectoral fins in order to cool their bodies in warmer waters or during times of exertion. The most common cause of death of a stranded whale is actually heat exhaustion. Once out of the water they cannot dissipate their body heat.[16]

Most of Ishmael's expository chapters on natural history or whale anatomy in *Moby-Dick* move far beyond the descriptive information of Beale and Bennett, and more toward the deep strivings of Emerson. As scholar Jennifer Baker explored, Ishmael often ends with a consideration of how we might assess our own lives in relation to some "honest wonder" that he's learned from one of God's ocean creatures. Emerson declared in his lecture that "the whole of Nature is a metaphor or image of the human Mind." Ahab echoes this concept in a soliloquy delivered to the sperm whale's head: "O Nature, and O soul of man! how far beyond all utterance are your linked analogies! not the smallest atom stirs or lives in matter, but has its cunning duplicate in mind."[17]

So, in turn, Ishmael ends "The Blanket," a biological chapter on the whale's skin, with what we humans can learn from the whale's blubber layer by keeping our own steady inner temperature: "Oh, man! admire and model thyself after the whale!"[18] As Ishmael begins the literal and metaphysical dissection of the whale, this is also his first suggestion in the novel that his analysis of whales might shake the very chain of being: perhaps a whale is even of a higher order than humans?

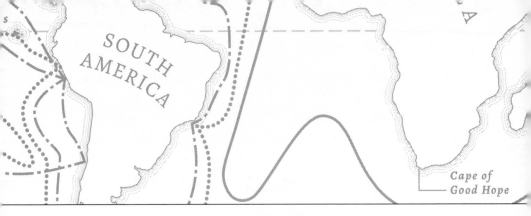

Ch. 5

WHALE MIGRATION

For with the charts of all four oceans before him, Ahab was threading a maze
of currents and eddies, with a view to the more certain accomplishment of that
monomaniac thought of his soul.
    Ishmael, "The Chart"

Mystic Seaport, the museum that maintains the *Charles W. Morgan*,
holds also a particular paper chart of the South Pacific that is a rare
treasure for an understanding of *Moby-Dick*. This chart, a map for navigation, is over 170 years old, nearly six feet wide, and it smells like an
attic. Folded hard in the middle, likely untouched for decades, the chart
was damaged in storage, so that on each end it has a broad brown stripe
the width of a book. The chart is sprinkled with water stains and foxing.
It's torn in places, ripped in others, and has been patched and repaired
at various stages. Partly obscured by more recent repair cloth is a note
at the bottom that it was printed by J. W. Norie in 1825 in London. But
this is a revised edition, since, for example, beside four neat little dots
near the equator it says in printed script: "Islands, seen 1832."[1]

In addition to the depths and coastlines, this Norie chart of the
South Pacific has information and misinformation to occupy several
afternoons of reading, especially with its antiquated names and the car-
tographer's candor. The islands of the Pacific Ocean were the last bits of

habitable land to be charted by Western mariners. On this chart, much of the southern coast of Papua New Guinea is simply left blank. Off Ecuador, lovely printed script with long serifs cautions: "The Galapagos Islands are said by Captains Hall, Krusenstern, and others, to lie from 14 to 30 miles more to the eastward, than they are placed in this Chart."

The curators at Mystic Seaport know very little about this copy of the chart. It was an accidental discovery by the director of the library who pulled it randomly out of a pile one morning. Mystic Seaport has another that's exactly the same, and several copies of this exact South Pacific chart are surely preserved in a few other collections around the world. What's special about this one is that it still has the working track lines of the courses of at least two different whaling voyages of the whaleship *Commodore Morris*, a vessel that was the same size as the *Acushnet* and the *Charles W. Morgan* and was built and launched for a voyage to the South Pacific in the same year of 1841. The logbooks from these two different four-year voyages of the *Commodore Morris* correspond to the track lines on this chart. One of the logbooks was kept by Captain Lewis H. Lawrence and the other was by an unnamed watch officer. Both are preserved at the Falmouth Historical Society near Woods Hole. It was common practice at the time on a chart of this scale to mark the ship's position every twenty-four hours or so, and then connect these dots with straight lines to mark where the ship had traveled. The ship's officer measured the distance with a set of dividers. The puncture marks from the steel points of the dividers are still visible on the paper. In pencil and in pen are a variety of notes, corrections, dots, dotted lines, straight lines, sketched lines, circles, boxes, recommendations, and even a couple of dates. The captains and mates occasionally printed "correct" or "incorrect" beside little islands and rocks, or they crossed out islands, or added their own.[2]

This aged scroll, with its beauty and detail and opportunity for interpretation and history, feels indeed, as Melville would have it in "The Chart," evocative of the wrinkled, complex, and often unreadable old brow of a grizzled sperm whale or of mad Ahab himself. Still more significantly, beyond the track lines and mariner's notes, the captains drew several clumps of whale flukes, representing whales sighted. Appropriately, the *Commodore Morris* sailed out of Woods Hole, the region at the base of Cape Cod, which beginning in the 1870s became the epicenter of marine biology, fisheries research, and oceanography in America. This began with a coastal field station nearby, started by Louis Agassiz. Captain Lawrence, who commanded the 1849–1853 voyage of the *Commodore Morris*, flogged his men, presided over a range of violence on board, and watched more than a dozen of his sailors desert at different times on various islands. But when it came to tracking whales, he was a model captain. Lawrence was especially thoughtful and quantitative as to where and when he saw whales, recording them in his logbook and on the chart. He also kept an additional section in his log that he titled "When we saw Whales Where we Saw Whales Etc. Etc." The penciled dots and fluke marks that Captain Lawrence drew on this chart are centered around a group of islets to the northeast of New Zealand, where on a few days in 1851 they caught whales in sight of the *Charles W. Morgan*. Lawrence also drew dense concentrations of flukes along the equator within what is now the vast island nation of Kiribati (pronounced Kiri-bas), whose islands are scattered like seeds across thousands of miles to the east and west of where the equator intersects the dateline of 180° longitude. Captain Lawrence's largest patch of over two dozen drawn flukes was penciled directly around the equator. He drew these tightly together and among a maze of track lines that go in so many different directions and among so many position dots that it seems as if he must be joking. It is a bird's nest of lines in and across and around the flukes. (See fig. 15.)

WHERE DOES *MOBY-DICK* END?

The clump of hand-drawn flukes in the equatorial Pacific on this old treasure of a chart is in one of the prime regions where American whalers hunted sperm whales in the mid-nineteenth century, along a vast whaling ground that Ishmael calls the "Season-on-the-Line." Ishmael never

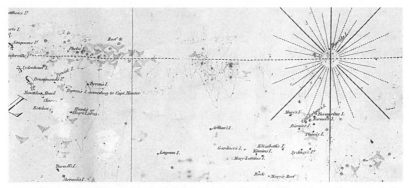

FIG. 15. Detail of a Norie Chart of the South Pacific, including the whale fluke marking and cruise track of the voyages of the whaleship *Commodore Morris* in 1852. The compass rose, just above the Phoenix Islands, intersects the equator and the 170° West meridian.

specifies, however, the exact longitude along the equator where Ahab has his final reunion with the White Whale.

It's reasonable to imagine Ahab meeting Moby Dick in the eastern equatorial Pacific, somewhere around 120° west. In the true story of the loss of the whaleship *Essex*, the bull sperm whale smashed the hull and sank this ship forty or so miles south of the equator at about 119° west in November 1820. Melville had sailed in this region himself, and to some whalemen this longitude was considered the beginning of the "On-the-Line" grounds, to the west of what was known as the "Off-shore Grounds." Toward the end of *Moby-Dick*, Starbuck says their intended direction is to the eastward from the Sea of Japan. Later Ahab gives the order to sail east-southeast for a time.[3]

Yet Melville knew that the Pacific is a very big place. It took months to sail across the entirety of the Pacific at this widest transect. If you wanted to cruise along the equator, a captain would've sailed from off Japan via the most direct route and then likely just proceeded eastward along The Line. Although Ishmael's *exact* "technical phrase" of "season-on-the-line" is hard to find anywhere in the period literature, for the captains of the *Commodore Morris* the "on-the-line grounds" were much farther beyond the sight of the sinking of the *Essex*, farther to west in the central and western equatorial Pacific within modern-day Kiribati.[4]

So for the geography of the end of Ahab and the *Pequod*, you could argue for really anywhere along the equator in the Pacific. But I imagine the meeting with the White Whale near the region of Kiribati on the basis of the period information that Melville read, the historic abundance of sperm whales in the mid-nineteenth century, the *Pequod*'s approach from the Sea of Japan, and the amount of time that transpires in the novel.

AHAB'S NEEDLE IN THE HAYSTACK

As the *Pequod* sails outbound in the North Atlantic, Ishmael buttresses his yarn by helping his reader understand that a single, violent mostly-white sperm whale could and did exist, and that sperm whales—white, named, or otherwise—could and did destroy whalemen, whaleboats, and whaleships. In "The Chart," Ishmael seeks to make certain that the reader can accept arguably the least believable plot element of the novel, which happens to be the essential driver to the suspense of his story: that one hell-bent man aboard one ship at his command can track down one single animal who was rarely at the surface and who could range throughout all of the oceans on Earth.[5]

In "The Chart," Ishmael explains and defends this possibility. The scene begins with Ahab in his cabin brooding over these scrolls:

> Now, to any one not fully acquainted with the ways of the leviathans, it might seem an absurdly hopeless task thus to seek out one solitary creature in the unhooped oceans of this planet. But not so did it seem to Ahab, who knew the sets of all tides and currents; and thereby calculating the driftings of the sperm whale's food; and, also, calling to mind the regular, ascertained seasons for hunting him in particular latitudes; could arrive at reasonable surmises, almost approaching to certainties, concerning the timeliest day to be upon this or that ground in search of his prey ... Were the logs for one voyage of the entire whale fleet carefully collated, then the migrations of the sperm whale would be found to

correspond in invariability to those of the herring-shoals or the flights of swallows.[6]

The *Pequod* leaves Nantucket on an icy Christmas Day, which has both poetic resonance as well as autobiographical connections for Melville. For American whalemen, beginning a voyage in the wintertime was as common as any other time of the year. When young Melville signed aboard the *Acushnet* in New Bedford on Christmas Day of 1840, eight other whaleships had sailed from this port for the Pacific Ocean in just the previous few weeks. The master of the *Acushnet*, Captain Valentine Pease Jr., set sail with his crew on the third of January. He presumably had the same intention as the others: to arrive off Cape Horn before the southern winter and when the opposing currents were reportedly less strong. This still gave his ship a few months to cruise south and perhaps stop briefly along the way to provision and pick up more crew in the Azores or the Cape Verde Islands, then do some whaling in the Atlantic before heading for the Pacific Ocean. We know that the crew of the *Acushnet* caught at least a few sperm whales in the Atlantic since Captain Pease shipped home from Rio de Janeiro some casks of sperm whale oil, but even by the 1820s the populations of sperm and right whales in the Atlantic and even off the southwest coast of South America had been so depleted that despite the distance and expense, the far offshore South Pacific provided the most potential for a profitable voyage.[7]

In *Moby-Dick* the *Pequod* crosses the Gulf Stream and sails southeasterly away from the New England winter. Instead of rounding Cape Horn, Ahab sails around the Cape of Good Hope, the southern tip of Africa, and then aims for the waters off Japan by way of the Indian Ocean and Indonesia. Prevailing winds in the far southern latitudes, below the capes, blow from west to east. The *Pequod* takes the path of least resistance. Part of the reason that Cape Horn was so dreaded by mariners was because to try to "double it," to attempt to sail around this far southern peninsula from the Atlantic, meant sailing into a freezing, contrary wind and current. Though often a necessity for most Pacific-bound routes before the completion of trans-isthmus railroad service in the 1850s and then the Panama Canal in 1914, Cape Horn represented some of the worst and most unpredictable waters for any sailor. A trip around Cape Horn was the mariner's badge of honor earned from hard labor amidst nature's harshest extremes. Melville once commented smugly in the margin of an essay by Emerson that spoke of how a sailor feels no terror in a storm: "To one who has weathered Cape Horne as a common sailor what stuff all this is."[8]

The owners of the *Pequod* anticipate the voyage of their whaleship to be about three years from Nantucket: one to get there, a year to catch whales, and a year or so to get back. Peleg and Bildad expect their ship to sail "beyond both stormy Capes." They've planned for Ahab to navigate around the world. Steering via Good Hope was not the most common direction, but American whalemen certainly sailed this way regularly, about 25 percent of the time. Traveling up toward the Japan Grounds instead of heading toward New Zealand below Australia was more rare, but not unprecedented.[9]

Ahab intends to meet the White Whale in the Pacific by the following winter, the season-on-the-line. Ishmael says that the expected time for encountering Moby Dick, this prime sperm whaling season, begins in January. The *Pequod* has neatly one year to get there.

Ishmael explains that Ahab chooses the eastbound route around Good Hope to match where Moby Dick had previously been sighted at other times, such as near the Seychelles in the Indian Ocean. Perhaps Melville chose to send the *Pequod* this way to distance the story from "Mocha Dick" and to create his first piece of fiction that had no direct connections to his own travels. He had just written about Cape Horn in his most recent novel *White-Jacket*—as had seemingly every other sailor-writer who had described this rite-of-passage in a literary arms race of stormier and icier waves. Going via the Cape of Good Hope also served the reader some variety and fresh foreign seas, while providing the additional poetry of Ahab sailing along with the rotation of the Earth as he chases the rising sun and the White Whale.

What Ishmael explains in "The Chart"—about navigation, geography, and whale migration—represents well the knowledge and practices of whalemen in the mid-nineteenth century. Although current research confirms some of their logic and experience, other aspects remain unknown or unproven, especially regarding the migrations of sperm whales. We can safely put in the fiction department, for example, Ishmael's notion that sperm whales migrate in consistent, predictable highways, "veins," that are the width of the sailor's view at the masthead—but this was an idea he learned from his fish documents, not one that he entirely fabricated for the sake of the novel.[10]

AMERICA'S FIRST OCEANOGRAPHER

In 1839, Lieutenant Matthew Fontaine Maury traveled from his family's home in Tennessee on his way to board a naval vessel in New York. In Ohio he gave up his seat to a woman boarding, choosing to sit on top of the coach with the driver. On the way up the Appalachians in the middle of the night, the vehicle rolled over. Maury was left crippled in his right leg for life. The recovery was grueling. It forever altered his career—as well as the history of oceanography. (Maury did not begin a deranged hunt for revenge against the particular horses involved or the stagecoach driver, but he did get a financial settlement. No peg leg was necessary.) Maury had enjoyed a promising career at sea before this accident, which began with a circumnavigation as a midshipman aboard the USS *Vincennes*. Next Maury was the sailing master on a naval voyage rounding Cape Horn in 1831. Maury returned to publish a study on the best route and season to weather Cape Horn. Now limited to shoreside service after his accident, he was assigned to be the superintendent of what would become the US Naval Observatory and Hydrographical Office in Washington, DC. He and his staff oversaw the Navy's navigational equipment, chronometers, and charts.[11]

Maury is in Ishmael's footnote to "The Chart." The note begins: "Since the above was written, the statement [about whales having predictable seasonal migrations like fish and birds] is happily borne out by an official circular, issued by Lieutenant Maury, of the National Observatory, Washington, April 16th, 1851. By that circular, it appears that precisely such a chart is in course of completion; and portions of it are presented in the circular." Ishmael quotes Maury directly about the details of this map.[12]

Maury expanded the role of his department to include the creation of deep ocean charts focused on winds and currents—as well as whale distribution. Maury and his staff did all this by collecting mariners' logbooks or the extracted data from their voyages, known as the "abstract logs." The captains, in exchange for sharing, got free copies of Maury's charts. For example, a senior captain in New Bedford took Captain Pease's log of the *Acushnet* and wrote out an abstract log to send to Maury (see fig. 16). So it's a lovely and accurate detail in "The Chart" when Ishmael says that "at intervals, [Ahab] would refer to piles of old log-books beside him, wherein were set down the seasons and places in which, on various former voyages of various ships, sperm whales had been captured or seen."[13]

Maury took over this government agency at a time when the American merchant marine was rapidly expanding. The number of vessels and men involved increased by four hundred percent between 1815 and 1860. Whaleships were an enormous part of this growth, both in tonnage and men. At its height in the 1840s, the American whaling fleet represented about 735 active ships. Meanwhile, the army and navy had surplus revenue, which filtered down to scientific activities, beginning the connection between American scientists and political and financial forces. This in turn led to the government institutions that would evolve into the United States Geological Survey (USGS) and the National Oceanographic and Atmospheric Administration (NOAA).[14]

With institutional and financial backing, between 1847 and 1860, Maury's office produced dozens of charts within six series: track charts, trade winds charts, pilot charts, thermal charts, storm and rain charts, and the final series, composed of six whale charts: two global maps of right and sperm whale distribution and four regional graph-style charts, focused on time of year for these two types of whales. (See fig. 17.) Melville almost certainly didn't see any of these charts for himself because of the timing of publication. He likely read the announcement, the "circular," in the newspaper, such as in the *New York Herald*. The graph-style whale chart that Ishmael mentions in his note to "The Chart" and that Melville read about in the announcement looks like a giant piece of graph paper or a seismograph.[15]

According to Maury's charts, sperm whales were far more commonly reported along the equator in warmer waters, while right whales were more common at more southern and more coastal regions, such as off Chile.

Limited by the number of logbooks and other hurdles, Maury's whale charts certainly had scientific shortcomings, but they were enormous early steps toward group data collection. Maury popularized and compiled information that was often held privately and tightly, just as that information on the chart of the Pacific with the drawn whale-flukes was proprietary information for the captains of the *Commodore Morris*.

The merit of Maury's research in science circles has been debated, but historian Graham Burnett wrote that Maury was "probably the single most decorated American man of learning in the nineteenth century," a scientist widely recognized and respected abroad. Maury's research stood piously, too, in the tradition of natural theology, occasionally even beyond the comfort of his scientific contemporaries. In 1855 Maury published *The Physical Geography of the Sea*, an immensely influential and popular volume. It's often considered the first American book focused entirely on what we now call oceanography. He wrote of the "grand machinery of the universe," a concept popularized by Isaac Newton, which Maury broadened to the ocean: God had created His clockwork universe, which also must apply to global meteorology and life under the sea. When considering ocean salinity, for example, Maury consulted the Bible as a factual source along with his experiments and mariners' reports. In advancing theories of ocean currents, trying to find the laws that govern their movements, Maury wrote, "They no doubt, therefore, maintain the order and preserve the harmony which characterize every department of God's handiwork."[16]

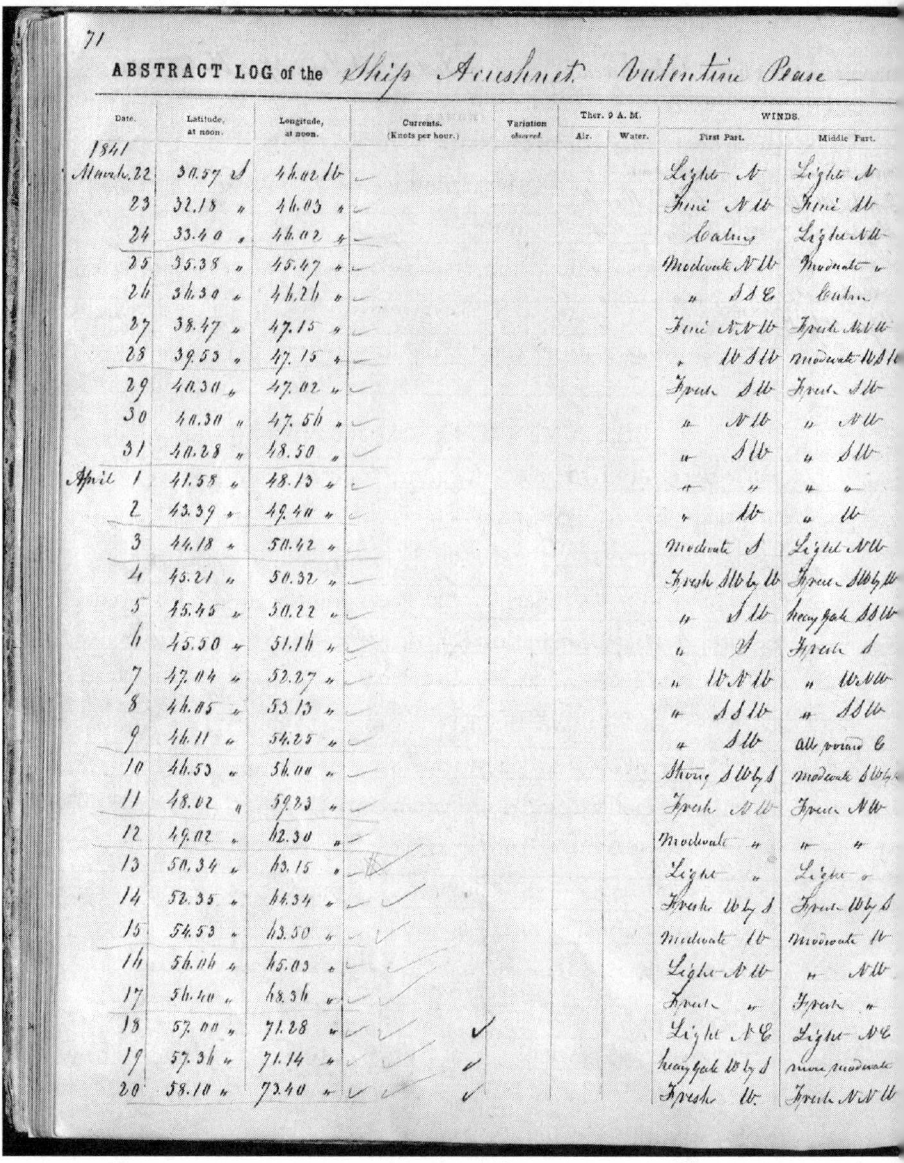

71

ABSTRACT LOG of the *Ship Acushnet. Valentine Pease*

Date.	Latitude, at noon.	Longitude, at noon.	Currents. (Knots per hour.)	Variation observed.	Ther. 9 A.M. Air.	Ther. 9 A.M. Water.	WINDS. First Part.	WINDS. Middle Part.
1841								
March 22	30.57 S	44.42 W					Light N	Light N
23	32.18 "	44.03 "					Fine NW	Fine SW
24	33.40 "	44.02 "					Calms	Light NW
25	35.38 "	45.47 "					Moderate NW	Moderate
26	36.30 "	46.26 "					" SSE	Calm
27	38.47 "	47.15 "					Fine NNW	Fresh NNW
28	39.53 "	47.15 "					" WSW	Moderate WSW
29	40.30 "	47.02 "					Fresh SW	Fresh SW
30	40.30 "	47.56 "					" NW	" NW
31	40.28 "	48.50 "					" SW	" SW
April 1	41.58 "	48.13 "					" "	" "
2	43.39 "	49.40 "					" W	" W
3	44.18 "	50.42 "					Moderate S	Light NW
4	45.21 "	50.32 "					Fresh SWbW	Fresh SWbW
5	45.45 "	50.22 "					" SW	heavy gale SSW
6	45.50 "	51.16 "					" S	Fresh S
7	47.04 "	52.27 "					" WNW	" WNW
8	46.05 "	53.13 "					" SSW	" SSW
9	46.11 "	54.25 "					" SW	all round C
10	46.53 "	56.00 "					Strong SWbS	moderate SWbS
11	48.02 "	59.23 "					Fresh NW	Fresh NW
12	49.02 "	62.30 "					Moderate "	" "
13	50.34 "	63.15 "					Light "	Light "
14	52.35 "	64.34 "					Fresh WbS	Fresh WbS
15	54.53 "	63.50 "					moderate W	moderate W
16	56.06 "	65.03 "					Light NW	" NW
17	56.40 "	68.36 "					Fresh "	Fresh "
18	57.00 "	71.28 "					Light NE	Light NE
19	57.36 "	71.14 "					heavy gale WbS	more moderate
20	58.10 "	72.40 "					Fresh W	Fresh NNW

FIG. 16. Two pages from the abstract log of Melville's first whaleship, the *Acushnet*, as it rounds Cape Horn into the Pacific (1841).

WINDS. Latter Part.	REMARKS.
Fresh N	Pleasant thro the day
Light W	squally " " "
" NW	
Fresh "	rugged
Calm	
moderate WSW	squally
" "	Fine weather thro the day. Latter heavy squall
Fresh W	" " " "
" WSW	squally
moderate SW	Bad sea.
Fresh "	Fine weather thro the day.
" WSW	
" NW	Latter Squally saw Sperm Whales.
" SW by W	squally
" gale SSW	Hail Squall Barometer fell below 29½.
Fresh NW	heavy squalls of rain " stands 28.50
" SSE	" " " "
toward SW	" "
strong SE	squally with rain
fresh W.NW	Fine weather thro the day
variable NW	" " " "
" "	" " " "
moderate W	" " " "
moderate W by S	" " " "
" W	thick heavy weather saw Staten Land bearg S. dist. 2 leagues.
" NW	heavy squall
more "	Diego Ramirez bearg N.N.E. dist. 3 leagues.
moderate N.N.W.	heavy squall of wind & rain
W	
light SE.	squalls of rain - thick weather

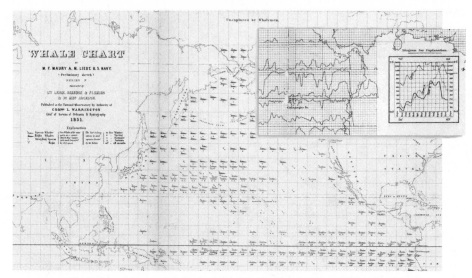

FIG. 17. Detail of the world map of whaling grounds by Matthew Fontaine Maury and his colleagues, with the sightings of sperm whales (lighter gray) and right whales (darker gray) as reported by whalemen (1851). Maury created four other regional graph-style maps, which Ishmael writes of in "The Chart," such as F:3 (detail inset). The graphs quantify past records of sperm or right whales by location and month.

It was Lieutenant Maury who appropriated the funds to commission the *Taney*, that schooner in 1850 that was the first American ship with the primary mission to sample the ocean itself, including the acquisition of deep-sea soundings. In that same notice Ishmael cites in *Moby-Dick* about the whale charts, Maury wrote both on the potential split between right whale species and a recommendation to whaleship owners that they equip all their ships with twine and scrap iron to allow their men to record depth each day. Maury wrote: "I am sure that the whalemen, from the great philosophical interest which many of them manifest with regard to my researches, would in calms get deep sea soundings for me."[17]

THE COMMANDER OF AMERICA'S FIRST VOYAGE OF DISCOVERY

Yet Maury was not Melville's most significant reference for "The Chart." The footnote seems to genuinely have been a late addition. For the bulk of the material, Melville had been reading the narrative reports of Charles Wilkes, another naval officer, who was the commander of the foundational US Exploring Expedition from 1838 to 1842.

By nearly every account, Charles Wilkes was, at best, an enormously capable, prolific, pompous prick. As Wilkes and his superiors planned the expedition, even Lieutenant Maury, who was not without his own obstinacies, pulled out of any involvement with the endeavor. Wilkes had been the superintendent of the chart and instrument department before him.[18]

Under Wilkes, the US Exploring Expedition, also known as the US Ex. Ex., surveyed coastlines, collected scientific and anthropological information, and represented the new United States as a growing imperial force. The four-year circumnavigation included six vessels, three hundred fifty sailors and officers, and six full-time naturalists known as the "scientifics." As would be true for most of American history, military and commercial interests provided motivations, platforms, and funding to advance sampling technologies and the collection of oceanographic and biological data. They brought home some sixty thousand natural history objects and specimens which would later form the core collection for the Smithsonian Institution in Washington—which John James Audubon had offered to direct. Wilkes claimed to be the first to prove that Antarctica was a continent.

When Wilkes returned from the expedition, he was court martialed for his abuses of power. He had an intensity of drive that has led to a theory that he was the model for Ahab. After he survived what he saw as bureaucratic badgering, Wilkes fervently published his five-volume *Narrative of the United States Exploring Expedition* in 1844, which included an atlas. Wilkes oversaw the production of some 180 charts, most of which were of the South Pacific, rendering my treasured chart of the *Commodore Morris* completely out of date.[19]

Melville purchased Wilkes's entire narrative in 1847. Wilkes's final chapter is titled "Currents and Whaling," from which Melville quoted in his "Extracts" and drew upon directly for "The Chart" and "Does

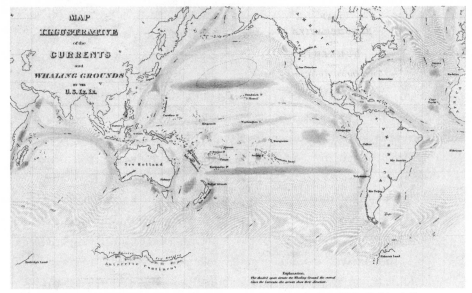

FIG. 18. Detail of "Map Illustrative of the Currents and Whaling Grounds of the U.S. Ex. Ex." (1845). The shaded areas are whaling grounds and the lines and arrows represent currents and their direction.

the Whale Diminish?" Wilkes opened "Currents and Whaling" with a two-page map with whaling grounds hatched in gray. Lines of currents point in various directions across every ocean (see fig. 18). Wilkes began:

> It may at first sight appear singular that subjects apparently so dissimilar as currents and whaling should be united to form the subject of one chapter. Before its conclusion, however, we trust to establish satisfactorily that the course of the great currents of the ocean, sweeping with them the proper food of the great cetaceous animals, determines not only the places to which they are in the habit of resorting, but the seasons at which they are to be found frequenting them.[20]

So this is where Melville got his ideas about currents. From the beginning of the voyage, Wilkes and his team tried to quantify and trace currents. Their research even included recording the direction of floating casks from a shipwrecked whaleship near Fiji. The US Ex. Ex. regularly took deep sea soundings and surface and subsurface water temperature. Wilkes mapped currents and eddies, and put forth theories on global circulation. His connections and ideas on water temperature, currents, trade winds, ocean bottom topography, and the effect of the coastlines portended future knowledge about physical oceanography, including what would later be called upwelling and downwelling. Ahab looks to currents to track sperm whale food, just as Wilkes theorized about sperm whale movements based on jellyfish or small squid being swept along by currents, although these organisms actually do not make up much of the sperm whale's diet.[21]

Today sperm whale biologists look more toward factors of high phytoplankton density, variations in underwater topography, and pro-ductive upwelling systems in order to predict sperm whale movements, since there are no data or methods even today to track the biomass of deep sea squids that make up the majority of their food.[22]

Melville cherry-picked from Wilkes what he wanted for *Moby-Dick*, more so than from Maury, but he used research from both to build the believability of Ahab's quest to find one single individual sperm whale in all the world's oceans.

THE WORLD WHALING HISTORY PROJECT

Tim Smith is one of those rare people who actually pauses and thinks before he speaks. He leans back in his chair in northern California and says: "Most of the people in my world had very little understanding about what happened before the industrial whaling of the twentieth century. It takes a couple hundred years to explain how we got to where we are today."[23]

Smith heads The World Whaling History Project, an endeavor that first began in Woods Hole. The team, co-led by cetacean expert Randall Reeves, digitized the distribution and seasonal data directly from whal-ing logbooks. They reassembled the data for sperm and right whales compiled by Maury in the 1850s, along with the data compiled in the 1920s by an American naturalist named Charles Townsend, who had included additional species, such as humpback whales. Smith and his team then added some of their own data through the Census of Marine

Life. After several years of work, Smith and his colleagues published in 2012 a brand-new set of maps that far more accurately visualize and record the locations of whale sightings from 1780 through 1920 (see plate 7).[24]

Smith's involvement in all this began in the 1970s when he was working for NOAA. He was advising the International Whaling Commission (IWC) on what was known about whale populations in order to help them make more informed management decisions. This inevitably led to questions about how many whales there used to be. Smith dove into the old logbooks and Maury's work.

He explained that, ironically, in many ways we know less about the status of whale distribution now than we did in the 1800s. American whaleships went everywhere. Today, with a few notable exceptions, large-scale industrial whaling is over. It's difficult to have any global census of whale populations in the twenty-first century. We have excellent local records from whale-watching boats and from aerial surveys in waters, say, off the coasts off South Africa, and in recent decades, scientists have attached ever-advancing lightweight tags on the animals that relay short-term diving and ecological information. But we still have few modern deep ocean records.

The migration patterns of most of the baleen whale species, such as humpbacks and gray whales, are fairly well-known today. Native Americans, surely, and then the whalemen of Melville's time built the foundations of this understanding. The movement of sperm whales is another story, however. Ishmael suggests in "Schools & Schoolmasters" that sperm whales have a predictable seasonal migration relative to temperature and food. Yet this was not known then—just as it is not known today.

"We still really don't understand the movement of sperm whales," Smith explains. In the 1800s whalemen reported very few sperm whales in the latitudes of the North Pacific and the Southern Ocean, above or below 60° north and south, but this is where twentieth-century whalers caught them. Did the whales move or are these data more about where people went to hunt them?

Nineteenth-century mariners understood that sperm whales traveled according to gender and age groups for most of the year. Whalemen could tell sex by size, since females are so much smaller. The men also observed the behaviors of family and age groupings. Captain Lawrence of the *Commodore Morris* regularly wrote in his logbook when he observed schools of "cows and calves" around the equator. Marine biologists have since confirmed the differences in the range of sexually mature male and female sperm whales, which are more segregated by distance than mates of any other mammal that has sex in the sea. Females and juvenile sperm whales spend their lives in the tropical and temperate zones. Modern experts believe the roaming of the units of female and juvenile sperm whales is likely nomadic, based on food abundance and the traditions of their social groups which were, and still are, recovering from the impacts of human hunting.[25]

How and why *male* sperm whales move around the world's oceans is even less understood. Melville was correct in imagining his old male, Moby Dick, to be solitary and capable of traveling tens of thousands of miles. Older male sperm whales tend to spend more and more time at higher, subpolar latitudes as they age, and then only occasionally migrate down toward equatorial feeding grounds and swim within a breeding area. Among these groups of females, these older males make just short visits, sometimes again and again, but often just for a matter of hours, presumably to mate. Their movements, true to Moby Dick, are highly individualistic. They do not, as Ishmael claims, act as lords of harems, an idea he learned from his fish documents and probably his own time at sea.[26]

"It's funny," Smith says, "because sperm whales are the image for protecting all whales, but they're still the ones we know the least about. And yet today we have the least interest in getting an overview."

No organizations or institutions today have the economic and political motivations to mount the kind of large studies necessary to really understand how sperm whales migrate. From the 1920s to 1930s scientists in the Southern Ocean collected data cards from Norwegian whaling captains. In 1925 British oceanographers launched a new ship named the *William Scoresby* for the express purpose of shooting tags into whales in the same way that other ecologists at the time had been developing systematic programs to band birds and mark fish. Tagging whales at that point was helpful for learning about the rorquals, but the program did not last very long.[27]

Part of the challenge is that sperm whales are especially hard to track. They dive so deeply and for so long that it's hard to count and identify them. Tagging devices are still limited even today because of the life of batteries and the pressure conditions that sperm whales subject them to. Researchers, such as those off Dominica and New Zealand, have used data loggers with suction cups, but those usually last for only a matter of hours. The technology is improving quickly, however; scientists working in the Gulf of California, for example, attached an Advanced Dive Behavior tag that remained on a sperm whale for over a month.[28]

When I ask Smith his thoughts on "The Chart" and the possibility of Ahab's quest, he pauses. He knows the novel well. He leans back in his chair. "I think the idea that Ahab was chasing a single whale and hoped to find him again was justified by stories of multiple strikes in the same animal. They knew that whales could survive a harpoon, so a given whale had a long-term identity. But the idea that you could look for that one single whale at sea—I think that is totally fiction."

The maps created by Smith's team suggest that the hunters were catching whales along the equator year-round. Not more so in January or February. So was there no specific time of year for the on-the-line grounds?

"No. There doesn't seem to have been any real season," Smith says. "That was one of the problems. There was this thing about seasonal changes, that sperm whales migrated predictably like fish or birds. We tried to do some analysis of that. What does the data actually say about seasonal distribution? In turns out there is little evidence that the sperm whales left the equator during certain seasons. These are mostly groups of females and juveniles. It was the *whalemen* that left the line during some seasons. But this just gets into the complexity of sperm whale life history. And how little we still know."

For Melville in that 1851 "Notice to Whalemen" by Maury, the lieutenant included a table that showed by equatorial region the times of year when sperm whales were cited. It shows seasonal differences, but also suggests year-round presence of sperm whales around the equator to the west of the dateline. Ahab's date to meet the White Whale in January specifically, the "season on-the-line," seems more for fiction than fact, both in his time and by what we know today.[29]

In the summer of 2017, a Spanish organization named CIRCE (Conservation, Information and Research on Cetaceans) reported that they had identified a single sperm whale returning to the Straits of Gibraltar for at least thirteen years, going back to 1998. They had photographed the distinctive fluke marks of a female sperm whale they've named "Amanita." The observers identify her by three evenly spaced scratches in a row on one fluke, below a little notch at the tail.[30]

*ATLANTIC
OCEAN*

*Cape of
Good H*

Cape Horn

Ch. 6

WIND

> For as in this world, head winds are far more prevalent than winds from astern
> (that is if you never violate the Pythagorean maxim).
> Fart joke by Ishmael, "Loomings"[1]

Melville wrote little about wind in "The Chart." His reader at the time understood this to be essential. Until quite recently, for the entire history of human passages across oceans and along coasts, the wind has been the beginning, the wind has been the end, and the wind has been everything in between. Today, the wind has some impact on modern shipping and air travel, but rarely anything significant to the general consumer or traveler. Unless you commute on a ferry to an offshore island, anything below a hurricane is usually at most an annoyance or a call to the barf bag compared to the way the wind impacted the lives of mariners and travelers in Melville's mid-nineteenth century and millennia before. Maury and Wilkes spent far more time researching, theorizing, and planning around wind than they did on whale abundance and migration.

Ahab's *Pequod* is powered entirely by wind. The ship does not have an engine. No characters in *Moby-Dick* encounter any boats or ships at sea that have an engine to propel their vessels. Before the development of commercial aircraft and underwater photography, the greatest single influence on our relationship to the ocean is likely the development of the marine engine.

Melville read of the well-publicized introduction of trans-Atlantic steam travel even before he went on his first voyage to Liverpool in 1839. Louis Agassiz, for example, arrived in Boston Harbor from across the Atlantic aboard a steamship in 1846. By 1856, Melville's third and final trip across the Atlantic, he traveled on an ocean steamer to Scotland. Although whaleships in Melville's day often got steamboat assistance to get in and out of harbors, it was not until the late nineteenth century that the technology was available for ocean-crossing cargo ships or for whaleships to have their own engines. It was rarely cost-effective or safe for these types of vessels, nor was it logistically practical to store or find the amount of coal necessary for deep-ocean meandering voyages.

The hull of a whaleship like the *Acushnet* is shaped above the water like a bathtub, so historians always assumed the ship sailed like one. Then in the summer of 2014 the sailing performance of an American wood whaleship was tested. Much to the nail-nibbling of curators and insurers and museum management—who all dreaded sending the museum's most important artifact out into harm's way—the *Charles W. Morgan* spent a summer touring southern New England for the first time under sail since 1921. (See plate 1.) It was far faster and far more maneuverable under sail than anyone expected. Yet regardless of her agility, the *Charles W. Morgan*, like most square-riggers, still could not get closer than sixty degrees to the wind, even in the most favorable conditions. In other words, if the wind was coming from due west, and that's the direction a whaleship captain wanted to go, then he'd have to zigzag roughly north-northwest and then roughly south-southwest, back and forth, to make a ship like the *Pequod* travel toward its destination.

Captain Lawrence of the *Commodore Morris* wrote of how they steered by the wind outbound on their way to the island of Flores in the Azores in September 1849:

Monday the 17th first part fresh gales from the Eastward and cloudy
weather tacked ship at 4 oclock and at 6 at sunset Flores bore about ESE
distant 30 miles middle part stood to the NE until 2 oclock Pm tacked
ship latter part headed S by W Flores and Corvo [Island] in sight to the
SSW.[2]

Lawrence wanted to sail directly to the anchorage, but with the wind
from the east he couldn't. So he steered the *Commodore Morris* as close
to the wind as possible, northeasterly and then just to the west of south,
far enough off the wind that the ship could keep up speed as it zig-
zagged back down toward its destination.

Tacking a big ship with three masts and more than a dozen potential
sails is an operation requiring a lot of people hauling and easing lines in
order, especially with sails suspended from heavy horizontal yards that
must be shifted from one angle to another with some precision.

During the New England tour in 2014, the sailors of the *Charles W.
Morgan* soon appreciated the whalemen still more as mariners, who in
historical discussions and comments by authors such as Richard Henry
Dana Jr. were usually denigrated as careless and lazy in managing their
ship. In *Moby-Dick*, Ishmael tries to defend the whalemen as sailors and
navigators in the chapter "The Advocate." Whalemen sailed a compli-
cated rig, they sailed back and forth around Cape Horn, and they often
sailed farther and to new places before any European or American explorers had even properly charted the area.[3]

In the days before engines, the inability to go in every direction affected ships in other ways, too—especially a whaleship. A whaleman
of the mid-nineteenth century, for example, would have been perfectly
happy to harpoon his prey from the safety of his deck. Larger vessels
were not fast or nimble enough or could not follow if a whale swam off
directly to windward. So at great risk to the men, they deployed small
rowboats, often equipped with their own sails, to try to get close enough
to a whale to harpoon it.

When the *Pequod* first sails from Nantucket, the ship is almost immediately hit with a contrary wind, one of so many omens. Melville

used this in his first truly dramatic chapter, "The Lee Shore," which is his definitive declaration that this story about a white whale is bound for far deeper water than the traditional yarn. A lee shore is a coastline upon which a ship is driven by wind, current, waves, or a combination of all these. A ship might not have enough speed or room to safely tack off the beach or rocks. This situation sets up Melville's metaphor—the irony of how much we might cherish a port, value the safety of land— and yet once a ship is underway and in rough weather, land is actually the terror of navigators, whether that be a harbor or an uncharted or misidentified rock in the middle of the Pacific. Melville used "The Lee Shore" to set up one of his favorite big themes of *Moby-Dick*, put well by scholar Howard Vincent, who wrote of Melville's "favorite antithesis of the sea, symbol of the half-known life, against the land, symbol of the known, the secure."[4]

Because of the limitations of a sailing ship, finding the global trade winds was essential to whalemen. Trade winds are bands of gentle, consistent breezes in different regions on Earth. With some seasonal variability, trade winds blow in the same direction with remarkable consistency. Trade winds have been well-known for centuries. Columbus sailed the trade winds and ocean currents to and from the Caribbean four times. Every sailing vessel crossing oceans looks for trade winds. Like you do in your car to get on a highway, captains under sail travel far beyond the most direct route in order to utilize these consistent winds. Ships today are not as concerned with these winds, but in order to save fuel, modern mariners still cruise on ocean currents, which are often driven by and thus coincide with these winds.

The locations and directions of some of the global trade winds were still being compiled as Melville wrote *Moby-Dick*, and the meteorological and physical factors that created these winds were also still very much an active debate. Maury had helped publish maps of global trade winds, and tried to explain them in his *The Physical Geography of the Sea*, but his theory on trade winds was possibly his most misguided in the entire text. To be fair, though, scientists struggled for centuries to fully understand what we do today: that global wind circulation is caused

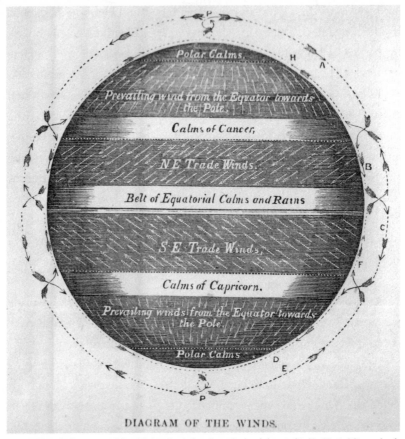

DIAGRAM OF THE WINDS.

FIG. 19. Maury's illustration of the global trade winds as he understood them in his *The Physical Geography of the Sea* (1855/1880).

by shifts in air pressure due to the sun's heat and what we now call the Coriolis effect, the shifts due to the speed and direction of the Earth's rotation.[5] (See fig. 19.)

In *Moby-Dick* Ishmael is not as concerned with the scientific causes of the trade winds, but more with them as a metaphor as he tells of how this wind, especially its consistency, steady force, and invisibility, drives the *Pequod*. By the final hunt at the climax of the novel, Ishmael uses wind to indicate the hand of fate or chance or God. On the second day of the chase, a favorable wind from astern rushes the *Pequod* in the direction of the White Whale. "The wind that made great bellies of their sails," Ishmael says, "and rushed the vessel on by arms invisible as irresistible; this seemed the symbol of that unseen agency which so enslaved them to the race." Captain Ahab continues beating upwind to try to capture Moby Dick. "Against the wind he now steers for the open jaw," Starbuck says as he's coiling down line after the *Pequod* tacked. "I misdoubt me that I disobey my God in obeying him." Starbuck believes as they steer upwind, they steer against God's wishes.[6]

Meanwhile, the wind inspires Ahab to deliver one of his most moving and powerful soliloquys. On that final day of the chase, Ahab tries to understand what drives him so relentlessly onward to avenge the loss of his leg. He gives human traits to the wind as he does the White Whale, arguing about its courage and questioning whether it is indeed a creation, an agent of God:

Were I the wind, I'd blow no more on such a wicked, miserable world. I'd crawl somewhere to a cave, and slink there. And yet, 'tis a noble and heroic thing, the wind! who ever conquered it? ... Would now the wind but had a body; but all things that most exasperate and outrage mortal man, all these things are bodiless, but only bodiless as objects, not as agents. There's a most special, a most cunning, oh, a most malicious difference! And yet, I say again, and swear it now, that there's something all glorious and gracious in the wind. These warm Trade Winds, at least, that in the clear heavens blow straight on, in strong and steadfast, vigorous mildness; and veer not from their mark, however the baser currents of the sea may turn and tack, and mightiest Mississippies of the land swift and swerve about, uncertain where to go at last. And by the eternal Poles! these same Trades that so directly blow my good ship on; these Trades, or something like them—something so unchangeable, and full as strong, blow my keeled soul along![7]

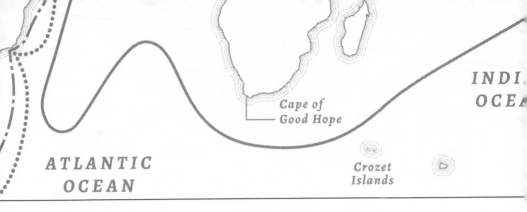

ATLANTIC
OCEAN

Cape of
Good Hope

Crozet
Islands

INDI.
OCE/

Ch. **7**

GULLS, SEA-RAVENS,
AND ALBATROSSES

> I saw a regal, feathery thing of unspotted whiteness, and with a hooked Roman
> bill sublime . . . Through its inexpressible, strange eyes, methought I peeped to
> secrets which took hold of God.
>   Ishmael, "The Whiteness of the Whale"

The *Pequod* approaches the Cape of Good Hope. Ishmael has established Ahab's mad mission and constructed the historical, cetological, and oceanographic details to establish the story's realism. After their first dramatic taste of action fails in a flurry of a squall during "The First Lowering," trying to kill their first sperm whale, the next close encounters with marine life are of a far different tone. These are with seabirds.

A discussion of birds on the ocean in any Anglophone literature must begin with Samuel Taylor Coleridge's "The Rime of the Ancient Mariner" (1798). It's a poem that's crucial for *Moby-Dick* because of its connection to the sailor's oral tradition and to understand several of Ishmael's references to the ballad. How we read an environmental message in "The Ancient Mariner" today also aligns and informs how we interpret environmental messages in *Moby-Dick*.

Coleridge wrote "The Ancient Mariner" in a verse form that evokes the Scottish narrative songs of earlier centuries, the ones passed down in the oral tradition. Coleridge was among a posse of English nature writers that included Wordsworth, Byron, Keats, and Percy and Mary Shelley—the collection of young literary idealists now known as the British Romantics who exulted in the experience of the outdoors and the sublimity of mountains and oceans, while also embracing darker and more spiritual worlds. In many ways, Melville seized the tiller of sea writing from the British Romantics and infused it with marine biology and period nautical realism in order to create a unique style of sea story.[1]

"The Rime of the Ancient Mariner" is told through the eyes of an unnamed lone survivor of a disaster, an old sailor with "a long grey beard and glittering eye" who accosts the best man on his way into a wedding. The wedding guest is spellbound, forced to sit and listen. The mariner begins his yarn explaining how his ship sailed south all the way to Cape Horn, where among stunning icebergs, the crew is terrified—until:

> At length did cross an Albatross,
> Through the fog it came;
> As if it had been a Christian soul,
> We hailed it in God's name.[2]

For over a week, the bird follows the ship, which, now with a favorable wind, makes progress through the ice. The mariner is suddenly struck with horror by the memory as he tells the story, startling the wedding guest, who says:

> "God save thee, ancient Mariner!
> From the fiends, that plague thee thus!—
> Why look'st thou so?"—With my crossbow [says the mariner]
> I shot the ALBATROSS.

The mariner gives no reason for why he shot the seabird. As the ship sails north toward the equator, the ocean and all its spirits begin to avenge the gratuitous killing. In the doldrums, an area of no wind between the trades, the sailors nearly die of thirst: "Water, water, everywhere,

Nor any drop to drink." The mariner's shipmates blame the whole thing on the mariner, hanging the dead bird around his neck. Soon a pair of ghosts cast a spell to convert all the sailors into what we'd now call zombies. The mariner is left to suffer by himself: "Alone, alone, all, all alone/ Alone on a wide, wide sea." He is wretched. He wants to die. He knows he should pray, but he cannot. Finally, after several days, he peers over the rail and sees some water snakes. The mariner is struck with a love and beauty for all living things. Unaware of what he's doing, he blesses these snakes. This is the climax of "The Rime of the Ancient Mariner."[3]

After the blessing of the sea snakes, the dead albatross falls off the mariner's neck into the water. A set of good spirits takes over and the zombie sailors work the ship as it's rushed home. Within view of his home port something comes up from beneath the surface and cracks the hull. The ship goes down in a vortex. A couple men in a rowboat rescue him. Then for the rest of his life the Ancient Mariner is doomed to travel the world and find those that need to hear his story.

For good reason did Coleridge spark the disaster in his "Ancient Mariner" with a sailor's interaction with a seabird. A mariner on open water, in Ahab's time or today, sees birds more than any other type of animal. Captain Lawrence aboard the *Commodore Morris*, for example, often wrote about birds. In his logbook in 1849, while southbound to Cape Horn, he wrote: "First part pleasant gales from the NW and fair weather ... gony[e?]s and speckled haglets [petrels] first made their appearance." Several weeks later just after making it around the Horn, he wrote: "Thick and foggy ... at about 10 oclock run into water very green many birds indications of land at no very great distance tacked ship." Prudent navigators, especially before the age of GPS, used seabirds to help them find their location. They understood which bird species were more coastal and which were more common over water far out to sea. They watched the direction of their flight. Seabirds are regularly found around whales, too, feeding off a carcass or on the same concentrations of plankton and smaller fish that attract the whales. Experienced whalemen and fishermen knew to look for birds on the water as a sign of activity below the surface. At the end of *Moby-Dick* in "The Chase—

First Day," Tashtego spots a line of seabirds flying toward Ahab's whale-boat—"their vision was keener than man's," warning them that the White Whale is ascending.[4]

So it makes sense that over the centuries birds at sea have been weighted with meaning by mariners and writers. Ishmael explains how after spending so many months and years alone on blue water, whale-men are the most superstitious of any sailors. Melville was aware, too, of the long-held conceit that birds hold the souls of drowned sailors, a superstition that Coleridge played into as well: "As if it had been a Christian soul," says the Ancient Mariner. Early in *Typee*, Melville wrote of a captain who killed seabirds for sport. The sailors were "struck aghast at his impiety" and believed their long passage around Cape Horn was due to "his sacrilegious slaughter of these inoffensive birds."[5]

Ishmael speaks generically of "gulls," "sea-fowls," and "seabirds" in *Moby-Dick*. Beyond these, he specifies four distinct seabirds in the novel, using the sailors' common names: cormorants, which he calls "sea-ravens"; frigatebirds, which he calls the "sky-hawk" or "sea-hawk"; albatrosses, known also as "goneys"; and storm petrels, which Ahab calls "Mother Carey's chickens." All of these birds have meaningful cameos in the drama and tone of *Moby-Dick*. At this point along the path of the *Pequod*, we'll discuss the gulls, the sea-ravens, and the albatrosses.

GULLS AND SEA-FOWLS

This might have been an accident, but it's a lovely bit of accurate orni-thology that after comparing the heroic Nantucketer to a "landless gull" and inserting a "screaming gull" as the *Pequod* left the island, Ish-mael does not again refer to seeing gulls specifically while at sea. Gulls (family Laridae) are coastal birds that tend to stay within a dozen or so miles of land. Their populations along the US East Coast are larger today than they were in Melville's time, due to factors such as the reduc-tion of hunting for eggs and feathers and the increase in human garbage and discards from fishing vessels. In various later scenes in the novel, Ishmael speaks of "white sea-fowls" or "sea-fowls" or just "fowls" on the water, usually involving a dead whale. The birds that Melville would've seen in southern latitudes, opportunistically snapping up bits of whale-meat, could be from several families, all in the order Procellariiformes, which are the truly pelagic seabirds, including albatrosses, petrels, shearwaters, and prions—the latter known by sailors at the time as "whale-birds." Dr. Bennett did some identification of offshore seabirds, but then, as now, offshore ornithology is a difficult field of study because the birds are usually seen far in the distance and their plumages can be so similar and variable, even within species. Melville often dabs generic birds, "sea-fowls," into his ocean scenes in the way a painter strokes a few v-shapes into a seascape to bring some activity and variety.[6]

SEA-RAVENS

Melville's "sea-ravens" set a different mood. These birds have a literary tradition of evoking gloom and death. Melville placed his sea-ravens in "The Spirit-Spout," immediately before the *Pequod*'s meeting with the *Goney*. Just as the departure from port in "The Lee Shore" signals a shift to a deeper, more dangerous story, "The Spirit-Spout" flags that the entire yarn is moving toward a more fantastical and spiritual tone. The sea state is changing. They are leaving the Atlantic. The ghostly white spout lures them ominously in the distance as the ship turns east into rough seas to round the Cape of Good Hope. Evoking Coleridge's water snakes, Ishmael says: "Strange forms in the water darted hither and thither before us; while thick in our rear flew the inscrutable sea-ravens."[7]

I read the black "sea-ravens" as cormorants (*Phalacrocorax* sp.), a family of dark deep-diving birds found all over the world. (See fig. 20.) In "The Spirit-Spout" they perch in the *Pequod*'s rigging where they "clung to the hemp." In *Paradise Lost* (1667), the epic poem that had great influence on *Moby-Dick*, John Milton compares Satan to a cormorant sitting, looming, looking down from the limbs of the tree of life, "devising death/To them who liv'd." The word *cormorant* likely derives from the Latin *corvus marinus*, which translates to sea-raven. Their long

THE CORMORANT.

" Not far from thence is seen a lake, the haunt
Of coots and of the fishing cormorant."

SOME persons, with Johnson, derive the name of
this bird from " *Corvus Marinus*," or the Sea
Crow. There is little resemblance to the crow,
however, except in its colour. The cormorant is

FIG. 20. The cormorant, "corvus marinus," which was likely Ishmael's "sea-raven" in *Moby-Dick*, as published in Reverend W. Tiler's *The Natural History of Birds, Beasts, Fishes, Serpents, Insects, &c.* (1862).

thin neck over the surface is snakelike. In "The Spirit-Spout" Ishmael describes these dark sea-ravens up in the rigging as the *Pequod* sails into the black waters off the Cape, "as if its vast tides were a conscience; and the great mundane soul were in anguish and remorse for the long sin and suffering it had bred." The *Pequod* is in tormented seas, "where guilty beings transformed into those fowls and these fish." What the sailors believe to be the White Whale's eternal spout is off in the distance, a "snow-white . . . fountain of feathers."[8]

ALBATROSSES

In the next scene, in "The Pequod meets the Albatross," Ishmael moves from this black bird of ill omen, to a white one that is far more famous. A few of Melville's connections to "The Ancient Mariner" hit you over the head with an oar. The vessel's name in *Moby-Dick*, unlike the chapter title, is the *Goney*—the sailor's name for the albatross (and today a common name for the smaller, northern albatrosses). This whaleship of "spectral appearance" that approaches the *Pequod*, the first whaleship with whom they gam at sea, is "bleached like the skeleton of a walrus" and manned by old, forlorn, "long-bearded" sailors at each masthead. The schools of fish that had been following under the hull of the *Pequod* ominously swim away to the safety of the *Goney*.

Just as in "The Ancient Mariner," the narrator of *Moby-Dick* is a sole survivor, a wanderer, telling his story. Ishmael is also rescued in a rowboat after watching his ship go down. His ship is also split and sunk by an aquatic nonhuman force. All of Ishmael's shipmates also die, and also, arguably, because of their passive aid in the attack of a single, magnificent animal on the open ocean. In the ballad the Ancient Mariner's shipmates physically rise up to heaven as angels; in *Moby-Dick* all Ishmael's shipmates sink to the bottom of the sea.

Ishmael's most direct engagement with the albatross and the ballad is in the footnote to "The Whiteness of the Whale," which sets up this later scene. In the note, Ishmael explains that before this voyage on the *Pequod* he'd seen an albatross on a ship's deck. He was on another ship in sub-Antarctic waters when he came up on deck to find an albatross, "a regal, feathery thing of unspotted whiteness, and with a hooked, Roman bill sublime." It was only after that experience, Ishmael says, that he read "Coleridge's wild Rhyme." Significantly, Ishmael explains that it was not Coleridge that "first threw that spell; but God's great, unflattering laureate, Nature." In the other words, the beauty, the sublimity is intrinsic to the animal. Yet seeing it for himself in life did "burnish a little brighter the noble merit of the poem and the poet."[9] This is exactly what that sperm whale I saw from the royal yard of my first ship, the *Concordia*, did for me for *Moby-Dick*.

If Melville saw an albatross for himself in the Southern Ocean, which is almost certain, he observed one of the six species of the wandering albatross group (*Diomedea* spp.). These are by far the largest of the approximately twenty-one albatross species globally. Albatrosses can live to be sixty years old in the wild. The male wandering or snowy albatross (*Diomedea exulans*) has been reliably measured to have a wing span of up to 11.5 feet. Ishmael describes in the note that the albatross had "vast archangel wings." After explaining the emotional effect the bird had on him, he emphasizes the importance of the white plumage, because he had *not* been as spiritually moved by "grey albatrosses"—which could have been juvenile wandering albatrosses, a few other albatross species, such as the sooty albatross (*Phoebetria fusca*), or even one of the large petrels.[10]

William Wordsworth said he originally gave Coleridge the idea for "The Ancient Mariner" after he read George Shelvocke's *A Voyage Round the World by the Way of the Great South Sea* (1726), in which a "disconsolate black *Albitross*" followed their ship off Cape Horn for several days. An officer "imagin'd, from his colour, that it might be some ill omen." So the officer shot the albatross, hoping the weather would improve. It didn't.[11]

In the ballad, Coleridge never actually states the color of the bird, although most readers today imagine a white wandering albatross. Although mostly white, the wandering albatrosses have at least some gray and black on the wings. As they age and molt, albatross plumage grows whiter. Ishmael assumes that the Ancient Mariner's albatross is a white wandering giant, too. This serves his narrative and metaphoric purposes in *Moby-Dick*, so it's not without meaning that in the last scene of the novel Melville placed a *black* bird going down with the ship as the white whale swims off. If trying to ascribe a feather of realism to Coleridge's wildly fantastical ballad, it does seem a stretch that the Ancient Mariner not only recovered the bird that he shot, but that the man can walk around the deck for a couple weeks with, hanging from his neck, a decaying wandering albatross with wings nearly as long as a person is tall. Even Nathaniel Hawthorne commented on his surprise as to the bird's size in relation to the ballad when he was in England and first saw a taxidermed albatross. Then again, after all, it was Coleridge himself who when describing his creation of the poem, coined the phrase "the willing *suspension* of disbelief."[12]

In "The Whiteness of the Whale," Ishmael continues in the footnote, which is by far the longest note in the novel, explaining that after his fellow whalemen fished up the albatross, the captain tied a note around its neck. "But I doubt not, that leathern tally, meant for man, was taken off in Heaven, when the white fowl flew to join the wing-folding, the invoking, and adoring cherubim!" This was not Melville's fabrication. Nineteenth-century sailors did indeed catch albatrosses, both before and after the publishing of "The Ancient Mariner." For example, whaleman-artist John F. Martin, on his third voyage into the Pacific, wrote in 1842 when sailing toward the Cape of Good Hope:

> This morning we caught an Albatross with a Dolphin [mahi-mahi] line baited with a piece of fat pork. It was a considerable of job [*sic*] to haul him on board. We put a leather label around his neck with the Ship's latitude & longitude, with the number of days from home marked upon it & let him go. He measured upwards of 12 feet from tip to tip. It's a beautiful sight to see them skimming over the water without apparently moving their wings. In the afternoon the mate shot one in the head. He was brought on board & skinned for next days dinner. They are webbed footed & make very handsome reticules [purses] for Ladies.[13]

Several other firsthand accounts show that sailors, even in the twentieth century, shot or hooked albatrosses occasionally for food, but more commonly just for the entertainment of hunting and hoisting the birds on board. Few referenced Coleridge's poem. Albatrosses have a keen sense of smell, allowing them to regularly detect fishy scents or a dead whale from over three miles away, perhaps even much farther. Sailors not only ate the birds, but they made wallets and tobacco pouches from albatross feet, rugs from their pelts, and needle-cases and pipe stems from their beaks. In *Omoo*, Melville's sailors use an albatross feather for an enormous mutinous quill.[14]

Nor was it without precedent that the crew of Ishmael's merchant ship and John Martin's whaleship sent an albatross off like a carrier pigeon. In 1847, for example, a captain named Hiram Luther shot an albatross off the coast of Chile. When he pulled it in he found the bird had a vial tied around its neck. It held a note from another captain, who complained in the message, "I have not seen a whale for 4 months." Based on the position and the date of the note, that albatross had flown over 3,150 nautical miles in twelve days.[15]

Melville's fish documents contain regular reference to catching albatrosses in the Southern Ocean. In *Two Years before the Mast*, Dana wrote of catching a couple albatrosses with a hook, the birds "which had been our companions a great part of the time off the Cape [Horn]." In a revised edition Dana tucked in a reference to Coleridge in this scene, yet with no irony about catching them. Browne wrote that at the latitude of Good Hope they caught an albatross with a twelve-foot wingspan. They set it loose with a note that had the ship's name and the date. Francis Allyn Olmsted, a naturalist-observer who sailed as a passenger on a whaleship and came home to publish *Incidents of a Whaling Voyage* in 1841, described catching seven albatrosses off Cape Horn. Olmsted described how the crew ate a "young albatross," which was likely a smaller species. It tasted like veal, he said, served in an "excellent 'sea pie'"—although some sailors would not eat it because "this bird has no gizzard."[16]

So, some sailors certainly did look to birds as omens and perceive killing or eating certain seabirds as bad luck, but the prevalence of this concern for these animals has been overstated. Albatrosses did not seem to be singled out as all that more significant, and the influence of Coleridge's ballad to those at sea seems fairly minimal—even among classically educated author-sailors. Albatrosses were, however, recognized for their stunning size and their ability to glide so effortlessly in the heaviest of weathers. Yet some nineteenth-century observers found themselves disappointed when they saw the albatrosses acting like scavengers or when witnessing the birds' awkwardness while waddling on deck—hence the "goney" name, derived from goon, a fool. Sailors often played cruel tricks on these birds, such as pitting one against the ship's dog or giving a piece of pork to two albatrosses on either side of a string.[17]

Today, our first association with albatrosses is usually an image that evokes sympathy bordering on pathos: red plastic gun shells, blue fishing line, and yellow plastic bottle caps are in the nests and stomachs of thousands of these seabirds who have ingested the objects, mistaking them for food and even feeding them to their young. Now about 175 years since Melville stood up at the masthead, the IUCN (International Union for Conservation of Nature) Red List classifies three albatross species as critically endangered, six as endangered, and six as officially vulnerable. Like nearly all seabirds, albatrosses are struggling due to habitat loss and a difficulty in raising chicks because of introduced mammals on their island rookeries. Since the 1960s and '70s, albatrosses are also drowned by the hooks of longlines. Though this problem seems to be improving somewhat due to adjusted fisheries practices and materials, a study in 2013 estimated that at least 100,000 albatrosses had been dying each year due to longline fisheries.[18]

For my part, I first saw wandering and black-browed albatrosses (*Thalassarche melanophrys*) off Cape Horn. It was difficult for me to get a sense of their size, even though they occasionally soared quite close to the ship. Nevertheless, I was awestruck. This is a description in *Moby-Dick* where I do not think Melville went overboard with his extravagance about an animal's glory. The albatrosses flashed pure white plumage on the underside when they wheeled perpendicular to the sea, like the blades of a modern windmill. I watched them glide astern of our ship, which rolled on towering slate blue swells. The wings of these albatross remained motionless regardless of the cold and wind. No photograph or video I've seen has matched the experience. Poetry comes closer.

Recent biophysics research on these enormous birds has only increased their sublimity. Albatrosses evolved specialized tendons to lock their wings in place. Modern racing ocean yachts now have enormously tall and thin sails, like albatross wings. Albatrosses soar and glide by "dynamic soaring," turning and dropping to use their own momentum and gradients of wind closer to the water, as well as "slope soaring" in which they use slight breezes by skimming just above waves and swells.[19]

The Ancient Mariner's final words to his wedding guest are:

> He prayeth best, who loveth best
> All things both great and small;
> For the dear God who loveth us,
> He made and loveth all.

For twenty-first-century readers, "The Rime of the Ancient Mariner" is a clear appeal to value the rights of animals, or, taken more broadly, to not heedlessly destroy our natural environment. In 1798 Coleridge did not think of himself or his Ancient Mariner as a proto-environmentalist. His mariner preaches about the need to be kind to animals, but more within the thinking of natural theology, rather than twentieth- and twenty-first-century ideas about conservation. The Ancient Mariner's albatross is an expression, even an agent, of God. The mariner's arrow at this seabird was a shot at Faith, aimed at His good works. In *The Island World of the Pacific* (1851), for example, Reverend Henry T. Cheever wrote a lengthy description of albatross, in which he discusses the moral and Christian lessons to be learned from Coleridge's poem.[20] (See fig. 21.)

This doesn't mean it's unreasonable to read "The Ancient Mariner"

FIG. 21. Illustration for "The Rime of the Ancient Mariner," paralleling the albatross with an angel (1857).

or *Moby-Dick* as an environmentalist work of literature today. In 2016, scholar Robert Louise Chianese wrote of the guest in the ballad in a way that we might also consider Ishmael: "Apparently he needs time to recover before he resumes his participation in society and its cultural traditions, such as marriage. He may not go off on an expedition to save albatrosses, but he may ask himself if he loves all creatures great and small. Such is the effect of powerful stories of personal transgression and transformation."[21]

In 1849 when Melville first left New York on that passage to London before composing *Moby-Dick*, the day after he climbed up to the masthead to remember the old feelings and the very same day that a passenger jumped overboard to drown himself, he wrote in this extraordinary journal of walking the decks with a new friend named George Adler, an accomplished German linguist and scholar. Melville wrote that night, seemingly encapsulating his own natural theological perspective that he would infuse into *Moby-Dick* as he diverged from the likes of Agassiz and even Maury: "[Adler's] philosophy is *Colredegian*: he accepts the Scriptures as divine, & yet leaves himself free to inquire into Nature. He does not take it, that the Bible is absolutely infallible, & that anything opposed to it in Science must be wrong."[22]

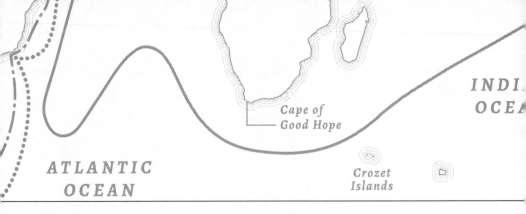

Cape of
Good Hope

Crozet
Islands

INDI.
OCE/

Ch. 8

SMALL HARMLESS FISH

> Shoals of small harmless fish that for some days before had been placidly swim-
> ming by our side, darted away with what seemed shuddering fins, and ranged
> themselves fore and aft with the stranger's flanks. Though in the course of his
> continual voyagings Ahab must often before have noticed a similar sight, yet,
> to any monomaniac man, the veriest trifles capriciously carry meanings.
>
> Ishmael, "The Pequod meets the Albatross"

In the previous chapter I mentioned the small harmless fish that depart
Ahab's ship in "The Pequod meets the Albatross." It's reasonable to
imagine these as pilot fish (*Naucrates ductor*), a silvery blue-striped
species well-known to whalemen. (See fig. 22.)

Melville had written about pilot fish as omens before he published
Moby-Dick. In *Mardi* his narrator describes the men killing an enormous hammerhead shark. They lance the shark in the head. As it sinks
down through its own blood to die, the "nimble Pilot fish" drop their
allegiance and swim over beside the men's boat. "A good omen," the
sailor says: "no harm will befall us so long as they stay."

In addition to coming from his experiences at sea, some of his ideas
on these fish likely derived from, or at least were reinforced by, Francis
Allyn Olmsted, who caught and described pilot fish carefully. In 1841
Olmsted explained that pilot fish commonly seek shelter and food by
living in the close vicinity of large sharks — "side by side with his fero-

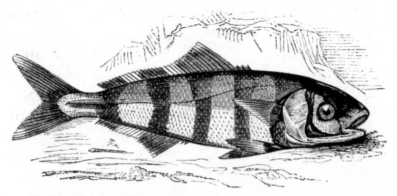

FIG. 22. "The Pilot-fish" in Frederick Bennett's *Narrative of a Whaling Voyage Round the Globe* (1840).

cious mate" — as well as for days sometimes beneath the hull of slow-moving sailing ships. Pilot fish are not to be confused with remora, the fish that suction themselves to sharks and whales (family Echeneidae). Olmsted wrote that it's "a trick sometimes practised by a brother whaler" to sail beside another's hull to get a school of fish to leave and join their ship. Perhaps this is where Melville got the idea for this scene, which he turned into an omen: the first hint that sharkish Ahab is not going to follow the redemptive path of the Ancient Mariner. The captain of the *Pequod* does not bless these small creatures of the sea that are named after guides to mariners.

Ahab murmurs sadly: "Swim away from me, do ye?"[1]

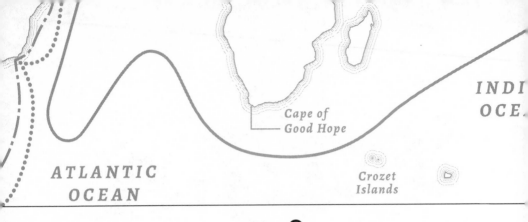

Ch. **9**

PHOSPHORESCENCE

A mariner sat on the shrouds one night,
The wind was piping free;
Now bright, now dimmed, was the moonlight pale,
And the phosher gleamed in the wake of the whale,
As it floundered in the sea.
> Elizabeth Oakes Smith, "The Drowned Mariner," *Western Literary Messenger: A Family Magazine of Literature, Science, Art, Morality, and General Intelligence*, 1846 (in "Extracts")[1]

Bioluminescence, known more often in Melville's day as "phosphorescence," is one of the most compelling and memorable natural phenomena at sea. It's a magic irresistible to ocean writers and artists. Yet in a novel so probing of the contrast between light and dark and so profoundly interested in the otherworldliness of the ocean environment compared to that of the land, how and why did Melville, other than the citation of the poem above in his "Extracts," write only two brief images in *Moby-Dick* that barely evoke bioluminescence? Did Melville somehow never see it during his years at sea? Or if he did, was it to him not as stunning as it is to us today?

For his part, Charles Darwin saw plenty of bioluminescence and wrote of it with that honest wonder identified in Melville by scholar Jennifer Baker. In December 1833, about two years into his voyage, the

Beagle cruised along the coast of Patagonia. Darwin wrote in his narrative of a makeshift plankton net that he made of bunting, the material they used for signal flags. He towed the net astern and "caught many curious animals." Of one dark night Darwin wrote:

> The sea presented a wonderful and most beautiful spectacle. There was a fresh breeze, and every part of the surface, which during the day is seen as foam, now glowed with a pale light. The vessel drove before her bows two billows of liquid phosphorus, and in her wake she was followed by a milky train. As far as the eye reached, the crest of every wave was bright, and the sky above the horizon, from the reflected glare of these livid flames, was not so utterly obscure as over the vault of the heavens.[2]

Bioluminescence is light created by living things. It is caused by chemical reactions occurring within the body of the organism, whether that's inside a firefly, a lantern fish, or a speck of bacteria. Most of the flashing visible in the wake of boats on the surface or in the crashing waves on the coast is produced by dinoflagellates, which are their own phylum of phytoplankton. One especially common species of dinoflagellate, found coastally worldwide, is known as "sea sparkle" (*Noctiluca scintillans*). About the size of a pen point, these gas-filled orbs each have a tiny tentacle for gathering food. Within this dinoflagellate's cytoplasm are speckled *scintillons* that make brief pulses of light when agitated.[3]

Many of the naturalists in the age of Darwin and Melville, those who were *not* focused on putrefaction or electricity as the cause of bioluminescence, correctly theorized that a large percentage of ocean organisms could make their own light. With the recent improvements in microscopes, nineteenth-century naturalists classified and illustrated these bioluminescent "animacules," notably the dinoflagellates. Experts such as C. G. Ehrenberg in Berlin did so from samples of sea water sent over land, sloshing about in horse-drawn carriages. Today, biologists estimate that not only are the producers of bioluminescence extremely diverse, but some three-quarters of all species in the ocean can make light.

And 90 percent of those organisms that live in the aphotic zone, meaning beyond the reach of solar light and deeper than about 650 feet, are capable of bioluminescence.[4]

Most of Melville's fish documents described this sparkling in the ocean, including those narratives of Surgeon Beale and Francis Allyn Olmsted. Dr. Bennett devoted an entire section to its analysis, considering its potential utility to the animals themselves. He also told a story in which the blubber from a sperm whale held a bioluminescent glow below decks for two full nights, something that even the "oldest whalers on board had never before witnessed."[5]

Yet even before consulting his books to write *Moby-Dick*, Melville saw marine bioluminescence himself. He saw it during his long night watches. He saw it over the gunwale when in small boats at night. He surely saw bioluminescence around sharks as they scavenged the flesh of a dead whale alongside. Aboard the *Acushnet* and the *United States*, Melville sailed through the waters off the coast of Patagonia that Darwin described, as well as through other areas known for bioluminescence.

In *Moby-Dick*, the first and most likely nod to bioluminescence is in "The Whiteness of the Whale." Ishmael muses as to why sailors are so afraid of "a midnight sea of milky whiteness—as if from encircling headlands shoals of combed white bears were swimming round him." Over the centuries, mariners have reported this sort of sea state, in which the surface is so densely full of opaque bioluminescence that the ocean has a consistent glow for miles, a milky cloud from horizon to horizon.[6]

For example, a few years after the publication of *Moby-Dick*, Lieutenant Maury received a letter from Captain W. E. Kingman of the American clipper ship *Shooting Star*, which the captain included when he mailed in his abstract log. When approaching Java from the southwest—a route similar to the *Pequod*'s—the sea was so white that Captain Kingman felt compelled to slow the ship and take a sounding, to make sure there was no hidden reef. But they had plenty of water beneath them. He wrote to Maury that the surface appeared "like a plain covered with snow." As he sailed along for hours, he ultimately measured the milky sea to be some twenty-three nautical miles wide with only a half-mile dark strip interrupting the center. In all his years Captain Kingman had seen "nothing that would compare with this in extent or whiteness." He ordered his crew to fill a sixty-gallon tub with glowing seawater, from which he examined several species of gelatinous zooplankton. Although he peered into the tub with the little telescope on his sextant, Kingman did not have on board the magnification nor the late twenty-first-century knowledge to understand that the zooplankton he saw likely swam among a vast bloom of bioluminescent bacteria. Captain Kingman wrote to Maury: "The scene was one of awful grandeur, the sea having turned to phosphorous, and the heavens being hung in blackness, and the stars going out, seemed to indicate that all nature was preparing for that last grand conflagration which we are taught to believe is to annihilate this material world." For the captain, as for Ishmael, bioluminescence held a dangerous apocalyptic portent.[7]

The second mention in *Moby-Dick* that just might be Ishmael describing bioluminescence is in "The Spirit-Spout," in that same dismal, foreboding scene with the sea-ravens as his whaleship approaches the Cape of Good Hope. Ishmael says: "When the ivory-tusked Pequod sharply bowed to the blast, and gored the dark waves in her madness, till, like showers of silver chips, the foam-flakes flew over her bulwarks." Perhaps Melville meant to evoke more than foam, but also bioluminescence? This would be another nod to "The Ancient Mariner," since the *Pequod* encounters the *Goney* in the next scene. Coleridge had written of bioluminescence in his ballad, describing "elfish light" and "hoary flakes" falling off the water snakes. And Coleridge's Pacific seas burn green, blue, and white.[8]

Yet even if these two scenes in *Moby-Dick* were meant to describe bioluminescence, Melville hardly wrote of it in the fashion or length that we expect for a novel so dripping with the unique mysteries of the ocean. We know Melville saw bioluminescence. He read about it in narratives and in fictional works. And the scientists, sailors, and readers of his time were fascinated by bioluminescence.

So, the answer might be, as with Melville's avoidance of Cape Horn in *Moby-Dick*, that he simply felt he had too recently and regularly drawn from this quiver of marine metaphors. A year earlier, in his novel *White-Jacket*, he had his fictional crew caught in a squall off Cape Horn. The men struggle to hold on as the "phosphorescence of the yeasting sea cast a glare" into their terror-stricken faces. In *Redburn* Melville wrote bioluminescence into a scene as gruesome and Gothic as anything conjured by Edgar Allan Poe: a dead sailor's flesh begins to glow as it rots in the dark of the forecastle. And then in *Mardi*, Melville wrote about bioluminescence to the extent that a modern reader expects in *Moby-Dick*—a mixture of dark magic and awestruck analysis. In a chapter of *Mardi* titled "The Sea on Fire," Melville's narrator and his fellows are in a small open boat traveling through the waters of the South Pacific when "we beheld the ocean of a pallid white color, corruscating all over with tiny golden sparkles." Sperm whales nearby spout "a bushy jet of flashes." (This, er, isn't possible; Melville corrects this in *Moby-Dick*.) As the sperm whale swims off, the narrator considers the various theories on the sources of bioluminescence, including electrical impulses from the atmosphere, decomposition of organic matter, its connection to small organisms, its possibility as a tool of communication among living "fire-fish," and even the sailor superstition that it was composed of strands of golden hair from mermaids. In this scene in *Mardi*, the milky sea lasts for hours. Of previous experiences with this phenomena, which illuminate our reading of "The Whiteness of the Whale" and perhaps "The Funeral," Melville wrote in *Mardi*:

> Whereas, in the Pacific, all instances of the sort, previously coming under my notice, had been marked by patches of greenish light, unattended with any pallidness of sea. Save twice on the coast of Peru, where I was summoned from my hammock to the alarming midnight cry of "All hands ahoy! tack ship!" And rushing on deck, beheld the sea white as a shroud; for which reason it was feared we were on soundings.
>
> Now, sailors love marvels, and love to repeat them. And from many an old shipmate have I heard various sage opinings, concerning the phenomenon in question.[9]

In short, Melville likely wrote so little of bioluminescence in *Moby-Dick* because he had done so recently in his other fiction. For Melville and other mariners and authors of his time, "phosphorescence" and milky seas were often a Gothic sign or symbol of danger, decay, and black magic—and even one of Judgment.

INDIAN
OCEAN

Cape of
Good Hope

ANTIC
CEAN

Crozet
Islands

Ch. 10

SWORD-FISH AND
LIVELY GROUNDS

They supposed a sword-fish had stabbed her, gentlemen.
Ishmael, "The Town Ho's Story"

After the *Pequod* rounds the Cape of Good Hope, Ishmael begins the long yarn "The Town Ho's Story" with a whaleship leaking: "they supposed a sword-fish had stabbed her."[1] Melville was fascinated with swordfish (*Xiphias gladius*). In *Mardi*, he wrote an entire chapter about this animal, extolling its solitary warrior characteristics and satirizing scientific attempts at classification, working through some ideas he'd later use to explore the cetology of the sperm whale.

To explore Ishmael's yarn, I turned to Linda Greenlaw, swordfisherman and author, to find out what she thinks of actual nineteenth-century reports of swordfish stabbing holes into wood hulls.

"It's hard to believe," she says. "And I'll tell you why. Swordfish go after prey with their bill. They do use it as a weapon. But they don't *stab*. They *slash* with it. Right? They're not going to swim at something and skewer it. When they swim through a school of fish, they slash, and then they come back and eat what they've injured. I've seen many swordfish fighting on a hook, and: *it's this*." She motioned a flat hand side to side, as if aggressively dusting off the table. "When you land them on the deck of a boat: *it's this*."[2]

Greenlaw goes back to holding her mug of tea. We're in her dining room in Maine.

"It's like a shark when they latch onto something, and they shake their head," she continues. "Now, could a swordfish slash at a wooden boat and happen to pierce it? I mean, I guess if the boat were really tender. I was never ever, ever nervous about a boat being hurt, damaged, by a fish—even when I was on my first boat, which was wood. Obviously, I can't say it's never happened. 'Course, maybe it did."

And, amazingly, it did.

Before sitting down to write *Moby-Dick*, Melville had heard of swordfish or other billfish stabbing into the hulls of wood vessels. By the nineteenth century, some sailors and naturalists had differentiated between swordfish and the billfish, the latter being the marlin and sailfish now in the family Istiophoridae. But for Melville and other fellow whalemen "sword-fish" could really have referred to any of these large predatory fish with a long bill. (See earlier fig. 3.) Baron Cuvier, in his volume on fish that Melville owned and marked up, mentions swordfish, sailfish, and sawfish all recorded leaving a bill or two in wood hulls. Melville might have read about such an instance in Liverpool in 1839 when he was ashore during his first voyage. A piece of wood from the ship *Priscilla* that had a swordfish bill rammed through the outer planking was on display. In 1840 Dr. Bennett wrote an account of the *Foxhound*, a whaleship that had a swordfish bill stuck through the protective copper underneath the ship's water line. As with the *Priscilla*, the fish's bill broke off like a bee's stinger and remained in the ship's plank from the South Seas all the way home. Two years later, when Melville was ashore in the Marquesas, a whaleship named the *London Packet* had to haul out in that harbor to make repairs to a leak that they believed was caused by a billfish. At the Natural History Museum in London is a hunk of ship's timber from about 1832, near the width and length of your shin, which had been stabbed by *two* separate marlin right next to each other—and the pieces of bill are still stuck in the wood. It's almost

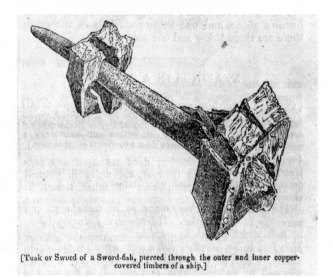

[Tusk or Sword of a Sword-fish, pierced through the outer and inner copper-covered timbers of a ship.]

FIG. 23. A swordfish bill speared through ship timbers, as illustrated in the *Penny Magazine* (1835).

certain that Melville saw this exact piece of wood when he toured the British Museum on his trip to London just before he began writing *Moby-Dick*. Perhaps the image *stuck* with him? I've held that chunk of wood myself. The bills are wedged into the fibers as if driven with a sledgehammer. The rostra come in at different angles. I could still see where the copper sheathing had once been nailed over—which the bills had pierced straight through.[3] (See fig. 23.)

So the part about a swordfish causing a leak in a wooden whaleship is not a stretch in Ishmael's yarn "The Town Ho's Story," although it was a matter of fascination among mariners and the public at the time. We know of no actual *sinkings* by swordfish, marlin, or sailfish bills, but they definitely pierced the wood hulls of sailing ships.

Why these fish stick into the hulls of ships remains a mystery. Even mariners and biologists today still do not have a good understanding of how swordfish might use their bills—if the extended rostra have evolved for anything other than for slashing smaller fish for food or as a deterrence from predation at the mouths of sharks and killer whales. Swordfish have relatively small mouths. Greenlaw explains that they barely have teeth, so they gulp up the chunks of fish they've slashed up. Their bills, on both males and females, can grow to a third of their total length. Swordfish can weigh over 1,100 pounds and, according to a recent study by a team of scientists in Taiwan who study propulsion dynamics, swordfish can swim nearly eighty miles per hour, faster than any other fish in the world.[4]

In "Cetology," Ishmael implies a knowledge of the swordfish's use of its bill, but he never follows through with what he thinks that is exactly. Dr. Bennett observed from his ship's rail in the 1830s, however, exactly what Greenlaw explains: swordfish slashing back and forth through a school of smaller fish. Thus, might a reasonable explanation be that they stab a hull accidentally when slashing through a school of fish congregating under a slow-moving ship in the open ocean?

Greenlaw isn't sure. She says that swordfish are always aware of a boat's hull. And when she's examined the stomachs of swordfish, it seems their diet is primarily deepwater fish and squid. Swordfish usually are not interested in a few surface schooling fish.

American whalemen sometimes rowed up and harpooned swordfish from their boats for food or entertainment, or to practice their harpooning skills. The largest of fish were even sliced up and put in the try works to yield a couple gallons of oil. Baron Cuvier in 1834 wrote that harpooning for swordfish was "precisely the whale fishery in miniature." In January 1844, for example, the men of the *Charles W. Morgan* lowered after swordfish and caught one off the coast of Peru—while Melville was sailing homeward bound in the same region aboard the *United States*. Two years later, also off the Peruvian coast, the men of the *Commodore Morris* harpooned three swordfish and were able to capture two of them. Beside this entry the logkeeper drew the swordfish and marked that they got two barrels of oil from each (see fig. 24).[5]

Whalemen could row fast enough to catch a swordfish?

"Sure," Greenlaw says. "Now *that* I believe. If you know what you're doing. In most parts of the world it's rare for swordfish to hang out at the surface. But if you're in an area like Georges Bank or off Nova Sco-

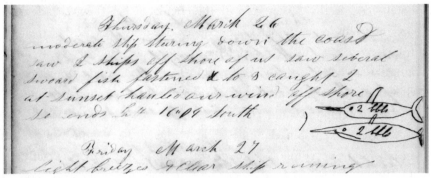

FIG. 24. Logbook entry from the whaleship *Commodore Morris*, recording the capture of two swordfish, which they boiled out for oil (1846).

tia where they do surface, and you go on them the right way, it's not that difficult to harpoon them. When they're on the surface, they're kind of lolly-gagging along."

At the time, Surgeon Beale and his fellow naturalists, along with the whalemen, believed that swordfish were predators on great whales, or at the very least an annoyance. Beale wrote of part of a swordfish bill impaled in the body of a beached whale in Yorkshire, England, and he also discussed a theory that sperm whales might breach to escape these fish. In 1856 artist-whaleman Robert Weir wrote in his journal: "Saw a very large sword fish this morning, that deadly enemy of the Sperm Whale. That is excepting ourselves and other whalemen."[6]

Twenty-first-century biologists do not believe that swordfish are predators of whales, or even of dolphins or small porpoises, although they concede that very occasionally a billfish might injure a marine mammal for reasons still unknown. It's more likely the other way around. Sperm whales, or more commonly orcas, will occasionally eat a swordfish or two. Yet swordfish interactions with whales and ship hulls are still not explained. Recent underwater video has recorded swordfish and marlin attacking oil rigs and getting their bills stuck in pipelines. In 1967, at the depth of 2,000 feet, a large swordfish mortally impaled itself into an edge of the steel submersible *Alvin*, and it remains a mystery as to why.[7]

LIVELY GROUNDS AND SHIFTING BASELINES

American whalemen of the nineteenth century observed fish at sea regularly, especially in prolific areas like the Peru Current off South America. Here the abundance of plankton supports baleen whales directly and sperm whales indirectly by feeding squid and swordfish, marlin, and shark populations. Highly productive waters like those off the west coast of South America also provide food for enormous schools of anchovetas (*Engraulis ringens*), which in turn feed baleen whales and enormous populations of seabirds, a range of medium-sized fish, such as mahi-mahi, several species of tuna, and then the billfish.

In "Stubb kills a Whale," Ishmael identifies the coast off Peru as a "lively ground," an ocean region loaded with flying fish, mahi-mahi, and "other vivacious denizens." Scientists today believe that these regions were once even far more lively than they are now, in part because of the effects of American whaling, but more so due to the large-scale industrial fishing of the second half of the twentieth century. Only in the 1850s, largely begun by zoologist Karl Ernst von Baer in Russia, did naturalists begin to examine and encourage the keeping of fisheries data in terms of catches, location, season, and fishing gear. Baer combed historical resources from monasteries to the archives of local governments in order to understand what was actually present in the past. He pushed for centralized, long-term data for the future. Even in 1854, Baer described what we now refer to as the shifting baseline: that past natural environments always seem more prolific—the fish "back then" were always more numerous and larger.[8]

The anecdotal reports of the Pacific from trusted sources, like Surgeon Beale, certainly suggest a more populous ocean in the southeastern Pacific. Beale wrote of schools of hundreds of swordfish. Once off the coast of Chile he described just how prolific the waters were in which his whaleship sailed:

The irregular and desert shore hemmed in the great ocean, which now swarmed with living creatures. The humpbacked whale sported in the smooth water, his polished skin glistening in the rays of the scorching sun; seals also, at a small distance from the shore, were lying as if asleep upon the surface, basking in the heat. Hundreds of large albacore and bonito [two species of tuna] now crowded our vessel, and gave employment to those who could be spared from the duties of the ship, in catching them with the hook. The ugly sun-fish [*Mola* sp.] now and then came floating by, and gave the young harpooner a chance of shewing his newly-acquired dexterity, by plunging the barbed iron into their grisly bodies. The ferocious sword-fish frequently shewed himself, much to the terror of the bonito and albacore, which shot through the fluid element with wonderful velocity, to escape from their voracious pursuers. The varieties of polypi [squid/octopus] and medusæ [jellyfish] which abound here are immense, and would find the naturalist with employment for a century.[9]

From his natural theological ship's rail, Beale recorded miles of seabirds flying over the sea and a variety of species along the coast—flamingoes, cormorants, penguins, and pelicans—all of which, he said, reflect on "the wisdom and greatness of Him . . . to animate the wilderness."[10]

The perspective and knowledge of Linda Greenlaw, one of the most successful fishermen in modern day New England, has much to teach us about Ishmael's views in *Moby-Dick*. Greenlaw wrote in *Seaworthy* that harpooning swordfish from the bow of a boat is "the most primitive and fundamental" way to kill these animals. She thought the "frenzied sensation" might be a feeling she shared with "the whalers of old." Although Greenlaw received a liberal arts college education, she sees her real education in life at sea. Greenlaw simply loves to fish—swordfish, tuna, lobster, halibut. As a twenty-first-century commercial hunter, she has little spiritual or cultural connection to fish. She has fished in some of the roughest weather to support herself and the families of the men whom she leads. Her self-respect, her identity, is in putting food on people's tables. Greenlaw tells me she learns about the biology of marine animals to help her catch more fish and to do so before the other fishermen can. Modern fisheries biologists have been of little help to her, she says. Similar to Ishmael's perspective on the naturalists ashore in 1851, Greenlaw wrote in 2010: "I had read most of the small amount of literature available on the biology of swordfish, and some of it is contradictory, indicating how little is actually known. Personal experience and observations through the course of my twenty-year work-study had taught me most of what I know about the behavior of swordfish."[11]

In this way, little has changed since the nineteenth century in terms of a divide between the fishermen who are out there every day and the scientists studying the matter with more limited experience on the water. The more significant difference, which Melville would never have considered, is that scientists now have a voice in managing fisheries—an idea that grates on Greenlaw.

Melville would, however, have been able to identify with how in her books Greenlaw wants the reader to know the blood, sweat, and tears that go into the fish that you're plopping on your grill. In 1816 Sir Walter Scott famously wrote a character in *The Antiquary* who said: "It's no fish ye're buying—it's men's lives." Melville followed in kind when in "The Affidavit" Ishmael says: "For God's sake, be economical with your lamps and candles! not a gallon you burn, but at least one drop of man's blood was spilled for it."[12]

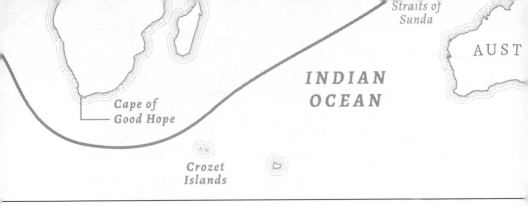

Straits of Sunda

INDIAN
OCEAN

AUST

Cape of
Good Hope

Crozet
Islands

Ch. 11

BRIT AND BALEEN

> Numbers of Right Whales were seen, who, secure from the attack of a Sperm Whaler like the Pequod, with open jaws sluggishly swam through the brit, which, adhering to the fringing fibres of that wondrous Venetian blind in their mouths, was in that manner separated from the water that escaped at the lip.
> Ishmael, "Brit"

The *Pequod* leaves the Crozet Islands astern, sailing eastward in the Indian Ocean. When they come upon right whales, Ahab and the mates do not lower their boats to try to catch these whales because the quality of their oil for human use was inferior to that of the sperm whale's. Actual whalemen in the mid-nineteenth century did tend to focus on either sperm whales or right whales, but they would certainly lower for the other if within reach, or captains would switch their focus based on the region or time of year. For example, after Melville jumped ship in 1842, the sperm-whaler *Acushnet* sailed up into the North Pacific for right whales, eventually returning to New Bedford with a full cargo of whale oil, spermaceti oil, and 13,500 pounds of baleen.[1]

Ishmael opens "Brit" this way: "Steering north-eastward from the Crozetts, we fell in with vast meadows of brit, the minute, yellow substance, upon which the Right Whale largely feeds. For leagues and leagues it undulated round us, so that we seemed to be sailing through boundless fields of ripe and golden wheat."[2]

WHAT IS BRIT?

Various scholars over nearly a century have defined the word *brit* as used in *Moby-Dick* as crustaceans, krill (shrimplike aquatic crustaceans), copepods (flea-like aquatic crustaceans), pteropods (tiny marine snails), or simply plankton. Plankton is a catchall term widely used today for any small pelagic animal species and their eggs and larvae (zooplankton), and photosynthetic microscopic organisms (phytoplankton), all of which are largely subject to tides, currents, and winds. It's worth dipping briefly into the etymological seaweeds on "brit" in order to learn more about the nineteenth-century understanding of oceanic food webs and marine ecology. Exploring this word also reveals the subtle ways that Melville again placed his sailors on a higher stage than the scientists ashore.

In *Moby-Dick* Ishmael does not define brit as krill—of the order *Euphausiacea*—or copepods, or even solely Crustacea. Beyond the "Brit" chapter, Ishmael uses the word four more times in the novel. In "The Affidavit," he refers to the right whale's food as "that peculiar substance called *brit* . . . the aliment of the right whale." As the novel progresses, he uses the words *meadow* and *wheat* and *grassy* in association with brit, connoting a more plant or algae-like substance. In "The Right Whale's Head," Ishmael implies brit is made up of fish: "The edges of these bones [baleen] are fringed with hairy fibres, through which the Right Whale strains the water, and in whose intricacies he retains the small fish, when open-mouthed he goes through the seas of brit in feeding time." Even with the caveat that Ishmael has a broad colloquial usage of "fish," it's clear that in *Moby-Dick* Ishmael does not have a specific single type of plankton (or small fish) in mind with the word brit. In Melville's copy of his "ark," Noah Webster defines brit as "fish of the herring kind." Melville's *Johnson's English Dictionary* defined brit as "the name of a fish."[3]

In *Moby-Dick* Melville never used the words shrimp, prawn, lobster, crayfish, or crustacean, all of which were available to him. But he did not have at his disposal in the mid-nineteenth century several other words in common usage today that might have been handy in describing these small organisms at sea. He would not have known the word krill, for example. Krill is from the Norwegian, meaning "very small fry of fish," and was not in regular English usage until the twentieth century. Melville did not have access to the word plankton either. Plankton wasn't coined until the 1890s. In short, Melville did not have the specific vocabulary to describe small pelagic organisms in the same language we do today.[4]

Curiously, the authors in Melville's fish documents did not use the word *brit*. In 1852 Lieutenant Maury published a letter in his *Explanations and Sailing Directions* from a captain who used the word, but even Maury never used it himself. So *brit* was a term Melville surely learned from his own experience. It was kicking around on ships off New England as early as the 1720s, and American whalemen in the 1800s used the word *brit* often. For example, in 1840 Captain Sanford aboard the New Bedford whaleship *Jasper* in the southwestern Indian Ocean, not far from the Crozets, reported in his logbook that he saw "quantityes Brit" for three straight days. Captain Elihu Gifford of the New Bedford whaleship *America* wrote while to the east of the Cape of Good Hope in 1836: "Fine weather steering by ESE with all sail set-saw plenty of Brit." The word *brit* is repeated in the margins of Gifford's entry with a box drawn around it. The captain did this to highlight the oceanographic information for Maury.[5]

Thus *brit*, synonymous with plankton aggregations upon which whales feed, was surely a term aboard the *Acushnet* and among the other whalemen that Melville sailed with or met. Whalemen such as Captain Daniel McKenzie of New Bedford, the man who helped assemble abstract logbooks for Maury, wrote in a personal letter to Maury in 1849 one of the most detailed accounts of an understanding of brit—as gelatinous, globular, and "about the size of a small pea"—including its relationship to right whale ecology. Melville himself likely saw brit on the surface, notably on the Brazil Banks or the Coast of Chile Ground. For *Moby-Dick* he chose an accurate part of the ocean where brit and right whales had been found.[6]

RIGHT WHALE ECOLOGY AND THE COPEPOD

Baleen whales, such as right whales, do swim open-mouthed through patches of zooplankton, as depicted by Ishmael in "Brit." These whales swallow millions to hundreds of millions of individuals in an hour. One way to differentiate baleen whales is by two primary feeding strategies: the "gulp feeders" and the "continuous ram feeders." Gulp feeders, such as most of the rorquals, open their massive jaws to engulf an enormous volume of ocean. With their tongue and abdominal muscles they push out the water in order to trap swarms of krill and schools of small fish within their baleen. Gulp feeders have pleats in the skin and muscles on the lower jaw and belly so they can stretch their mouths out wide. In contrast, the right whales and bowheads are continuous ram feeders, also known as skim feeders when they do so at the surface. Ram feeders swim open-mouthed through clouds of zooplankton. Ram feeders have a big gap in the front of their mouths to let water in and then spill out through their filtering baleen fibers. Ram feeders have finer filaments and longer baleen plates than the gulpers, seemingly in order to open wide and keep their jaws agape while feeding.[7] (See plate 4.)

In "The Right Whale's Head" Ishmael declares:

> Over this lip, as over a slippery threshold, we now slide into the mouth. Upon my word were I at Mackinaw, I should take this to be the inside of an Indian wigwam. Good lord! is this the road that Jonah went? The roof is about twelve feet high, and runs to a pretty sharp angle, as if there were a regular ridge-pole there; while these ribbed, arched, hairy sides, present us with those wondrous, half vertical, scimetar-shaped slats of whalebone, say three hundred on a side, which depending from the upper part of the head or crown bone, form those Venetian blinds which have elsewhere been cursorily mentioned.[8]

All that Melville described here is anatomically correct. The right whales and their close cousins, the bowhead whale, have the largest

FIG. 25. A view of the right whale's baleen and tongue on a beach of Cape Cod (c. 1880).

mouths on Earth, even bigger than blue whales. Ishmael and his reader could never stand inside a right whale's mouth, however, because the massive tongue fills up most of the space. (See fig. 25.) If you removed that tongue, Abraham Lincoln—who had not yet decided to run for president when *Moby-Dick* was published—could have stood inside of a right whale's mouth, even while wearing his top hat. The right whale's head, which is nearly all mouth, is indeed nearly triangular. The upper jaw bones curve narrowly beside each other. Southern right whales have between 220 and 260 baleen plates on each side of their mouth, some of which can extend to nine feet long. Bowhead whales of the Arctic can have as many as 360 plates per side, with the longest at least four-teen feet long.[9]

In 1855 a greenhand named Robert Weir, who I mentioned earlier regarding his observations of swordfish, began keeping a journal of a whaling voyage out of New Bedford aboard the *Clara Bell*. Weir was a prolific reader, earnestly religious, and a formally trained artist. (See figs. 14, 33, and 51.) A few months into the voyage he wrote of his first impression of the right whale's head:

Got the head on deck before breakfast. And what a looking thing it was—a person could very easily stand upright in its mouth—then what

a tongue—it would weigh about a ton & a quarter who can imagine what
such a mass as that ever had life & animation—Oh! how wonderful—
The right whale has whalebone [baleen] in his mouth in lieu of teeth—
the longest length of bone from this whale is 6 ft.; from that tapering to
two or three inches at the end of his nose.[10]

Before spring steel and plastics, the baleen, about the thickness and
flexibility of your average frisbee, was cut into thin strips and used to
support umbrellas, corsets, and hoop skirts, and to make tools, shoe-
horns, horse whips, and home crafts.

In "The Right Whale's Head" Ishmael describes a technique by
which whalemen use baleen to "calculate the creature's age, as the age
of an oak by its circular rings." Ishmael approaches the idea, which is in
his fish documents, with some caution, but he declares it has "the savor
of analogical probability." It suggests to him an exceptionally long life
span for right whales. Like fingernails, baleen grows continuously. But
it wears down at the ends due to normal feeding—so you can't use it
for age. Thus a slat of baleen in these whales can be at most from ten
to twenty years old. I have a piece of baleen beside me as I write this.
At a certain angle I can see concentric treelike growth lines, but more
prominent are lines that are more straight and vertical, extending along
the length of the plate, as if each line leads to a hair at the end. Twenty-
first-century researchers have been analyzing baleen plates not so much
for age, but to provide long-term samples of hormonal changes over a
period of the whale's life. The baleen plates reveal gestation periods,
mating periods, and even the presence of stress in possible relation to
injury, changes in food supply, and climatic shifts.[11]

We know now that all of the three right whale species around the
world feed on planktonic crustaceans, especially copepods and, less
so, krill. Robert Kenney, an ecologist with an Ahabian four decades
of studying North Atlantic right whales, sums up for me the current
knowledge of the right whale's diet one afternoon while sinking his
teeth into a beef burger at a pub in Rhode Island: "All three species of
right whales feed primarily on larger copepods," he says. "Their baleen

will not efficiently filter organisms much smaller than those, although they'll feed on smaller copepods and other organisms like barnacle larvae if those are abundant enough."[12]

Phytoplankton species are too small to get caught in right whale baleen. Right whales do feed on krill at times, but to capture this large, more mobile food, they have to swim faster than normal, unlike the rorquals, which can catch krill more easily since they are faster swimmers. It becomes a question of energetic trade-offs.

"In other words," Kenney says, "right whales will probably eat anything that aggregates into sufficiently dense patches, can be effectively filtered by their baleen, and cannot swim well enough to escape."

In "Brit," Ishmael recognizes the poetic irony and the ecological miracle that the largest animals on the planet survive on some of the smallest prey. Marine invertebrates such as copepods and krill reproduce by the trillions globally. Copepods—of which there are hundreds of marine planktonic species that make up the majority of all global zooplankton biomass—nearly all look like tiny teardrops, each with antennae, paddle feet, and an exoskeleton. Copepods are likely the most numerous multicelled organisms in the sea, perhaps even on Earth. Krill, which are usually much larger individually than copepods, congregate in swarms so vast and dense that in the Southern Ocean mariners have sailed through clouds at the surface that covered an area of 175 square miles. The larger rorquals, the gulpers, more commonly feed on these swarms of krill.[13]

In *Moby-Dick* Ishmael's brit is yellow. Both copepod and krill species are more commonly pinkish-brown in their clouds at the surface, but historic and current accounts do attest they can appear yellow, green, or brown, aligning with Ishmael's description. "Yellow or green-colored water is more likely to be caused by concentrations of phytoplankton," Kenney says, "rather than zooplankton—which tends to be on the yellow to red spectrum." During his years at sea, Melville certainly could have seen patches of green or brown or yellow blooms of phytoplankton—but, again, baleen whales rarely if ever skim feed through these plant-like patches for sustenance. Then again, Christy Hudak, who samples zooplankton in Cape Cod Bay in relation to the diet of right whales, told me that she has seen clouds of copepods of the genus *Centropages* that appear more greenish-brown at the surface.[14]

With our understanding that the way Ishmael describes baleen and right whales skim feeding through large patches of organic material is remarkably accurate for his time, and we know that many zooplanktonic species would've composed those large patches, especially copepods and occasionally krill, did the author or the naturalists and whalers of his time recognize these specific organisms *within* brit?

In the 1830s, when Darwin sailed aboard the HMS *Beagle*, he observed aggregations of tiny species at the surface. Darwin collected samples to consider the exact nature of the organic slick at the surface, the color of the water, and how patches of organisms stayed together under the influence of wind and current. Darwin witnessed clouds of krill when cruising near Tierra del Fuego. He wrote in *The Voyage of the Beagle*: "I have seen narrow lines of water of a bright red colour, from the number of crustacea, which somewhat resemble in form large prawns. The sealers call them whale-food."[15]

Surgeon Beale in his *The Natural History of the Sperm Whale* had a lot to say about the food of the right whale. Beale included swaths of material from Sir William Jardine, who wrote of "the medusæ and minute fish" that supply the food for these whales. Beale wrote of "discolorations of the water caused by myriads of animalculæ which perhaps form the common black [right] whale's food, that consists of 'squillæ' and other small animals." (Melville wrote "Barley Banks" in the margin of his copy of Beale's book.) By *squillae*, Beale was likely referring to shrimp, and *animalculae* probably meant any collection of tiny animals. Beale cited these areas as "submarine pastures" and "fields of spawn." Unlike the deep water whaling grounds along the equatorial Pacific that attracted sperm whales, the whaling grounds around the world that were rich in zooplankton and attracted right whales were often called "banks." These are shallower, often more coastal areas, which because of upwelling and sunlight provide exceptionally productive habitat for planktonic organisms. One of these areas was, and is, the Brazil Banks off the Rio del Plata, and another was and is, as Ishmael describes, near the Crozet Islands. These right whale regions were described and depicted by Beale, as well as by Maury in his maps of whale abundance. Ishmael references to these banks in his note to "Brit."[16] (See earlier fig. 9.)

To learn about baleen whales in particular, Melville often turned to the writings of Williams Scoresby Jr., his most trusted expert on right whales, especially because that man had actually thrown the harpoon. Although Melville poked fun at Scoresby multiple times throughout the novel, recreating him as fictional experts and mariners by the name of "Fogo Von Slack," "Dr. Snodhead," and "Captain Sleet," he clearly had enormous respect for Scoresby as a hunter and a naturalist. From Yorkshire, England, and the son of a successful whaleman, William Jr. had been to sea aboard his father's ships beginning when he was only ten. In between voyages Scoresby studied at the University of Edinburgh—more than fifteen years before Darwin did—and then he commanded his own whaleships hunting bowheads up in northern waters and into the Arctic. Scoresby wrote two popular books about his experiences, in which he included his scientific findings on whales, magnetism, and oceanography, both often cited or pirated verbatim. Scoresby was a friend of the famous naturalist Joseph Banks and named a Fellow of the Royal Society.

In his *An Account of the Arctic Regions* (1820), Scoresby observed and collected several species of plankton in the Arctic seas, some of which were illustrated with a plate labeled: "Medusae and other animals, constituting the principle food of the whale," an image we know Melville saw while writing *Moby-Dick* (see fig. 26).[17]

Scoresby described a range of species at the surface, although he usually found only "squillae, or shrimps," presumably what we now call krill, in the whales' stomachs.[18]

Elsewhere in *Arctic Regions* and in his later *Journal of a Voyage to the Northern Whale Fishery* (1823)—both of which Melville checked out of the New York Society Library while writing *Moby-Dick*—Scoresby de-

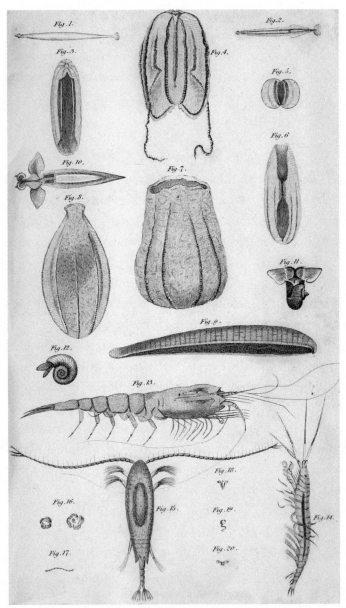

FIG. 26. "Medusæ and other animals, constituting the principle food of the whale," in William Scoresby Jr.'s *An Account of the Arctic Regions* (1820). Figs 4–7 are comb jellies; 10–12 are pteropods; 14 is an amphipod; 15 is a copepod; and 13 seems to be a drawing of either a shrimp or a krill.

scribed patches of Arctic water in several shades of green, brown, and yellow. One day Scoresby scooped up a bucketful and looked at a sample under the microscope. He wrote: "These were evidently the remains of animalcules . . . I make no doubt, but they are of a kind similar to that which gives the yellowish-green colour to the sea."[19]

ISHMAEL'S MORNING MOWERS

Ishmael's description of the *sound* of the right whales feeding in "Brit," the "strange, grassy cutting sound," also suggests that he infused his own time at sea into this scene, because I haven't read of anything anyone wrote in the nineteenth century about *listening* to whales feeding. As far as I've found so far, it wasn't until a 1976 scientific paper by Bill Watkins and Bill Schevill, in which someone reported a sound they called "baleen rattle." Audible from above the water and with the aid of hydrophones beneath, it was believed to be the result of the clattering of the baleen plates together due to small waves entering the mouth while skim feeding, potentially even as a way to communicate feeding success to others. The sound is not, as Melville implies in his description, the crunching through the brit like "morning mowers." Melville likely wrote this description of the sound either purely from his imagination for fictional purposes, from a description he heard from sailors' yarns, or, I like to think, from his own direct, close experience in the seat of an engine-less whaleboat watching and listening to right whales feeding on the surface.[20]

PLANKTON AS THE BOUNTY OF PROVIDENCE

Scoresby wrote quantitatively and philosophically about the enormous number of these small individual "animalcules" that inhabit the ocean and serve as whale food. In Arctic Regions, after trying to calculate the number of these individuals, Scoresby expounded from his own natural theological masthead:

What a stupendous idea this fact gives of the immensity of creation, and of the bounty of Divine Providence, in furnishing such a profusion of life in a region so remote from the habitations of men! ... These animals are not without their evident economy, as on their existence possibly depends the beginning of the whole race of mysticete, and some other species of cetaceous animals ... thus producing a dependant chain of animal life, one particular link of which being destroyed, the whole must necessarily perish.[21]

Scoresby's commentary on whale food surely influenced Melville's writing in "Brit," especially in this chapter when Ishmael expands out into the significance of the sea to all global life. Scoresby, for his part, revealed an early understanding of food webs and the critical importance of what we now call primary and secondary productivity in the ocean. Two decades later, Dr. Bennett described the ocean food system in terms of a column, in which the tiny "mollusks" formed the base, while at the top of the column were the whales.[22]

Melville, of course, in addition perhaps to the sounds of the whales feeding, added a few splashes of fictional license to make the "Brit" chapter whole and dripping with metaphor as his *Pequod* sails peacefully in the Indian Ocean beside these right whales placidly feeding. Ishmael opens the scene with a view from the masthead: a vision of grassy yellow: the pastoral, the peaceful. Ishmael recognizes that brit is the base of Scoresby's pelagic food chain, just as bread is the human staple or grass is the primary food for herds of herbivorous land animals. Melville likely wrote this scene while working on his small farm. Historians estimate that in the mid-nineteenth century about 80 percent of the American population lived on farms. Whalemen, in turn, often perceived their work at sea in agrarian terms. In Obed Macy's *The History of Nantucket* (1835), the "worthy Obed," as Ishmael calls him, explained: "In the year 1690 ... some persons were on a high hill ... observing the whales spouting and sporting with each other, when one observed '*there*,' pointing to the sea, '*is a green pasture where our children's grand-*

children will go for bread.'" In the "Nantucket" chapter Ishmael declares how the island's whaleman hero is the most comfortable and suited to the sea, "he alone, in Bible language, goes down to it in ships; to and fro ploughing it as his own special plantation ... He lives on the sea, as prairie cocks in the prairie." Less than a year before beginning *Moby-Dick*, Melville wrote a review of Francis Parkman's *California and Oregon Trail* (1849), a narrative that helped inspire much of the prairie and grasslands imagery that Melville wrote into his novel. Melville was well aware, even in his time, of a sad shrinking and taming of the American West, which Ishmael contrasts with a boundless, ferocious, and forever unspoiled ocean.[23]

The right whales in "Brit" graze through "golden wheat," creating a visual image of wide blue stripes expanding through yellow. Metaphorically, the land (and thus sanity: consider the phrase *to be grounded*) is disappearing as the *Pequod* and her crew progress eastward, deeper into the Indian Ocean. Melville's baleen whales are removing the brit, like "long wet grass," leaving only deep blue, slowly wiping clean all traces of the color that represents the safety of the shore. In "Brit," Melville advanced that master theme of *Moby-Dick*, first emphasized in "The Lee Shore," that the sea is a profoundly different place than the land. The sea is immortal and vicious and one should never be fooled by what appears to be a calm scene of whales grazing on crustacea. The sailors from the mastheads first mistake the whales feeding as stationary rocks, looking like elephants: the slow-moving, grazing, vegetarian, largest of living land animals. Quickly Ishmael pivots to state that this comparison to elephants that he just offered is misleading, even false. No. Animals at sea are entirely different from those on land. There are no direct analogies. Here Ishmael alludes to his skepticism of the "old naturalists" by first pointing out that, despite what they say, there is no comparing land animals to those of the sea. Ishmael declares that under the surface are "numberless unknown worlds" that humankind has yet to discover. His lack of clarity or specificity about what is the actual makeup of "the peculiar substance called brit" is part of the point. Our society's knowledge of the ocean is enormously and inevitably limited—both about the sea's inhabitants and its overall dangers—expressed in that single spiked club of a sentence with which I ended the introduction to this natural history. By choosing the sailor's word *brit*, instead of Darwin's or Beale's or Scoresby's *crustacea*, *animaculae*, or *medusae*—wonderful words that Melville resisted including—he once again sided with those who knew the vast wild ocean most empirically: the whalemen.[24]

To close "Brit," Melville doubled-down on the difference between land and sea. Just as in "The Lee Shore," Ishmael warns of a physical or existential departure from the safety of land and home. Melville's nineteenth-century sea is merciless, "savage," and "masterless." For his metaphysical end to the chapter, he returns his reader back to the pastoral, the "green, gentle, and most docile earth," the tiniest of creatures. The whales are still slowly eating away at the brit, erasing the ground, removing safety, and nibbling closer to the peaceful "insular Tahiti" in the soul of man. After the brit is chomped away, "thou canst never return."

Ishmael, the ancient mariner, blesses these tiny creatures anyway.

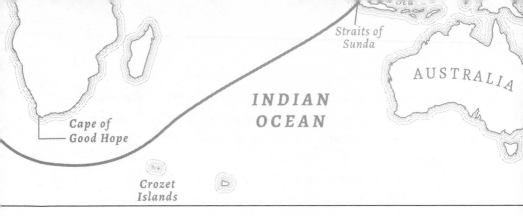

INDIAN
OCEAN

Straits of
Sunda

AUSTRALIA

Cape of
Good Hope

Crozet
Islands

Ch. <u>12</u>

GIANT SQUID

> The great live squid, which, they say, few whale-ships ever beheld, and returned
> to their ports to tell of it.
> Starbuck, "Squid"

> Leviathan is not the biggest fish; — I have heard of Krakens.
> Melville in a letter to Nathaniel Hawthorne, reflecting on his next book,
> after *Moby-Dick* bombed[1]

In "Squid," the chapter directly after "Brit," the whalemen of the *Pequod* come upon a giant squid on the surface of the Indian Ocean. It's an eerily calm day with a "languid breeze." From aloft, Daggoo, the African harpooner, mistakes it for the White Whale. As they lower the boats and row toward it, the creature continues to rise and sink beneath the surface. Ishmael says:

> Almost forgetting for the moment all thoughts of Moby Dick, we now gazed at the most wondrous phenomenon which the secret seas have hitherto revealed to mankind. A vast pulpy mass, furlongs in length and breadth, of a glancing cream-color, lay floating on the water, innumerable long arms radiating from its centre, and curling and twisting like a nest of anacondas, as if blindly to clutch at any hapless object within reach. No perceptible face or front did it have; no conceivable token of

either sensation or instinct; but undulated there on the billows, an unearthly, formless, chance-like apparition of life.[2]

The giant squid in *Moby-Dick*, as remains true today, is one of the
most elusive and unknown large animals on the planet. Ishmael ends
"Brit" warning that the sea's "most dreaded creatures glide under water,
unapparent for the most part, and treacherously hidden beneath the
loveliest tints of azure." As if conjured by this, the *Pequod* raises a giant
squid in the next scene. Thus Melville created one of his tidy chapter
pairs about the diet of the great whales: "Brit" is an essay-style discussion of the right whale's food, while "Squid," in a narrative form
through his characters, explores the sperm whale's food.[3]

Ishmael's giant squid is unknown to anyone ashore, enormously long
and wide, and witnessed by only a few sailors at sea. He explains that the
giant squid makes up the sole food of the sperm whale, which has huge
teeth in order to eat them. He says that naturalists place giant squid
in the same class as cuttlefish, that they cling to the seabed with their
arms, and that some of the legends of the Kraken and other sea monsters might reasonably be explained as encounters with the giant squid.
What of this is true to what we know today?

GIANT SQUID PICKLED, PART I

It's taken me two days of observing, sketching, photographing, lying
underneath the tank, and standing up on a ladder looking down on the
tank, to come to the realization that despite the enormity and rarity, I
am unexpectedly and entirely underwhelmed by Archie. The specimen
is so large that submerged in its olive-yellow fluid it looks more like a
sunken puppet or a faded beanbag. Tiny tendrils of cottony tissue twirl
out from the body, arms, and tentacles. It appears sewn of pink insu-
lation.

Preserved in a thick, clear acrylic coffin holding some 870 gallons of
a 4 percent formalin-saline solution, this giant squid (*Architeuthis dux*)
is twenty-eight feet long stretched flat from the tip of the mantle to the

end of the two tentacles. It is one of the largest complete specimens of its kind in the world, but not nearly the largest of its species. In 2004 the crew of a fishing boat off the Falkland Islands, just a few hundred cold hard miles from Cape Horn, unintentionally netted the animal in a trawl. They froze it. When the men returned to the dock they donated the squid to the biologists at the local fisheries department, who in turn shipped this specimen of the largest invertebrate on Earth up here to the Natural History Museum in London. Though the squid is a female—she does not have a three-foot-long penis sticking out of her mantle beside her head—the British press named her Archie, after the scientific name.[4]

My host here at the museum has been Jon Ablett, Senior Curator of Nonmarine Mollusca and Cephalopoda. Bookish and boyish, he wears thick glasses and an untucked flannel shirt. Ablett has been escorting me to the tank room each day, where I'm now alone and coming to grips with my Kraken crisis of unrealized expectations.

The tank room is a clinical, industrial space in the basement of the museum's Darwin Centre Spirit Building. The room is quiet and safe from the thousands of visitors on the floors above who are stomping around in wellies and swarming around the dinosaur skeletons, taxidermy African animals, and a great white statue of Charles Darwin presiding over the great hall, the same "cathedral to nature" that formed the center of the Natural History Museum after it branched out from the British Museum in 1881, led by Richard Owen, the same fossil expert that Ishmael refers to a couple times in *Moby-Dick* and whom we'll talk about more later. Ablett and his fellow curators have filled the shelves along the walls in this subterranean tank room with all manner of glass jars that preserve specimens of fish, reptiles, and a few small mammals, such as a desert fox. Some of the specimens date back to the early nineteenth century. These include, in the only locked cabinet, a collection of some of Darwin's pickled specimens from his expedition aboard the *Beagle*.

The floor of the tank room is filled with five rows of multiple stainless-steel tanks, each an enormous vault of alcohol that holds specimens too large for the bottles.

"They can be a little disturbing inside," Ablett said when he first showed me around, because these tanks preserve, among other specimens, a ten-foot-long Greenland shark, a thresher shark, a full-sized marlin, an eagle ray, and the head from a thirteen-foot Russian sturgeon.[5]

Archie's tank dominates the room. It fills half the central row. What I initially thought was her eye, is actually the funnel—the exhaust out which the squid flushes waste, ink, and the seawater sucked across its gills. Water thrust out of this funnel is the squid's jet propulsion, how it speeds through the sea. The giant squid's eyes are more than three times larger than those of an ostrich or a swordfish or even a blue whale. The diameter of her eye is nearly eleven inches, about the size of your head. These eyes enable her to see in the dark water of her preferred depth, about 400 to 3,300 feet. One theory is that giant squid eyes evolved to be so large in order to see the bioluminescence around a sperm whale, its predator, diving after them.[6]

Yet Archie's eyes are no longer visible in its coffin tank. After time in preservative, the eyes have sunk in and collapsed, as if gathered in fabric before being turned inside out. So maybe it's the lack of eye that makes the appearance of this specimen less powerful: thousands of eyes stare out from their bottles down here, but this giant squid stares at nothing.

What turns out then to be the more impressive aspect of this giant squid is her "clubs," which are still very much intact. Squids, by definition, have two tentacles and eight arms that extend out from their head and encircle their mouth. Each arm has two rows of suckers. The two tentacles, which are more than twice as long as the arms and seem impossibly delicate, have suckers only at the ends, on the clubs. These clubs, shaped like spatulas, hold a field of circular suckers similar to those on the arms, each inlaid with sharp-toothed rings, like miniature band saws. Scientists believe giant squid are active predators of midwater fish and other squid species, but their diet and feeding behavior, along with pretty much everything else about them, remains a mystery. The first real records of the stomach contents of mature giant squid, usually much masticated and barely identifiable, were not published until the late twentieth century.[7]

Archie is one of roughly fifty or so known complete specimens of giant squid in collections around the world. These are among some five hundred recorded ever, anywhere, found in a range of physical states—just the mantle here, only three arms washed up there—that were once measured and recorded carefully or whose length a biologist, beachcomber, fisherman, or whaleman estimated out in the field. Addison Verrill, a former student of Louis Agassiz and then a professor of zoology at Yale, was the first scientist to publish a trusted analysis and assemble the first set of illustrations and one photograph, gathered from several specimens of giant squid that began appearing in 1871, found floating at the surface off Newfoundland. One of the squid, still alive, reached into a small boat of fishermen, who brought back ashore a nineteen-foot tentacle that they chopped off with an axe. The first photograph of a live giant squid underwater was not captured until 2004, when a Japanese biologist named Tsunemi Kubodera used an automatic camera and a light placed on a line 2,950 feet deep. He and his colleagues used a bag of squid and shrimp as bait, which they fixed beneath a camera. When they brought the line back up, after the camera had snapped the photographs, the giant squid's eighteen-foot tentacle had torn off. Back up on the boat's deck, the tentacle still grasped at the scientists' fingers. Eight years later, in 2012, researchers for the first time captured a video of a live giant squid at depth.[8]

As I'm staring at Archie's clubs and trying to imagine her swimming and hunting in the dark, Jon Ablett walks in through the door from the dissection room. He tells me that he got bored halfway through his first read of *Moby-Dick*, but then he reread it in full a couple years later in the rain forests of Vietnam, while looking for new species of terrestrial snails.

He asks me how things are going.

I tell him I think I built up the squid too much in my imagination.

He points above Archie's clubs to a submerged plexiglas shelf that holds a scrap of an arm that I'd assumed was from another giant squid.

"This is from the colossal squid," he says. "*Mesonychoteuthis hamiltoni*. It's a relative of the giant squid but less closely related than you'd think. By weight, colossal squids get even larger than *Architeuthis*."

He explains that one colossal squid specimen caught by fishermen in Antarctic waters in 2007 weighed over one thousand pounds.

Could Melville have seen one of these?

"Not likely. Unless he sailed in Antarctic waters. But look more closely at its suckers."

Instead of the inset band saws, the colossal squid's suckers each have a menacing hook curling out of each one, creating a teeth-like row that looks like the inside of a mako shark's mouth.

"And each hook can flex and twist as much as one hundred-eighty degrees when attached to prey," Ablett says.

Next, in a tall jar on a table to the left, is another colossal squid tentacle. Ablett says this one was found in the belly of a sperm whale, which is how, in addition to specimens washed ashore, scientists have learned the most about giant and colossal squids—although bits of smaller specimens of giant squids have been recovered in the stomach and regurgitants of wandering albatross, sharks, toothfish, and swordfish.[9]

Aware that I'm starting to come around, he brings out juicier bait. Ablett walks me over to one of the stainless-steel tanks in another part of the basement.

A GREAT WHITE SQUID?

Nineteenth-century whalemen and the Polynesian voyagers centuries before them were the humans who would have had the best chance to see giant squids alive at sea, but no entirely reliable records are known, alive or dead. We don't know whether Melville ever saw a giant squid himself. Again, no museums had any collections back then, and no one in the 1850s could have seen a published illustration of a giant squid that was not an entirely fantastical sea monster. Nelson Cole Haley, a harpooner aboard the *Charles W. Morgan* in the early 1850s, wrote one of the few accounts of the sighting of an alleged giant squid, which recalls, albeit written three decades later, one morning north of New Zealand when he saw three enormous cylindrical forms, the largest about three hundred feet long. He adamantly swears it was three giant squids, but he saw no tentacles or arms, and no one else on the ship got to see them.[10]

Melville was ahead of his time in wondering whether various reports of sea monsters might have been in fact giant squid. While Melville was writing *Moby-Dick*, reports of a sea serpent off southern New England and New York Harbor were all over the newspapers, which included a range of theories as to what it actually was. Even Agassiz chimed in to state that he "can no longer doubt the existence of some large marine reptile," even though it had not yet been identified. We know the alleged serpent was even a topic of conversation one night when Melville was at a dinner party with Hawthorne in August 1850. Scholars and scientists continue to debate these sightings and similar historical accounts, with most analysts today agreeing with Melville in the conclusion that many of the early sea monster reports were probably cases of giant squid up on the surface. Others argue, however, that some or most of these sightings simply could not equate with the geography or behavior of the giant squid.[11]

In *Moby-Dick*, Ishmael declares that the squid that the men of the *Pequod* see on the surface is "furlongs in length and breadth." Clyde Roper, the senior statesman of all squid experts in the world and another Ahabian veteran of forty years of ocean research, told me his own first impression of this scene in *Moby-Dick*: "Dear old Ishmael certainly must have had a triple ration of grog that day."[12]

A furlong is an agricultural measurement, common in Melville's America. It's equivalent to over two football fields at 660 feet. Melville used a furlong for yarn-spinning purposes and perhaps to connect to the pastoral images with "Brit." Melville, of course, had no real idea of the size of giant squid. Throughout the novel, Ishmael actually casts doubt on previous, far larger mythical lengths of whales and other sea monsters. Perhaps ironically, considering his own vast exaggeration to the length of his giant squid, at the end of this scene Ishmael pleads prudence when he remarks on the Danish historian Pontoppidan's 1755 account of the Kraken: "But much abatement is necessary with respect to the incredible bulk he assigns it." This is because Pontoppidan declared the Kraken—which he thought could be squid-like or maybe even a massive sea star—to be more than a mile and a half in circumference. Pontoppidan, for his part, also claims that *he* is being conservative in the information *he* relays.[13]

Melville would be pleased to know that our understanding of how large the giant squid might grow remains debated even today. As he walks me over toward another, mystery tank, Ablett says that he, too, questioned the "furlongs" when he read this chapter in *Moby-Dick*. "The current thinking is that the giant squid might get to be between ten and thirteen meters [33′–43′] total length," he said, "far longer than Archie. Scientists know that females grow larger than males. And the colossal squid, a *guesstimate*, might grow to be eighteen meters [59′]. Although the current *current* thinking is that they both might be still bigger."

Ablett pictures the giant squid in *Moby-Dick* as unhealthy or dying, since that would be the only time a giant squid has been observed at the surface and would be so white.

"When Archie was caught," Ablett tells me, "she was in good health. She was a bright magenta. Squid and octopus have elastic chromatophores. These are cells that contain pigment, which under muscular control allow it to change color. Giant squid have them too, although we think they can only go to red-dy, silvery, or beige. In the description in *Moby-Dick* it's a creamy color. I know, I know, Melville was interested in the color white. But the image is completely at odds of how I know Archie. Sadly, their colors fade over time because of the preservative, but even when she arrived after being frozen for six months on a ship, she was a deep magenta, a deep red color on its dorsal side."

In thinking out loud, however, Ablett recognizes that few if any observers have ever seen a live giant squid at the surface in all its redness.

These animals simply do not come up unless they're floating dead or have been brought up by a sperm whale. Giant squid that people find washed up on the beach seem to always be white, because their pigment has disappeared in death and also the skin has been scraped off by the sand or the rocks. (See plate 9.) Ablett muses that maybe large squids might even be observed floating on the surface for some unknown behavior.

"Just so little is known," Ablett says.

For Melville's composition of "Squid," biological accuracy in this case seems a secondary interest to poetic effect. Ishmael describes an unknown white ghostly creature: a seemingly less-honest wonder, although based on Cole's account perhaps Melville would have genuinely believed in a giant squid that large. Regardless, in his novel, the sighting provides similar narrative effects as that of the spirit spout. The giant squid is yet another evil omen. Starbuck, the faithful rational Christian, is terrified of the thing.

Ishmael soon flips this, however, by explaining that the sighting of the vast white giant squid is a lucky harbinger of sperm whales—understood and introduced by the man on board whom the reader might assume to be far more superstitious than Starbuck. In "Stubb kills a Whale," two chapters after "Squid," Queequeg reveals himself to be the more experienced and more rational observer, because mariners since at least the late eighteenth century understood that squid was the primary food of the sperm whale.[14] Ishmael opens the scene:

> If to Starbuck the apparition of the Squid was a thing of portents, to Queequeg it was quite a different object. "When you see him 'quid," said the savage, honing his harpoon in the bow of his hoisted boat, "then you quick see him 'parm [spermaceti] whale."[15]

Just as actual whalemen tracked fields of brit to find right whales, the men at the masthead, like Queequeg, took care to look out for squid parts and smaller squid to try to locate sperm whales. Whalemen on the world's oceans often saw sperm whales vomiting up squid parts during the violence of being harpooned. In *The Whale and His Captors*, Reverend Cheever wrote that the sperm whale "is supported principally by the squid, otherwise called cuttle-fish, or Sepia Octopus, of which one sperm whale that we have lately captured disgorged pieces as long as the whale boat." Surgeon Beale explained that whenever he found large bitten-off squid limbs on the surface he inevitably saw sperm whales "within a few hours." Sperm whales eat fish of a range of sizes—from schooling groundfish to sharks, but squid make up the majority of their global diet. Sperm whales prey on giant and colossal squid, as well as rugose hooked squid (*Moroteuthis robsoni*), Dana octopus squid (*Taningia danae*), scaled squid (*Lepidoteuthis grimaldii*), and dozens of other species of deep ocean squids, most of which still do not have common names.[16]

Melville read still more in his fish documents. Dr. Bennett wrote of the unknown "gigantic squid" that has been mistaken as "a reef, or shoal." Olmsted wrote of a three-foot squid, "a flabby mass of a white color," found in the mouth of a captured pilot whale. In the full chapter on squid in his copy of Beale's *The Natural History of the Sperm Whale*, Melville commented in the margins and heavily underlined several passages, including paragraphs on flying squid and an eighteenth-century account of a "gigantic cephalopod" collected by Joseph Banks on Cook's first voyage, "found floating dead upon the sea." Banks brought back one of the arms with its suckers and talons, too. The afterguard ate most of the rest of the carcass. Banks wrote in his journal that night: "Of it was made one of the best soups I ever ate."[17]

Ishmael learned from his reading that the giant squid clings to the seabed with its arms, perhaps as a defense against sperm whales. Marine biologists today do not believe this to be true. Giant squid are midwater, deep sea inhabitants, catching prey actively by swimming after them.[18]

Ishmael describes the "spikes," the "long, sharp teeth of the sperm whale," claiming that the animal can tear apart human limbs and giant squid off the bottom. Sperm whales do have twenty to twenty-six pairs in the lower jaw that can grow to be over ten inches long, but sperm whale teeth are not well adapted for chewing or ripping squid or fish (or human limbs). They clamp down on slippery animals in order to swallow them whole with their backward-thrusting tongues. Although sperm whales have scars on their skin from squid suckers and hooks, the squid parts found inside sperm whale stomachs have only punctures from sperm whale teeth, never rips or lacerations. I once held a huge chunk of a warty squid (*Onykia ingens*) that had two perfect holes made by the teeth of a sperm whale. We found it on the surface after a male sperm whale had sounded.

For centuries whalemen and naturalists wondered how sperm whales in the dark depths of the ocean find and catch their prey. Beale thought the whales sunk down, opened their mouths and simply waited with their glittering mouths and teeth. Now scientists are confident that the hunting strategy of sperm whales involves echolocation. One theory, if highly unlikely, suggests that sperm whales actually stun the squid with a focused beam of sound, known as "acoustic debilitation" or a "sonic boom."[19]

GIANT SQUID PICKLED, PART II

Down in the basement of the Natural History Museum in London, Ablett pauses on the way to the mystery tank to show me a tall glass bottle of alcohol with what looks like a dark brown shell. It's the beak of a colossal squid.

At the center of the base of every squid's arms is a beak that is as sharp and strong as a parrot's. It looks like it's been made of black plastic. It's actually chitin, the stuff of lobster exoskeletons. This particular colossal squid beak was recovered from the stomach of a sperm whale. It was donated to the museum by Malcolm Clarke, the same giant squid specialist who once found a sperm whale beached in Cornwall that held in its stomach the beaks from forty-seven giant squid.[20]

Ablett puts on a white coat. He takes off his glasses and places them on another lid behind him. "I've had them drop into a tank of alcohol before," he says. "Very hard to get rid of the smell."

He and another curator wrestle the lid off to the side. Ablett pulls on thick rubber gloves that extend nearly to his shoulders. We peer into the dark rusty-colored fluid. I can just make out a sack of swirled gauze. The smell of alcohol is strong, but not overpowering.

"This is the *Architeuthis* that washed up on the beach of Scarborough in 1933," he says. He reaches his own arms all the way into the tank and begins fishing around for the giant squid's arms. He tries to disentangle the squid from a second specimen's arms, which are also wrapped loosely in the sunken twisted fabric.

I think I see a plate-sized eye peering out at me through the liquid.

"What? Oh, no, that's the other one," Ablett says, splashing around. "Where *is* the label? Right, okay, here's a nice arm then. From the Scarborough specimen. This whole squid is nearly complete, but I think the story is that when it arrived on the beach everyone in town started attacking it, and they ripped up the tentacles and messed it up pretty badly until someone could come down to stop them. Just look at this arm!"

I ask to touch it. He slides over a box of disposable gloves.

The arm feels far thicker, tougher than I'd expected. It's like a slippery hawser from a tugboat.

"You ever eat calamari?" Ablett says. "Or a big piece of octopus sushi? Think about it? This is all muscle."

I look over at Archie and my impression of her delicacy sloshes away. The eyeless face doesn't matter anymore. I appreciate now Melville's description of the arms like a "nest of anacondas," which is more apt than I suspect he realized. Still trying to find the label, Jon Ablett searches around, in past his elbows, trying to separate the other specimen. I'm still captivated by the arm. I'm still holding it, weighing the heft of it, considering what crushing constriction power she must have possessed. I feel around the suckers and rasp one of these saw rings against my glove. It splits the latex clean open to the skin.

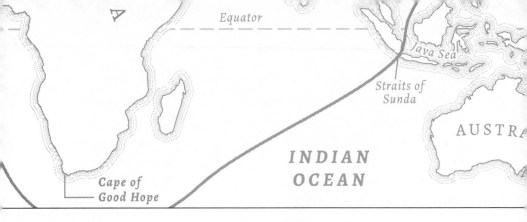

Equator

Java Sea

Straits of
Sunda

INDIAN
OCEAN

AUSTRA

Cape of
Good Hope

Ch. 13

SHARKS

You is sharks, sartin; but if you gobern de shark in you, why den you be angel;
for all angel is not'ing more dan de shark well goberned.
    Fleece, "Stubb's Supper"

After the sea-ravens and the brit and witnessing the giant squid in all its enormous, nebulous whiteness, the men of the *Pequod* continue eastbound in the Indian Ocean. As Queequeg predicted, they soon see sperm whales. Stubb kills the first one of the voyage. They lash it alongside. Night falls. Now enter the sharks.

Introductions to shark field guides and other quick studies of the American perception of sharks often explain that it was twentieth-century attacks on coastal swimmers and the stories about shipwrecked soldiers mauled and eaten by sharks while floating at sea in World War II that kicked off an American cultural fear and hatred of these animals— an automatic fear response when an author or filmmaker sends a dorsal fin gliding into a scene. The correlation with sharks was then expanded and cemented by the likes of Peter Benchley's novel *Jaws* (1974)—which has distinct nods to *Moby-Dick*—then far more so by Steven Spielberg's film version the following year. A flood of sensationalized documentaries followed, bleeding into today's "Shark Week." Our current fear of sharks, these authors suggest, is the result of a recently learned, mostly unreasonable cultural fear of these animals. This learned fear theory of the American perception of sharks might be valid for nearly all of us who never come into contact with these animals in our professions or recreational life, but the chronology of this cultural antagonism fails to account for the American whalemen in previous centuries. For those aboard whaleships, those who had regular relationships with deep ocean sharks, familiarity bred contempt. Whalemen hated sharks, reserving for them a level of cruelty that they inflicted on no other animal. And they brought their stories home.

ISHMAEL'S SHARKS AND SHARKISHNESS

Throughout *Moby-Dick*, Ishmael uses the word and image of sharks to evoke lawlessness, danger, and ferocity. Ishmael says that whalemen have explored "the heathenish sharked waters" of the Pacific, beyond where no other mariners had dared. Starbuck soliloquizes on his "heathen crew" that was "whelped somewhere by the sharkish sea." And along with swordfish and whales, sharks are part of the "murderous thinkings of the masculine sea."[1]

In other parts of his storytelling, Ishmael uses sharks to evoke nightmares and ghosts. The crunching teeth of sharks are the stuff of the sailors' bad dreams, evoked by Ahab's crunching the decks over their heads with his ivory leg. "Insatiate sharks" swim around a whale corpse floating off in the distance, becoming a ghostly myth and metaphor for the dangers of religious orthodoxy. Sharks swim around a dead whale that Fedallah and Ahab float beside one night: the sharks sounding, with biblical allusions, "like the moaning in squadrons . . . of unforgiven ghosts." Only in "The Whiteness of the Whale" does Melville specify a kind of shark, the great white shark (*Carcharodon carcharias*), with which he furthers the ghostly imagery and elevates the power of his White Whale. Ishmael's White Shark lurks in "white gliding ghostliness." He explains in a footnote that in French this shark is called *Requin*, sharing the name of the "funereal music" of the requiem at the Catholic Mass for the dead.[2]

Ishmael gives sharks their most significant role in the closely aligned chapters "Stubb's Supper," "The Shark Massacre," and "The Monkey-rope." The whalemen have killed their first sperm whale and lashed it alongside. Sharks scavenge and tear off pieces of the bloody carcass. Ishmael explains that if they were in the equatorial region where sharks were more common, the whale would be but a skeleton by morning, but since the ship is not, Stubb takes the first watch while the rest of the crew sleep so they may rest for the labor of processing the whale the next day. Still, the ship is surrounded by "thousands on thousands of sharks."[3]

In "Stubb's Supper," Ishmael describes the sights and sounds of sharks scavenging on a whale as he directly connects these sharks with Stubb, who has awoken the cook, Fleece, to prepare him a whale meat steak. Stubb prefers the whale especially rare, just as the sharks eat their rare meat in the water, "mingling their mumblings with his own mastications." Ishmael also associates these sharks with dogs around humans at a table, or a scene of battle, waiting in the shadows while men kill each other as warriors or slavers:

> Though amid all the smoking horror and diabolism of a sea-fight . . . the valiant butchers over the deck-table are thus cannibally carving each other's live meat with carving-knives all gilded and tasseled, the sharks, also, with their jewel-hilted mouths, are quarrelsomely carving away under the table at the dead meat; and though, were you to turn the whole affair upside down, it would still be pretty much the same thing, that is to say, a shocking sharkish business enough for all parties; and though sharks also are the invariable outriders of all slave ships crossing the Atlantic, systematically trotting alongside, to be handy in case a parcel is to be carried anywhere, or a dead slave to be decently buried.[4]

Ishmael continued that meandering, encircling sentence to explain that rarely do sharks collect in such numbers anywhere at sea than around a whaleship at night. Then he punctuates the paragraph with a snap: "If you have never seen that sight, then suspend your decision about the propriety of devil-worship, and the expediency of conciliating the devil."[5]

In "The Shark Massacre," Ishmael turns to the sharks themselves, down in the water. While Stubb works on his, the sharks work on their own meal. Queequeg and another sailor lower lanterns over the side and stand on the cutting stage to try to kill and hack away at the sharks in order to reduce the loss of blubber. Looking over the rail at night to watch this feeding is a scene of Gothic horror:

> These two mariners, darting their long whaling-spades, kept up an incessant murdering of the sharks, by striking the keen steel deep into their skulls, seemingly their only vital part. But in the foamy confusion of their mixed and struggling hosts, the marksmen could not always hit their mark; and this brought about new revelations of the incredible ferocity of the foe. They viciously snapped, not only at each other's disembowelments, but like flexible bows, bent round, and bit their own; till those entrails seemed swallowed over and over again by the same mouth, to be oppositely voided by the gaping wound.[6]

This is the most grotesque vision in all of *Moby-Dick*. It's more graphic than anything you'll read in *Frankenstein* (1818), *Dracula* (1897), or Poe's *The Narrative of Arthur Gordon Pym* (1838). Ishmael gives a dead shark the ability to clank its jaws closed, nearly chomping off Queequeg's hand. "It was unsafe to meddle with the corpses and ghosts of these creatures," Ishmael says as Queequeg wonders what sort of heathen God would create a shark.[7]

At the end of *Moby-Dick*, sharks have a different role in the drama. On the third day of the final chase, the "unpitying sharks" lurk and crunch at the oars as Ahab's boat tries to stroke after the White Whale. The sharks follow and snap at Ahab's boat alone. While they pursue Moby Dick, Ahab wonders aloud whether the sharks want to feast on the whale or on him. Eco-critically minded readers might interpret here that the sharks are instead trying to protect the whale, their own apex predator leader.

The final cameo of the sharks in *Moby-Dick* is in the final image of the novel. In "The Epilogue" Ishmael floats alone on his coffin: "the unharming sharks, they glided by as if with padlocks on their mouths."[8]

So Melville fed the period fear and contempt for sharks, writing of these fish as a ghastly, fierce, and cannibalistic metaphor and also as a very real masticating menace to the men of the *Pequod*. He saved his most gruesome, horrific imagery in the novel for sharks. Yet in other ways, Melville wrote of these predators in a more tempered manner than did his whalemen contemporaries and even the naturalists ashore, describing his sharks with a surprising, subtle degree of sympathy. Melville had Ishmael and Fleece explain that humankind is no more ethical than sharks, and perhaps even more cruel and insatiable. In *Moby-Dick*, humans are more sharkish than the sharks, first hinted by the departure of the pilot fish from sharkish Ahab. As ichthyologists today learn more and more about sharks, we see that once again, Melville was sneakily, surprisingly, accurate when read from a twenty-first-century biological perspective. For example, Ishmael's sharks will eat humans on purpose only if the people are already dead: "A thing altogether incredible were it not that attracted by such prey as a dead whale, the otherwise miscellaneously carnivorous shark will seldom touch a man."[9]

THE WHALEMAN'S PERCEPTION OF SHARKS
IN THE NINETEENTH CENTURY

Francis Allyn Olmsted, whom we've already turned to regarding his accounts of albatross, pilot fish, and squid, was the American answer to the British surgeon-naturalists Beale and Bennett. Only a few weeks older than Melville, he boarded a whaleship in 1839 out of New London, Connecticut. Similar to Dana, who in 1834 left Harvard to go to sea in part to cure his bad eyesight, Olmsted had just graduated Yale and boarded a ship for his own health. Olmsted was in much worse shape than Dana, however. In his *Incidents of a Whaling Voyage*, he refers to himself as an "invalid," but he doesn't specify his symptoms beyond "a chronic debility of the nervous system." Olmsted shipped at the age of twenty on the whaleship *North America* as a gentleman-naturalist. He returned in February 1841 from his voyage aboard a merchant ship, which dropped anchor off Sandy Hook, New Jersey, as Melville was barely a month outward bound from New Bedford aboard the *Acushnet*. Olmsted hustled home to New Haven and sent out his narrative, which was snatched up immediately and published the same year. Back at Yale, he earned a medical degree with a dissertation on the role of how narcotics helped to cure insanity. Then he got so sick again that he tried to go to sea once more, but this time he returned home to die less than a week after he turned twenty-five.[10]

Notably different from most other naturalists and whalemen authors at the time, Olmsted did his own drawings for his narrative. His images in *Incidents of a Whaling Voyage* were the first commercially published illustrations of American whaling (see later fig. 42).[11]

Olmsted was a broad-minded field naturalist who was equally comfortable with a gun, a dissecting kit, or a scientific paper in his hands. Olmsted's father was a professor at the University of North Carolina who moved his family to Connecticut to take a post at Yale in mathematics and natural philosophy. Also an author of text books and an inventor, he cultivated a wide range of interests that included studies of hail stones and shooting stars. So Francis grew up in a household that valued science and a careful, empirical eye. In contrast to the bowing formality of Surgeon Beale, Olmsted in his narrative often slipped in slivers of self-deprecatory, subtle humor. Although he had not yet had any medical training, the captain and men aboard the *North America* soon began referring to him as the "doctor." He was put in charge of the ship's medical chest, with which he did his best to care for his shipmates and the locals that they met who requested aid. When the ship made its first port stop in Ecuador after five months at sea, Olmsted wrote that since the locals wanted confirmation of their health: "It is a singular fact that in proportion to the beauty of the fair applicant, a longer time was required to count the pulsations of her arm." When crossing the equator on the way home, Olmsted put a piece of string across the spyglass so the passengers' kids could see "the *line*."[12]

Olmsted often described great abundance in the South Pacific: "Among the various amusements which make the time pass away pleasantly aboard ship, catching fish is one of the most agreeable. Vast schools of fish frequently accompany ships for several days in succession, and whalers are often surrounded for month after month by countless hosts of the finny tribe." Olmsted caught a range of fish himself. In order to catch sharks, in particular, he used a stout hook with chain.[13]

One April day to the west of the Galápagos Islands, the crew killed a large male sperm whale and rowed it back to the ship. Lashed to the side, the whale was left overnight. In order to rest, the crew tolerated a collection of sharks, just as does the crew of the *Pequod*. The next day Olmsted captured six or seven "Peaked-Nose Sharks," each about seven feet long. He said these were also known as the "Blue shark." He gave a detailed account of their fins, teeth, tail, and five "orifices" forward of the side fin. Olmsted summed up his shipmates' perspective:

> The shark in all his varieties, is regarded with inveterate hatred by the sailor, and is considered a legitimate subject for the exercise of his skill in darting a lance or spade, to which this savage animal is admirably adapted from his apparent insensibility to pain. At the repeated gashes he receives from these formidable instruments, he manifests the utmost indifference and calm composure, and even with a large hook in his mouth he still continues to exercise his voracious propensities. Aboard whale ships, sometimes, upon the capture of a shark during the process of trying out, he is drawn out of the water by two or three men, and a gallon or more of boiling oil is poured down his open mouth, a most cruel act, but defended on the ground that "nothing is too bad for a shark."[14]

In the same breath that he called them "ravenous monsters," Olmsted relayed, just as Ishmael does, that this shark rarely bites humans unless they put themselves in harm's way in the flurry of the bloody, blubbery water.

This does not mean that American whalemen did not worry that sharks would prey on them intentionally. When they stepped aboard their first ships, their perception, taken as a whole, was likely the same as how most of us feel today: intellectually we know the chance of shark attack is rare, but that doesn't mean we're not terrified by the prospect. And as we do today, when whalemen in the 1800s discussed drowning or death in the water it was often associated directly with sharks, whether they saw them alongside or imagined their immediate appearance. For example, greenhand-artist Robert Weir associated sharks with death in his journal at the start of his first voyage: "We are far very far out of sight of land—of sweet Ameriky. I was sent aloft to the lookout for whales + whatnots—And oh! how dreadfully [sea]sick I was. saw two sharks, one about 12 ft long + the other 5 or 6 ft. I felt very much tempted to throw myself to them for food." Later Weir drew in his journal a detailed illustration of sharks preying on a whale as the men cut into the blubber (see earlier fig. 14).[15]

Sometimes the whaleman's fear of sharks was warranted, even when they did not have a dead whale alongside. Owen Chase, for example, in his 1821 account of his small-boat journey after the wreck of the *Essex*, describes a shark chomping at the oars of a whaleboat. As the men were slowly dying of exposure and starvation, Chase said that one night "a very large shark was observed swimming about us in a most ravenous manner, making attempts every now and then upon different parts of the boat, as if he would devour the very wood with hunger; he came several times and snapped at the steering oar, and even the stern-post." Beale, Colnett, and others, like Ishmael in the final chase, also described sharks chomping at the whaleboat's oars.[16]

Ishmael's gruesome description in *Moby-Dick* of sharks cannibalizing each other and feeling no pain when hacked or sliced by the men's spades matches the descriptions of Melville's other contemporaries and likely his own observations when looking over the rail when processing whales at sea. J. Ross Browne led his *Etchings of a Whaling Cruise* with his own personal illustration of ferocious sharks feeding on a whale carcass alongside. Browne told the story of a shark that was coming up on a man trying to insert the hook into the dead whale while it was floating alongside, almost exactly as Melville would later write in "The

Monkey-rope." Browne described a shipmate from above lopping off the tail of the shark with one slice from his cutting spade. "Strange to say," Browne wrote, "the greedy monster did not appear to be particularly concerned at this indignity, but, sliding back into his native element, very leisurely swam off, to the great apparent amusement of his comrades, who pursued him with every variety of gyrations."[17]

A few years later, the Reverend Cheever described how he witnessed sailors skinning sharks alive in order to get the skin for sandpaper. Similar to Queequeg's experience, Cheever wrote: "One [shark] that we hauled upon deck, after it was cut open, and the heart and all the internal viscera were removed would still flap and thrash with its tail, and try to bite it off. The heart was contracting for twenty minutes after it was taken out and pierced with the knife." Cheever also relayed a story of a gutted shark with its tail chopped off that was still able to swim away. "Sailors . . . kill them whenever they can," Cheever said, "and there is little wonder, considering they are so likely to be themselves eaten by this greedy ranger." Mary Brewster, a captain's wife, recorded the whaleman's hatred for sharks, too, explaining in her journal that the men sometimes caught them to use the oil for their boots. On her second voyage in 1848, she wrote that one day off the Azores the crew caught a large shark, which they killed and threw right back over. She wrote: "The delight of the sailor is to kill every one they can get." Men on whaleships sometimes ripped out the teeth, jaws, and vertebrae of sharks for their craftwork on board. Some, as Olmsted described, wrote of torturing the animals, cutting off their tails and throwing them back alive, or even sticking a steel rod down a living shark's throat before tossing it back overboard.[18]

THE LANDLUBBER'S PERCEPTION OF SHARKS
IN THE NINETEENTH CENTURY

Images of cannibalistic, ravenous sharks that felt no pain made their way back to popular works ashore. Melville's *Penny Cyclopædia* simply, unemotionally listed the known species of sharks along with the rays, organized with morphological descriptions under the heading of their taxonomic family at the time, the Squalidæ. Other popular science works, however, such as Samuel Maunder's *A Treasury of Natural History* (1852), declared that sharks "devour with indiscriminating voracity almost every animal substance, whether living or dead." Good's *Book of Nature* said the shark was "the most dreadful tyrant of the ocean," capable of devouring everything. Good wrote that the "white shark" can be thirty feet long and 4,000 pounds, and can "swallow a man whole at a mouthful." (The longest trusted record of a great white is about 19.5 feet.) Thousands of people in Boston, likely including Melville, viewed John Singleton Copley's famous painting *Watson and the Shark* (1778), which hung on exhibit at the Boston Athenaeum in 1850. The painting depicts a naked boy in the water only seconds away from the jaws of an enormous great white shark-monster. This was based on a true story of an English teenager named Brook Watson who was bathing off a small boat in Havana Harbor. A shark bit off his leg beneath the knee. Watson, who would later be the Lord Mayor of London, had a peg leg for the rest of his life.[19] (See fig. 27.)

Even the preeminent scientists of the mid-nineteenth century were no less emotional or factually conservative about sharks. Baron Georges Cuvier, in the volume on fish that Melville owned, described the white shark as a man-eater and just beneath the sperm whale, which was the apex of all predators. As Ishmael described in "The Shark Massacre," Cuvier wrote in his work of science: "The white shark ... is so impatient to pass its half-digested food, to make room for more, that, as Commerson observed, the intestines are frequently forced out a considerable distance from the anus. So great indeed is the gluttony of this animal, that, as Vancouver relates, when harpooned, and no longer able to defend itself, it is sometimes torn to pieces by its companions."[20]

SHARK TAXONOMY

In his earlier novel *Mardi*, Melville not only wrote of the killing of a hammerhead from a small boat (the scene with the pilot fish), but also

FIG. 27. Detail of J. S. Copley's *Watson and the Shark* (1778), a painting scholars believe Melville saw while writing *Moby-Dick*.

he composed a lovely little chapter setting up that scene in the story. It's called "Of the Chondropterygii, and other uncouth Hordes infesting the South Seas." It has some ideas similar to those that he would express a few years later in "Cetology," "Brit," and other scenes in *Moby-Dick*. Here in *Mardi* Melville wrote of all the wonders of the ocean unknown to people ashore: "I commend the student of Ichthyology to an open boat, and the ocean moors of the Pacific. As your craft glides along, what strange monsters float by. Elsewhere, was never seen their like. And nowhere are they found in the books of the naturalists." Here Melville wrote brief, anthropomorphized descriptions of the different types of sharks, those identified by whalemen. Melville's common names of sharks align with the narratives of the time and the logbooks of experienced whalemen. Men on whaleships often identified different species of sharks that they saw scavenging on whale meat or that they saw from the rails of their boats and ships. The men at times just recorded them generically as "sharks," but typical common names included the white shark, the "bone shark," the "brown shark," the "shovel-nosed shark," and, as Olmsted wrote, the "Peak-nosed Shark," also known as the "Blue Shark."[21] (See earlier fig. 3.)

Today over five hundred species of sharks have been named and agreed upon by modern taxonomists, with more classified each year. By the time *Moby-Dick* was published, naturalists had named only a little over half of these sharks. About a dozen of the known species today are the large pelagic sharks. They are in the families Laminidae and Carcharhinidae, the latter of which are still known as the requiem sharks. All large pelagic sharks will opportunistically scavenge on a sick or recently dead marine mammal and might commonly follow a small boat inquisitively. What whalemen called the shovel-nosed was one of the hammerhead varieties (Family Sphyrinidae). The "bone" shark came from the same word that they used to describe baleen. These would be the big-mouthed plankton-filtering sharks, the whale shark (*Rhincodon typus*) and the basking shark (*Cetorhinus maximus*). Olmsted and Melville's peak-nosed/blue shark corresponds to what today's ichthyologists still call the blue shark (*Prionace glauca*), which does indeed have a longer, pointed snout. Mariners might have also used this common name when they saw a mako (*Isurus* sp.) or even a great white, which can be blue or black dorsally. The whalemen's "brown shark" is the trickiest in retrospect. Bennett described it around a whale carcass and said they never got larger than eight feet, with the scientific name *Squalus carcharias*. In *Mardi* Melville wrote that the "ordinary" Brown Sharks, "the vultures of the deep," were the most common around a whale carcass, and this shark would snap at their oars, and even swim in "herds." His mariners might've been speaking of a dozen different shark species here, perhaps most commonly the oceanic whitetip shark (*Carcharhinus longimanus*) or the great white.[22]

A SHARK EXPERT REVIEWS ISHMAEL'S SHARKS

At his lab beside Monterey Bay, I show shark expert David Ebert a sailor's illustration labeled "brown shark," but it doesn't have enough detail other than to suggest it likely wasn't a great white. Ebert says the most likely species to feed on carcasses mid-ocean and around the equator would be blue sharks, oceanic whitetips, occasionally great whites, tiger sharks (*Galeocordo cuvier*), and several *Carcharhinus* sharks, which include the silky sharks (*C. falciformis*), Galápagos sharks (*C. galapagensis*), dusky sharks (*C. obscurus*), bull sharks (*C. leuchas*), and bronze whaler sharks (*C. brachyrus*).[23]

Ebert is a native Californian who grew up in the water around great whites and then spent years studying large predator sharks off South Africa. He returned to Monterey Bay, in part because of the productivity from the ocean canyon and the large populations of marine mammals. The bay is a lively ground for sharks, especially great whites.

Ebert's lab at Moss Landing is in a building lined with stuffed marine mammals and skeletons of seabirds and whales, and even the library has an enormous shark mural. Ebert is the author of shark field guides and scientific papers, and he's a regular on Shark Week documentaries. He has seen sharks feeding on whales, including one time from the cliffs of South Africa when he watched a dozen or so great white sharks thrashing into a dead southern right whale. Ebert explains that bloody fish guts can work to attract sharks, but he's found that dolphin and whale blubber is an even better attractant. The sharks smell the scent from the slick of oil. He says, "You just see these fins pop up on the horizon in the slick."[24]

Ebert supports Ishmael's observations about shark cannibalism: "That's true. Definitely happens. The sharks are very precise when they're feeding on stuff, but sometimes if one gets wounded or winged or injured, then the others will pile on and start eating him or her." Ebert says there's a hierarchy among sharks at a carcass, within a species and among others, say, with a great white shark pushing out or even "snacking" on a blue shark that's in its way. "I've seen the events where the smaller sharks will come in initially and then the bigger sharks will come in with their teeth and graze them, along with other behaviors, to get them to leave, to say, *hey, the big boys are in now*. But usually they're pretty focused on the whale." Ebert compares the sharks to hyenas in the wild, in the same way that Ishmael compares them to dogs.[25]

Regarding Ishmael's depiction of sharks' indifference to pain, Ebert confirms that these animals can survive serious wounds and tremendous loss of tissue, and, yes, the musculature can still respond after a shark is technically dead. "People think, *My god, you can't kill this thing*, but it is dying. It's just an involuntary reflex—so you got to be careful because they still snap, but not consciously. Best thing to do is sever the vertebral column—or the brain if you can get into it."

Ebert explains that for all the negative press that came with *Jaws*, it also helped stimulate interest and funding for shark research and even some conservation, too. Unprompted, he makes an Ishmaelian point to emphasize the importance of learning from the commercial fishermen and the whale-watch captains who have the most experience out in Monterey Bay. Ebert says he's always prodding his students to go out and talk to these men and women.

I ask Ebert about the status of shark populations. Ishmael's "thousands upon thousands" of sharks around Stubb's sperm whale carcass is surely fictional hyperbole, but, is it likely that more large sharks did swim around the global ocean in the mid-nineteenth century, based on the accounts of early mariners?

Ebert explains that facts about past shark abundance are hard to come by, yet shark specialists and fisheries biologists who study present populations in relation to past or healthy ocean ecosystems are confident there have been significant losses in both the numbers and the size of sharks since Melville's mid-nineteenth century. For example, historians believe that human hunting practically eradicated the large plankton-eating basking sharks from Massachusetts Bay by the 1830s. Late eighteenth-century explorers such as James Colnett and George Vancouver reported enormous sharks in the Pacific, animals eighteen to twenty feet long, which is easy to discount as exaggerations, but the men could measure them well in comparison to their small boats. Part of the challenge is that few nineteenth-century or even twentieth-century scientists or fishermen ever actually counted sharks. They were only considered bycatch or a nuisance. And no recent scientists have yet done DNA work to estimate historical populations. Today, the shark-fin trade still thrives despite international regulations, and small sharks are a regular target for food and for bait for higher value fisheries. Sharks of all sizes are regularly killed as bycatch by the hundreds of thousands of tons each year, unmanaged—and particularly post-*Jaws*, the large sharks are now targeted in recreational fisheries, although there is more value now in cage diving and photography trips.[26]

Today's commercial fishermen, like their nineteenth-century whale-men counterparts, often compete with scavenging sharks. They experience the same frustrations with sharks eating their bait or their catch as did nineteenth-century whalemen. In 2010 Linda Greenlaw wrote about the destruction that blue and mako sharks inflicted on her swordfishing gear or the fish that they'd caught on the line. Over the years Greenlaw watched her shipmates pounding sharks with ice mallets, slicing them up in pieces, and setting the sharks on fire with lighter fluid, some of which she'd even participated herself when she was younger. She wrote that she'd grown out of that kind of hatred of sharks and discouraged it with her crew, but she expected that it continued on other boats: "Maiming, torturing, and killing sharks out of frustration and some weird sense of retaliation and revenge were bound to occur."[27]

A 2003 study in *Nature* found that more than ninety percent of the biomass of all predatory fish, sharks included, have been fished out of the global ocean since before the Industrial Age, when Ahab first rolled into the Pacific. In 2014 shark specialists for the IUCN estimated that over a quarter of all sharks and rays are currently in danger of extinction. The blue shark is "near threatened"; the great white shark is "vulnerable"; the great hammerhead shark is "endangered"; and scientists do not yet have enough information to assess the oceanic whitetip shark.[28]

On the other end of things, however, Ebert explains that the great white shark is exceptionally well-protected today. With the Marine

Mammal Protection Act of 1972, elephant seals, eared seals, and true seals are all making a comeback, especially in areas like Monterey Bay. This food source has brought back great white shark populations.

Motioning out toward the water, Ebert says: "You get tired of trying to count white sharks around here."

Only a week after I meet with Ebert a man is attacked by a great white shark while spearfishing with his father at the south end of Monterey Bay. As is commonly the case with these sorts of attacks, the shark bites a couple times at the legs and then backs off. The young man, named Grigor Azatian, bleeds profusely. Trained professionals just happen to be on the beach. They apply a tourniquet and save his life.[29]

A SECOND SHARK EXPERT REVIEWS ISHMAEL'S SHARKS

While sharks feeding on a whale carcass was a common vision for tens of thousands of whalemen in the nineteenth century, few people have witnessed this since. It's rarely described in any scientific literature.

Chris Fallows is a wildlife photographer and a guide for great white shark observing and cage diving off South Africa. He often works with field scientists, during which he is one of the few who has actually observed and photographed events where great whites scavenge dead whales (see plate 10). Fallows once observed sharks eating a near-term fetus that they'd bitten out of a dead Bryde's whale. He's seen firsthand the circular bites in whale meat from great whites, which, just as Ishmael says, appear as near-perfect countersunk circles.

One extraordinary behavior that Fallows and his colleagues have witnessed is that sharks regularly regurgitate chunks of whale blubber, seemingly, they theorize, to make room for other bites of higher calorie whale blubber or meat, with "higher energy yield chunks." Other researchers off South Africa have watched a thirteen-foot female white shark feed on its own regurgitated whale meat. Was this what Cuvier had heard about, and what Ishmael described in his gruesome scene of sharks eating their own innards?[30]

On the *sound* of sharks feeding on a carcass, Fallows says: "Yes, yes, yes. It is a sound that never leaves you. It sounds like the noise of a bellow used to fan a fire. Coupled with the fatty spray, blood, and smell—it is an experience that you never forget."[31]

SHARKS AND SLAVERY

Through a twenty-first-century lens, Melville's depiction of Fleece in "Stubb's Supper" seems mocking, if not hostile, toward racial equality. Fleece is a comic slave. As he gets relentlessly ordered about by Stubb, commanded to give a speech to animals, Fleece appears so foolish that he doesn't know his own age or that the United States is his country.

In 1973 a literary scholar named Robert Zoellner wrote of "Stubb's Supper": "Over a century after Melville wrote these lines, one can only feel embarrassment at such evidence—pervasive in nineteenth-century American literature from Fenimore Cooper on—that possession of moral sensitivity and high creative gifts does not necessarily arm one against the prejudices and stereotypical thinking of one's culture." In other words, we'd like Melville to have been an active abolitionist. He was not.[32]

Melville was, however, more enlightened on race than might first appear, especially by contrast to others of his day. Francis Allyn Olmsted, for example, wrote disdainfully of the native Hawaiians. The comments of 1940s whalemen in their journals about their African-American shipmates, such as those by John F. Martin, are often ugly and offensive. American whaleships were diverse islands and one of the rare workplaces where merit took precedence over skin color, but they were certainly not beyond the racism on land. As the new United States struggled to mend its divisions over slavery, Melville did see the institution as evil. Some today read *Moby-Dick* as prescient of the Civil War, with the *Pequod*, the ship of state, sunk because of its obsession with race and whiteness. Only four years after *Moby-Dick*, even closer to the Civil War, Melville engaged directly with the slave trade and its horrors with his novella *Benito Cereno*. Melville, along with his father-in-law, the Chief Justice in Massachusetts, seemed to have worried that to emancipate the millions of slaves in one sign of the pen would be too dangerous and chaotic for America. He hoped that the southern states would soon end slavery on their own. Within this historical context, Melville does not go as far as we might wish, but he does have something profound and progressive to say on both race and nature, which he centered, more safely, on sharks. He does not, unfortunately, directly address immoral hatred among humans, yet with these marine predators, in a proto-Darwinian way, Ishmael decenters man in *Moby-Dick*, regardless of race or ethnicity, and explores the similarities between the main drivers of human and nonhuman animals.[33]

It is Queequeg who first teaches Ishmael cultural empathy in the novel. The Polynesian hero saves at least two human lives over the course of the story. He then saves Ishmael's life indirectly with his coffin. Zoellner wrote that Melville wanted us to read Queequeg as the human embodiment of a shark. Queequeg is nomadic, a superior swimmer and hunter, and a cannibal with "filed and pointed teeth." A good harpooner needs to be "pretty sharkish," Peleg explains. In "Cetology" Ishmael lays the groundwork to sympathize with sharks and equate them to the dark-side of humans: "For we are all killers, on land and on sea; Bonapartes and Sharks included." Later in "The Monkey-rope," when Queequeg balances on a dead whale in order to insert the blubber hook, balancing between the sharks and the men, Ishmael says of Queequeg, as he is tied physically and metaphorically to the sharkish hero: "Are you not the precious image of each and all of us men in this whaling world? That unsounded ocean you gasp in, is Life; those sharks, your foes; those spades, your friends; and what between sharks and spades you are in a sad pickle and peril, poor lad."[34]

The two African-American characters in *Moby-Dick*, Pip and Fleece, are not treated as heroically as Queequeg. Nor is the African Daggoo, forced to be Flask's harpooner, treated altogether favorably. Dagoo is the one sent down to get the steak in "Stubb's Supper." Yet if you could remove the dialect, for example, Fleece's sermon is brilliant. He directs and extends this discussion on what is cruel and indifferent in nature, in comparison to what is more evil and dangerous in man. Fleece delivers, Zoellner argues, one of the major morals of *Moby-Dick* with vocabulary and rhetoric that is beyond most of the other crew members: embrace what is innate, Fleece says, what is natural. Fleece has a thorough understanding of Scripture, even though he claims to Stubb that he's never been in a church. Though he listens to Stubb out of necessity, Fleece is clearly the better person throughout this interaction, which I find hard to believe that this would have escaped a nineteenth-century reader. In direct contrast to the Calvinist sermon in New Bedford, which argues that people must bend to God's wishes, Fleece preaches the governance of our evil thoughts, universal equality, and charity for the weak. The cook shows sympathy for sharks in the way Ishmael has learned to revere the morality of Queequeg.[35] Fleece says:

> Your woraciousness, fellow-critters, I don't blame ye so much for; dat is natur, and can't be helped; but to gobern dat wicked natur, dat is de pint. You is sharks, sartin; but if you gobern de shark in you, why den you be angel; for all angel is not'ing more dan de shark well goberned. Now, look here, bred'ren, just try wonst to be cibil, a helping yourselbs from dat whale. Don't be tearin' de blubber out your neighbour's mout, I say. Is not one shark good right as toder to dat whale?[36]

Fleece then settles on the still more forgiving path, which accepts that sharks will be sharks, men will be men, and we must forgive that which is innate. Fleece understands that Stubb is more shark than the sharks—because he could know better, but does not act morally. The real "shocking sharkish business" is war and violence and hatred. Perhaps even whaling itself, or the broader hunting of marine animals. Melville certainly wanted to make clear in "Stubb's Supper" that by the shocking business, he also meant the American institution of slavery.

Even more subtle and more shocking to the twenty-first-century reader, is that there was nothing fictional about sharks in the 1800s scavenging on sick and dead men, women, and children who were thrown off slave ships by slave traders or on those African people who jumped off to commit suicide to escape the horror or to gain a final sliver of autonomy. The terror of sharks eating dead African people in the sea was used regularly by abolitionists to expose what was happening on the Middle Passage. This vision of sharks as man-eaters was in the nineteenth-century public consciousness. In 1840, for example, one of Melville's favorite artists, J. M. W. Turner, painted *The Slave Ship* with a nightmarish vision of blood, hands, shackles, and shark fins.[37]

With all that in mind, it's still a disappointment to the modern reader that Ishmael displays racial prejudice with his suggestion in "The Chase—Third Day" that the sharks might have been more attracted to the flesh of Fedallah and his crew, "infidel sharks" themselves, even if he pulled this directly from Baron Cuvier's comments that great white sharks used their keen sense of smell toward killing black men over white, because men with darker skin were "more odoriferous." Ishmael does not correct this or suggest it ironically.[38]

In "Stubb's Supper" Fleece has lived long enough to accept the evil in man. He has survived years of degradation. He does not rely on traditional religion, pleads ignorance of his connection to a country that would enslave people, and seeks solace in a peaceful afterlife, a deliverance from an angel. Fleece, like Pip, has been the victim of the sharkish evil in mankind. Fleece has made peace. Pip goes mad. They both know how this voyage in pursuit of the White Whale is going to end.

Starbuck, meanwhile, still wonders whether the good in man might prevail: that Ahab will not lead them all to their death. Even this is connected with sharks. Many scenes later in "The Gilder," gazing over the rail at a lovely Pacific sunset, Starbuck seems to heed Fleece's sermon and remain with Christianity in the hopes that everything will turn out all right with this hunt. Starbuck prays that humankind, as a group and as an individual, can overcome, can govern, our innate sharkishness, despite what has been proven to be true over and over again throughout history. Starbuck soliloquizes while looking into the sea: "Loveliness unfathomable, as ever lover saw in his young bride's eye! — Tell me not of thy teeth-tiered sharks, and thy kidnapping cannibal ways. Let faith oust fact; let fancy oust memory; I look deep down and do believe."[39]

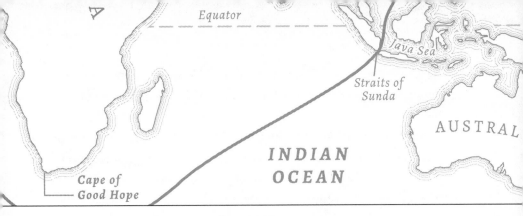

Equator

Java Sea

Straits of
Sunda

AUSTRAL

INDIAN
OCEAN

Cape of
Good Hope

Ch. 14

FRESH FARE

Yet now and then you find some of these Nantucketers who have a genuine
relish for that particular part of the Sperm Whale designated by Stubb; com-
prising the tapering extremity of the body.
  Ishmael, "Stubb's Supper"

While the sharks are eating whale meat above, sharkish Stubb is eating a whale steak, a piece from near the animal's tail. In "The Whale as
a Dish," Ishmael opens with the declaration: "That mortal man should
feed upon the creature that feeds his lamp, and, like Stubb, eat him by
his own light, as you may say; this seems so outlandish a thing that one
must needs go a little into the history and philosophy of it."[1]

THE WHALEMAN'S DIET

Whalemen under sail in the 1840s and '50s ate mostly salted beef and
pork, salted cod, hard biscuit, rice, and a regional collection of vegetables that would not spoil as quickly, such as potatoes, onions, squashes,
and beans. They had molasses as a sweetener for weekly "duff." They
drank water and coffee. By Melville's time, scurvy was largely a disease of the past, even on lengthy voyages with few port stops. In "The
Decanter," Ishmael jokes about scurvy when describing hardtack with

weevils: "The bread—but that couldn't be helped; besides, it was an anti-scorbutic; in short, the bread contained the only fresh fare they had."[2]

Moby-Dick is piled high with discussions of food, beginning immediately with "Loomings" and Ishmael's comic mumblings about broiled fowl, continuing all the way through to "The Decanter," in which he extols the virtues of the food on British whaleships (despite the rotten bread). How well the whalemen ate in terms of volume and quality on these ships is a matter for scholarly debate. One morning, as we toured the galley of the *Charles W. Morgan*, food historian Sandy Oliver tells me she thinks the records of the men served bread infested with cockroaches and weevils are true, but did not occur as often as it's represented today. It made no sense to have men weak and unable to hunt effectively. On the other hand, some scholars believe that whaleship owners were more often penny-pinchers, like Peleg and Bildad, so that the desire for more and better food was the primary motivation for desertion by the time whalemen arrived in the Pacific. In the first paragraph of Melville's very first book, the opening to *Typee* in which he brags about his six months at sea, he laments the loss of their fresh bananas, sweet potatoes, yams, and delicious oranges: "There is nothing left us but salt-horse and sea-biscuit." (Seaman told stories of occasionally being fed horsemeat—Ishmael jokes in *Moby-Dick* that it can be camel—but the salted meat they normally ate was cow beef.) Most whaleship captains by midcentury, however, planned enough to acquire fruits and vegetables to supplement the seamen's diet. They thought ahead along their network of suppliers for fresh food and water at a range of trading posts that had been established for whalemen all across the islands of the Pacific. For example, when Melville escaped from his Tahitian jail after his second whaleship, he worked on a farm in Moorea that cultivated sweet potatoes, taro, yams, and sugar cane.[3]

A half century earlier, when whalemen first began hunting in the Pacific, the food networks were far more sparse and scurvy was still a major concern. The most illuminating of Melville's fish documents about sailor-fare was that by James Colnett of the Royal Navy, the same

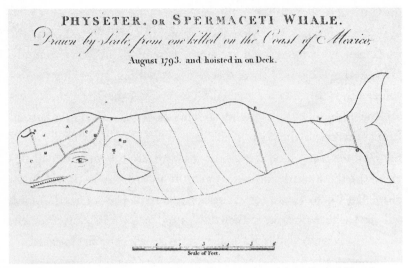

PHYSETER, or SPERMACETI WHALE.
_Drawn by Scale, from one killed on the Coast of Mexico,
August 1793. and hoisted in on Deck.

Scale of Feet.

FIG. 28. James Colnett's cub sperm whale and diagram for cutting into a whale (1798), which Ishmael ridicules because of the enormous eye.

captain who reported on eighteen- to twenty-foot sharks. In 1798, Colnett published the narrative of his Pacific voyage, funded in part by the Enderby whaling family in order to gather information for whaling voyages. It was Colnett's illustration of a sperm whale with an enormous eye that Ishmael ridicules in "Monstrous Pictures of Whales" (see fig. 28). Colnett had learned the importance of crew health as a midshipman on James Cook's second circumnavigation, and a significant part of the mission of Colnett's voyage in the 1790s seems to have been to plan for food for the future whaling fleet.[4]

Colnett sailed from England with two live cows and regularly kept pigs and chickens on board. When anchoring besides islands or along coasts in the Pacific, the men of the *Rattler* ate a variety of fish species. They also ate coconuts, fruit of the "molie tree," shellfish, a variety of sea and land birds, as well as the Ancient Mariner's sea snakes — in the belly of which they found more fish. They ate "alligator," which was more likely crocodile. They shot two or three monkeys, presumably for food. They made soup from turtles and from large flocks of sea ducks (*Anas* spp.). With a Spanish ship, Colnett traded for pumpkin, chickens, two sheep, two bags of bread, and twelve (presumably beef) tongues. On the Galápagos, they captured the immense Galápagos tortoises (*Chelonoidis* spp.), animals that were just beginning to be discovered by the whalemen, who would go on to capture these reptiles by the thousands, precipitating the loss of three species to extinction and leaving the rest vulnerable or critically endangered. Colnett wrote that the tortoise "was considered by all of us as the most delicious food we had ever tasted." Ishmael refers to them as "Gallipagos terrapin," and Melville likely heaved a few himself off the islands while aboard the *Acushnet*. Whalemen kept these tortoises alive for months aboard ship to provide fresh meat. On Cocos Island (now a national park of Costa Rica), Colnett ordered the introduction of a pair of pigs and goats. Rats, unfortunately, were already well established on Cocos. In another bay on Cocos Island he had his seamen plant "garden seeds, of every kind."[5]

While out whaling at sea, they ate "sun-fish" (*Mola* spp.). They ate "devil-fish," which were likely rays. The crew of the *Rattler* also "saw dolphins and porpoises in abundance, and took many of the latter, which we mixed with salt pork, and made excellent sausages." For men who had scurvy symptoms, Colnett gave them preserved fruits, pickles, fresh bread, and three times a day "twenty drops of elixir of vitriol, and half a pint of wine." The men recovered.[6]

A half century later, by Melville's day, whalemen like Captain Lawrence on the *Commodore Morris* had far more options for fresh food around the Pacific. After rounding Cape Horn, Lawrence sailed to Mocha Island in 1850 to load up two barrels of potatoes, five pigs (who would later have piglets), two roosters, and twenty-five pumpkins. At other islands throughout the voyage, Lawrence's crew captured, gathered, or traded for fish, peaches, yams, and the eggs of "mutton birds [likely shearwaters, *Puffinus* spp.], which are very good eating they resemble a hens egg." On the "Nancytucket Island" of the South Pacific, his seaman collected about a "half barrel of gulls eggs and eight curlews." At one point Lawrence flogged the cook for "wasting provision."[7]

In short, the diet of American whalemen in the mid-nineteenth century was mostly salted meat and bread, but it was also varied and regularly mixed with fresh meat, fruits, and vegetables whenever possible.

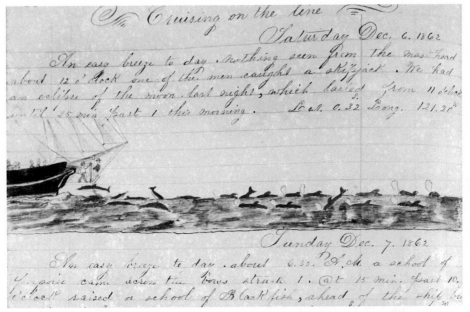

FIG 29. Whaleman Thomas White's watercolor of harpooning dolphins from the bowsprit aboard the *Sunbeam* (1862), standing on the part of the ship often known as the "dolphin striker" (although this name is likely more for the iron point that might hit a dolphin that was bow-riding). In color, this image shows a splash of red blood underneath the dolphin just beneath the men.

DID NINETEENTH-CENTURY WHALEMEN EAT WHALE?

Like Colnett's crew, the whalemen of the *Charles W. Morgan* and the other whaleships captured and ate a variety of smaller toothed whales, such as dolphins and pilot whales.[8] (See fig. 29.)

Reverend Cheever wrote that, "as every one knows," the whalemen harpooned and ate dolphin, known as "sea beef." The cook used the dolphin oil for cooking. Robert Weir wrote that dolphin meat tasted like veal. John Jones, the steward of the whaleship *Eliza Adams*, wrote in 1852: "Got orders to make 10 000 balls of it," referring to meat from seven pilot whales killed off Cape Horn. Thus true to culinary life at sea, Ishmael in *Moby-Dick* extols the flavor of dolphin meat and explains that many people, including the monks in Scotland, have enjoyed meat-balls made of porpoise.[9]

Yet for all of that diversity of diet, need for fresh food, and regular eating of smaller marine mammals, there is less evidence that whalemen like Stubb actually ate the meat from the larger cetaceans with a "genuine relish," especially meat from their regular catch of sperm or right whales. Ishmael puns accurately that Stubb's taste is rare, but it was not unprecedented. Ishmael says he liked to fry ship's biscuit in the oil of the tryworks. Melville likely buttressed his own experience and developed some of his foodie ideas—and the relish pun—from J. Ross Browne, who wrote of "delicious" whale oil biscuits during night-watches, as well as the frying up of sperm whale brains. Browne wrote, too, that "certain portions of the whale's flesh are also eaten with relish, though, to my thinking, not a very great luxury, being coarse and strong." Browne added that whale meat was better mixed with potatoes, like the porpoise balls, but was gobbled up by the men just for the variety, just as they'd eat "with as much relish" roast beef back on land.[10]

Cruising up in the Arctic for bowheads, Scoresby wrote that if whale meat was cleared of fat, broiled, and seasoned with salt and pepper, it was "not unlike coarse beef." Like other whalemen, Captain Scoresby recommended the meat from younger whales, or at least the more tender parts. Dr. Bennett, for his part, described the delicacy of meat from small humpback whales. Scoresby heard whale breast milk is "well-flavoured," and it was Bennett who wrote about sampling this milk: "it has a very rich taste." In turn, Ishmael mentions that whale milk is "very sweet and rich," adding that "it might do well with strawberries." Whaleman-artist John F. Martin wrote in his journal about frying pieces of the right whale's lip: "When eaten with pepper & vinegar it tastes very much like soused tripe."[11]

Aboard Colnett's *Rattler*, the men ate the heart of a young sperm whale—the very same one that was the model for the outline with the large eye. Colnett's cook baked the heart into a "sea-pye," which "afforded an excellent meal." Even as late as 1912, Robert Cushman Murphy, a naturalist aboard an American whaleship, reported their chef making a meal of dolphin fish, sperm whale balls, and baked beans.[12]

So why, if it was so available, was whale meat not an even more regular, even daily part of their diet? If they had chosen to, the whalemen could have surely smoked, dried, or salted the meat. A large portion of the men would have had experience with this from farm and coastal fishing work.

Ishmael takes up the question in "The Whale as a Dish," in which most of the history appears accurate. Ishmael explains that despite the long and royal tradition of eating whale meat, American whalemen did not eat it regularly, beyond dolphins and pilot whales, because there was just too much of it all at once during the butchering process, and that whale meat on the whole was just too fatty and rich for their taste. Sperm whale muscle is a dark, red meat filled with the myoglobin that enables the whale to dive so deeply. Even today, baleen whales, such as the minkes eaten in Iceland, are known to taste better. Only the Japanese and small pockets of Indonesians, historically and presently, have eaten sperm whale on a regular basis.[13]

Melville's use of whale meat for the novel works in part to tickle perceptions of the exotic. The author knew, as is still often the case today, that whale meat is stigmatized as for only the uncivilized, the sharkish. This Melville implies with Queequeg and subsequently with Stubb, as well as with his story of the British whalemen *forced* to eat whale in the polar regions—as if those men had lowered themselves to the level of the Eskimo. Environmental historian Nancy Shoemaker found that many of the sailors considered eating whale meat too savage, too otherly. They'd try it once or twice to have the story to tell back home, or even if they ate it often in some manner, they had no interest when they returned. Whale meat was considered food for Native Americans, Inuit, and Pacific Islanders. Ishmael explains early in the novel that even food cooked rare is heathenish and implies a taste for human flesh: "We will not speak of all Queequeg's peculiarities here; how he eschewed coffee and hot rolls, and applied his undivided attention to beefsteaks, done rare." Melville's perception that whale meat was for savage tribes only strengthens the image of Stubb eating rare whale meat as sharklike—no better or worse than the sea animals gnawing at the freshly killed meat beside the *Pequod*.[14]

At home in the United States in the nineteenth century, whale meat was considered food for the lower classes. In *Cape Cod*, published serially beginning in 1855, Thoreau learned that local boys ate sandwiches with blubber from pilot whales driven onto the beach. An older fisherman said he actually preferred pilot whale meat to beef. Thoreau didn't try it. In his narrative he quickly added that in previous decades in France "blackfish were used as food by the poor." Thoreau's implication here is that whale meat was a necessity of poverty. (It's a fun, if surely coincidental, detail that Stubb is from Cape Cod.[15])

As today, food avoidances and taste tended to be more often culturally driven social constructions, even if these choices might have had historical roots in logistics or health. In Melville's America, of course, vegetarianism and more ethical eating was not as significant a movement as it is today, but Ishmael's closing comments in "The Whale as a Dish," certainly show these ideas had begun to arise. In "Cetology" Ishmael shows that we can hold opposing ideas together at the same time. The anthropomorphized playfulness of the "huzzah porpoise" reaches even the most hard-hearted of mariners: "He always swims in hilarious shoals . . . They are accounted a lucky omen. If you yourself can withstand three cheers at beholding these vivacious fish, then heaven help ye." Then, immediately after this celebration of these animals, Ishmael explains that they make "good eating" and a plump porpoise will amount to about one gallon of "exceedingly valuable oil." The sailors held a hunter's reverence for a dolphin right alongside the desire to eat it or use it for daily living.[16]

Shoemaker explained that the modern debate over whaling is often about animal sentience and environmental ethics, but it's also tied intricately with food, with taste, which extends across and beyond indigenous cultures, such as the bowhead and beluga-eating Inuit, and the industrial, high tech countries such as Japan, which sends its hunters all the way to the Antarctic to capture the meat. Both the Japanese and the Inuit can trace their hunting and eating of whales back over one thousand years.

Ishmael has two closing points in his "history and philosophy" of eating whale meat. The first, as we discussed earlier when talking swordfish, is to again remind the reader sitting comfortably at home to re-spect and recognize the whalemen out at sea, men who are in the ugly business of bringing home the whale oil. The second is to point out the hypocrisy of those ashore who might condescendingly judge those who choose to eat meat, use animal products, or even hunt the whales in the first place. This is an argument used by representatives of the cur-rent whaling nations—Japan, Norway, and Iceland—against those that judge the practice most vehemently: the people in animal-advocacy groups who use the products of animals such as cows and pigs in their daily lives, even while arguing the very cases against killing animals. Ishmael chides a fictional "Society for the Suppression of Cruelty to Ganders." In 1840 Queen Victoria agreed to add "Royal" to the Society for the Prevention of Cruelty to Animals in England, and the SPCA had been active in England even then for about fifteen years, but there was yet no organized animal advocacy movement in the United States at the time of *Moby-Dick*. Whaling nations today ask why the United States can allow indigenous whaling from their own shores off Alaska, or, more significantly, why killing a "free range" minke whale, whose population seems stable and healthy in the North Atlantic according to both the IWC and the IUCN Red List of Threatened Species, is worse than tying up and fattening a baby cow for veal.[17]

Deeper still, as the *Pequod* continues eastward in the Indian Ocean, Ishmael in "The Whale as a Dish" subtly equalizes humans and non-human animals: "Go to the meat-market of a Saturday night and see the crowds of live bipeds staring up at the long rows of dead quadru-peds. Does not that sight take a tooth out of the cannibal's jaw? Canni-bals? Who is not a cannibal?"

One sub-Arctic night in Reykjavík, Iceland, I am on my way down along the wharf to go on a whale watch when I notice, opposite the whale watch boat, two black-hulled ships named the *Hvalur 8* and the *Hvalur 9*. The two whale hunting vessels, named for the whale, each have white masthead capsules for the lookouts, evolutions of Scoresby's crow's nests that Ishmael chides in "The Mast-Head." Even at midnight it is light outside, because of the high latitude. On the whale watch we see a few minke whales in the distance, but otherwise it is a quiet ride out in the fjord beyond the capital. After we return to the dock, I hear some other tourists talking about finding a restaurant to dine on whale meat, despite walking past a large blue sign at the head of the wharf with a smiling cartoon whale that says: "Meet Us, Don't Eat Us." The sign is sponsored by the International Fund for Animal Welfare and the Icelandic Whale Watching Association, the latter of which advertises a list of local "whale friendly" restaurants that will not serve whale meat.[18]

The next day a painter working on *Hvalur 8* tells me that the whale hunting boats are at the dock for repairs. They hunt on another part of the coast. According to the documentary *Breach* (2015), however, in previous years a whale watching vessel and a whale hunting ship were out on these waters of Hólmasund at the same time, within view of each other. Once the hunter boat steamed by with a couple dead whales dragged alongside in full view of the whale watchers. Tourists continue to eat whale meat in the restaurants in Reykjavík. Whale is eaten by some current Icelanders but it's apparently *not* a meal that has deep roots in Icelandic culture. According to a 2017 report in the *Iceland Review*, sixty-five percent of the minke whale meat is sold to restaurants, primarily to tourists.[19]

I go to a restaurant that serves whale. But I *chicken* out.

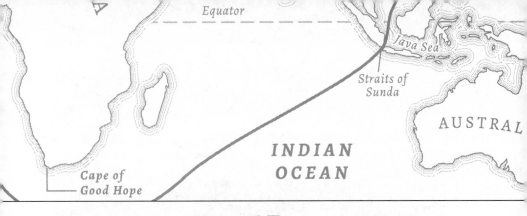

Equator

Java Sea

Straits of
Sunda

INDIAN
OCEAN

Cape of
Good Hope

AUSTRAL

Ch. 15

BARNACLES AND
SEA CANDIES

No great and enduring volume can ever be written on the flea, although many
there be who have tried it.
    Ishmael, "The Fossil Whale"

Melville represented mid-nineteenth-century knowledge of small pelagic organisms in "Brit" and likely in "The Whiteness of the Whale" with his bioluminescent milky sea. In a few other scenes in the novel, Melville wrote of still more kinds of small invertebrates, notably barnacles and whale lice.

In "Cetology" Ishmael writes of the *Pequod* rolling on the waves "side by side with the barnacled hulls of the leviathan." In "Less Erroneous Pictures of Whales," Ishmael extols the vivacity and accuracy of a French painting by Ambroise Louis Garnery, which he believes properly depicts a right whale hunt, including where "sea fowls are pecking at the small crabs, shell-fish, and other sea candies and maccaroni, which the Right Whale sometimes carries on his pestilent back." Ishmael gives far more detail, in fact, about the life on the whale's head than is in the actual painting (see fig. 30).[1]

Later in the novel in the middle of the Indian Ocean, Ahab orders

FIG 30. Detail of *Pêche de la Baleine* by Ambroise Louis Garnery (1835), engraved by Frédéric Martens, one of the paintings that Ishmael respects in "Less Erroneous Pictures of Whales."

the men to kill a right whale and hack off the head to lash to the side of the *Pequod*, opposite a sperm whale's head. Ishmael leads the reader over to the rail to look down on the little ecosystem on the right whale's skin:

> If you fix your eye upon this strange, crested, comb-like incrustation on the top of the mass — this green, barnacled thing, which the Greenlanders call the "crown," and the Southern fishers the "bonnet" of the Right Whale . . . when you watch those live crabs that nestle here on this bonnet, such an idea will be almost sure to occur to you; unless, indeed, your fancy has been fixed by the technical term "crown" also bestowed upon it; in which case you will take great interest in thinking how this mighty monster is actually a diademed king of the sea, whose green crown has been put together for him in this marvellous manner.[2]

Biologists today refer to the patches of hard skin tissue that Ishmael calls the "bonnet" or "crown" as the whale's callosities, a word less in use in Melville's day regarding marine mammals. Modern scientists, beginning with the Soviet ecologist S. K. Klumov in the early 1960s, identify individual right whales by photographing this pattern of growth on their heads. Within the first several months of life each right whale develops a tough keratinized tissue on his or her lower lip, chin, above the eyes, and from the tip of the upper lip to back around the blowhole—similar to the locations of human facial hair. (See plate 4.) The callosities usually have, in fact, as Ishmael notes, a few hairs sprouting out from within. Once formed, the whale callosities are then colonized, especially by cyamids and barnacles.[3]

The cyamids, which Ishmael refers to as crabs and maybe also sea candies, were known to naturalists as "whale lice" or "crab lice," a common name that's still in use today. Although they indeed look like little crabs or even our own hair-infesting insects, cyamids are flattened amphipods. There are about forty species. They can be so host specific that researchers have used them as "tags" to help identify the historical and geographic separation of right whale species. Cyamids are about the size of your thumbnail. They have three pairs of back legs evolved to dig and hook into the whale's skin.[4]

Barnacles also live on the right whale's callosities. Perhaps twenty or so species of "whale barnacles" (of the subclass Cirripedia) may live on a variety of marine mammals. In order to hang on, the *Tubicenella major* barnacle, which is common on the southern right whale, builds an accordion-style column of shell that burrows deeply into the callosity. The gorgeous *Coronula diadema* barnacle has evolved a distinct hexagonal opening. Linnaeus gave this species its scientific name in the eighteenth century. Darwin identified and named its close relative in 1854, *Coronula reginae*. If Melville knew the genus name, he worked it into his puns in "The Right Whale's Head," appreciating the Latin reference to crown.[5]

Cyamids and barnacles feed off scraps from their whale hosts as well as phytoplankton floating past. Cyamids also eat flakes of the whale's skin. Although there's little evidence that barnacles and cyamids do any real harm to the whale, either of these hitchhikers can introduce irritation and, just like on a boat, reduce the whale's swimming efficiency (which adds to Ishmael's analogy of the whaleship's fouled hull as compared to the whale's body). A compelling recent theory is that the hairs in the callosities might even sense the movement of cyamids, which in aquaria studies actually rise up when copepods are nearby. Maybe this helps the whale hosts identify passing schools of plankton?

In a vision similar to Ishmael's "Brit," Dr. Bennett in 1840 wrote that "the True-Whale [right whale] of the South ... has its body encrusted with barnacles and other parasites, often to the extent of resembling a rugged rock."[6] While sailing in the South Atlantic aboard the *Clara Bell*, Robert Weir wrote in his journal something similar about the invertebrate hitchhikers, although less affectionately than Ishmael:

> The right whale is a very dirty mamal compared to others of the same tribe. I have noticed they are covered with small insects very much resembling crabs—about half an inch in diameter. On the end of their nose is a bunch of barnacles about 18 inches wide this the whalemen call his bonnet. And when you see a whale just rising out of water it has the appearance of a rock, the barnacles are enormous—as much as two inches deep—the boys often roast them and eat them the same as oysters.[7]

Ishmael makes it clear, accurately, that the sperm whale has no callosities and barely any barnacles or any other fouling, likely due to their deep diving. The sperm whale's smooth skin, aside from the scars of battle, was simply another way to elevate the sperm whale as the new monarch of the seas, with Moby Dick as the clean, unencumbered high emperor.

Equator

Java Sea

Straits of
Sunda

AUSTRALIA

INDIAN
OCEAN

Cape of
Good Hope

Ch. 16

PRACTICAL CETOLOGY

Spout, Senses, and the Dissection of Heads

Down to this blessed minute (fifteen and a quarter minutes past one
o'clock P.M. of this sixteenth day of December, A.D. 1850), it should still
remain a problem, whether these spoutings are, after all, really water, or nothing but vapor—this is surely a noteworthy thing.
    Ishmael, "The Fountain"

"Where, I should like to know," Ishmael says as the *Pequod* continues
sailing east in the Indian Ocean, "will you obtain a better chance to
study practical cetology than here?"

Presumably for superstitious purposes, and conveniently for Ishmael's explanations of sea candies and whale anatomy, Ahab orders the
head of a right whale and the head of a sperm whale lashed along either
side of the hull of the *Pequod*. Ishmael devotes no fewer than six chapters to various considerations of the whale's head: its internal physiology, its sensory life, and then, as the ship nears Indonesia, the exact
composition of the whale's spout.

THE SPOUT

Tall, with a bowl-style haircut, Dr. Justin Richard is so genuinely enthusiastic about whales that you barely raise an eyebrow when he uses
"snot" or "bonkers" in his speech. He studies reproductive success in
beluga whales, which are also, if smaller, white toothed whales with oil-
filled heads. Richard travels up to the Arctic to study belugas in their
natural habitat, but he has spent far more time working with them at
Mystic Aquarium, just down the road from the *Charles W. Morgan* and
that chart of the *Commodore Morris*. Richard was a trainer at Mystic
Aquarium for a decade, and here he now continues to work with belugas in these controlled conditions for his research. He's been developing
tools to study the response of beluga whales to rapidly rising ocean temperatures and decreasing ice. He works in particular on ways to monitor
whale health by sampling their blow. In his scientific papers, Richard
refers to the whale's spout as "respiratory vapor" or "exhaled breath con-
densate."[1]

Ishmael concludes accurately in "The Fountain" that the spout is a
condensed mist that is often mixed with a bit of seawater resting around
the blowhole. This was not settled at the time. Dr. Bennett devoted
pages to both sides of the issue, explaining that "the entire question is
involved in much perplexity." Several others, such as Surgeon Beale, had
arrived at the conclusion that it is indeed condensed mist.[2]

"Melville gets the spout mostly right," Richard says as we look over
the rail at the beluga exhibit. "But it's not *just mist*. Like he said, part of
it is the seawater that's been sitting in the depression over the blowhole,
which combines with this super powerful exhale, filled with carbon dioxide. This condenses with the outside air."[3]

Kela, the female beluga whale at the aquarium, takes a quick breath,
which we *cannot* see, before she dives back under the water.

Richard continues: "The spout is also not just mist in the sense that
it's also a complex biological matrix that has lots of stuff in it. There's
mucus in there, snot, which lines the respiratory tract. And there's skin

cells and microorganisms that get carried up from their upper respiratory passages above the lungs, just like you have bacteria inside your nose. There's so much there in whale blow. It's bonkers. That's why it's such an exciting tool for research."

Working with other colleagues, Richard has been using whale blow to provide information about reproductive status and other hormone levels. They also use blow to sample DNA, microorganisms, and traces of other cellular debris. It's far less invasive than a blood sample and easier to collect. Scientists are even now exploring ways to collect whale blow in the wild with hovering drones.

"I really loved reading that passage in 'The Fountain,'" Richard says, "about how the whalers were too afraid to get close because their skin might melt off from the spout. I love this! Having had so much of it on me, I can attest my skin has *not* melted off! Was Melville kidding?"

I think so. But Bennett did write that it smelled fetid and felt acrid on the skin. And Ishmael quotes a source in the "Extracts" that whale blow smells and "brings on a disorder of the brain." I ask Richard if whale blow smells bad.[4]

"It can. It can. I find that different individuals tend to have a distinct smell, and I wonder if it has to do with their particular flora, like normal bacterial growth, just as some people have distinctly smelling breath that you can pick up on, probably dependent on what microorganisms are growing in there at the time. But remember, the whale's mouth, the trachea to the lungs, has no connection, no little valve, to this spout—despite what Melville says. So this scent in the blow is not from rotting food or anything like that."

We watch the male beluga named Juno surface, exhale, and submerge again. Juno is notorious here at the aquarium for playing with kids by spraying water over the glass. He even targets staff members when they're giving talks with their back to him. But visitors sometimes don't recognize that Juno does not, cannot, get them wet by way of his spout. The beluga whale *spits* up a gulp of water out of his mouth.

From a boat or from the shore, the first view of whales out in the wild could be a fin, an arching back, or a tail, but it's usually first the whale's

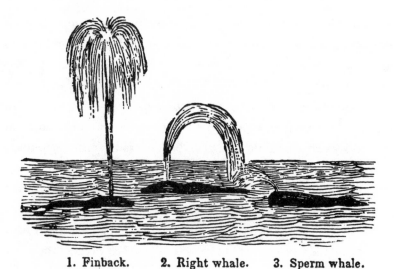

1. Finback.   2. Right whale.   3. Sperm whale.

FIG. 31. Spout key from J. Ross Browne's *Etchings of a Whaling Cruise* (1846).

blow. The spout was essential to whalemen, who could identify species, distance, and sometimes even the animal's swimming and diving behavior. The nineteenth-century whalemen who drew pictures in their journals more often illustrated spouts and fins rather than full profiles of whales or fish.[5] (See fig. 31.)

In "The First Lowering" Ishmael explains how Tashtego, the Native American harpooner, identifies the presence of a whale from the merest spray off in the distance, even when sitting in a whaleboat: "To a landsman, no whale, nor any sign of herring, would have been visible at that moment; nothing but a troubled bit of greenish white water, and thin scattered puffs of vapor hovering over it, and suffusingly blowing off to leeward, like the confused scud from white rolling billows."[6]

Indeed, nineteenth-century mariners hunting throughout the world's oceans could diagnose species of whales by their spout characteristics, just as whale watch captains and naturalists do today. In 1858, the Danish zoologist Daniel Fredrik Eschricht wrote about how impressed he was with the observational skills of these mariners, their ability to distinguish species only by "the shape of the vapor." In "The Grand Armada" Ishmael describes the difference between sperm and right whale spouts. Daggoo, the African harpooner, confirms with his captain that he's seen the White Whale by referring to the individual's blow, which is particularly "bushy" and "quick." Mariners knew that toothed whales, such as sperm whales and belugas, have a single, forward-spraying spout from a single blowhole, while baleen whales have double-spouts from a *pair* of nostrils. A right whale, for example, blows two puffs that make a "v" shape. A fin whale, as Ishmael says in "Cetology," exhales a single tall, thick, columnar blow that's a combination from its two blowholes, a "straight and single lofty jet rising like a tall misanthropic spear upon a barren plain."[7]

Once I went on a whale watch on Monterey Bay, California, with a captain named J. J. Rasler, who drives the only tourist whale-watch boat on the California coast that uses biodiesel. Although a young man, he is a bit of a local legend, having grown up in the area as a fisherman. His boss told me that Rasler turned down a law school scholarship to stay out on the water. Rasler has read *Moby-Dick* at least three times. He tells me a story of a humpback whale that came up under one of their boats and bent their propeller. During a whale watch he is on the radio most of the time with his fellow captains to find whales, but it really comes down to his ability to see and identify spouts off in the distance.

"Melville's actually not too far off in a lot of it," Rasler tells me while driving the boat and pausing to listen to the VHF radio. "Although it was kind of funny how Melville twisted certain things to be entertaining." Rasler explains how he also identifies different whales by their blow, but there's variability within species and situations. While we are out there, we see a shockingly high and tall spout of a blue whale far in the distance. "Sometimes when they're feeding, and staying in the same area, they don't breathe as hard. You just take little pieces and try to put it all together. I've been out here working for thirteen years, and I'm still seeing new things."[8]

With all this in mind, it's intriguing that Melville chose the whale's spout as the first instance in the novel that moves into the realm of the magical. Melville chose the spout as his primary symbol for the fantastical, the alluring, and the sometimes dangerous unknown. It is in "The Spirit-Spout" that Fedallah, the eerie, satanic fourth harpooner—

the only human character whom Ishmael pushes beyond human—first sees a "silvery, moon-lit jet." Fedallah believes this distant spout to be a sperm whale. When the whalemen try to lower after it, the blow disappears. And so this happens night after night until "no one heeded it but to wonder at it."[9]

THE WHALE'S SENSE OF SMELL

The whale's "sense of smell seems obliterated in him," declares Ishmael. Justin Richard confirms that any extreme halitosis from the spout would be unrecognizable to fellow whales—at least for toothed whales, such as sperm whales and belugas. Over millennia, toothed whales lost their olfactory bulbs and the nerves that anatomists believe to be necessary for smell. There is, however, an active line of thinking today that the baleen whales might have at least some ability to detect some scent. Maybe it helps them find food?[10]

THE PHYSIOLOGY OF THE DEEP-DIVING SPERM WHALE

In "The Fountain" Ishmael claims that a sperm whale spends "one seventh or Sunday of his time" breathing on the surface, while the rest of the animal's life is fully underwater with the blowhole shut tight. Ishmael declares: "Every one knows that by the peculiar cunning of their gills . . . a herring or a cod might live a century, and never once raise its head above the surface. But owing to his marked internal structure which gives him regular lungs, like a human being's, the whale can only live by inhaling the disengaged air in the open atmosphere."[11]

Surgeon Beale wrote that an experienced whalemen could predict "to a minute" when a given sperm whale, once observed for a while, will surface to breathe. Since humans live in the air, we breathe automatically. But living primarily underwater, whales need to be aware of their breathing. This means they can't sleep in the same way we do. Whales have adapted a way to sleep with only one side of their brain at a time. It's hard to calculate the fraction of one-seventh of their time at the surface, though, which Melville enjoys for its religious parallel. A study in 1995 tracked a single sperm whale in the southeastern Caribbean for over four days with a radio tag, revealing that this sperm whale, likely a male, spent a little over half of his time on deep dives, 23.4 percent on shallow dives, and 22.6 percent "in activities at or near the surface." A study published in 2017 found that sperm whales in the Gulf of California spent closer to 30 percent of their time at or near the surface. Sperm whale expert Hal Whitehead observed a ratio that was almost the same as Ishmael's one-seventh of the time. He described a "group of twenty or thirty sperm whales, each diving for forty minutes or so and then coming to the surface for about eight minutes at a time."[12]

Justin Richard and I can't see this as we watch the two belugas at the aquarium, but they expire and inspire when at the surface in a fraction of a second, far faster than land animals, since they need to keep water out of their respiratory tract. Scientists and poets, then and today, have been appropriately astounded at the sperm whale's ability to dive so deeply and for so long—which involves two factors that Ishmael understood: the efficient use of oxygen and the ability to withstand the pressure.

During his examinations of the sperm whale's head and when describing the old, sick, injured sperm whale in "The Pequod meets the Virgin," Ishmael recognizes the pressure that the sperm whale must withstand at depths of 1,200 feet, which they do seem to dive to on a regular basis. Sperm whales dive far deeper than anything a beluga whale or any baleen whale could withstand. Among the cetaceans, only the beaked whales can dive as deep and for as long. Ishmael calculates the water pressure the whale needed to withstand at that 1,200-foot depth to be about fifty atmospheres, which he overestimates at nearly the same ratio as did his source on this, whaleman William Scoresby in *Arctic Regions*. Naturalists in Melville's time wondered whether the rubberlike nature of the whale's skin might protect them during these dives. Scientists know now, that as all whales evolved, they lost their frontal cranial sinuses, so they wouldn't have that trapped air when they dove. In addition, the sperm whale's lungs likely collapse to expel any air, while other cavities, such as in the middle-ear, are lined with veins that fill the cavity with blood to expel any other trapped air that would be susceptible to pressure changes.[13]

When humans take a breath to dive underwater, we generally store about half of our oxygen in our lungs, and then store the rest of the oxygen in our blood and, in the smallest proportion, in our muscles. By contrast, in order to thrive and hunt in the deep ocean, sperm whales store the majority of their oxygen, about half, in their blood, distributing this in a manner more efficient than humans do. Ishmael explains that the sperm whale's ability to make air last is related to a "labyrinth of vermicelli-like vessels" between the animal's ribs and around the spine, which stores the "oxygenated blood." Melville's description of these spaghetti-like blood vessels is accurate in that cetaceans do indeed have specially adapted blood vessels that look like that, although physiologists are still not certain of their function even today. The blood vessels are known as the *retia mirabilia*, which means "wonderful nets," and they might help deep-diving mammals with thermoregulation, protecting against the effects of pressure at depth on the lungs, and/or nitrogen regulation to protect against possible decompression sickness—what scuba divers call "the bends." Long before Melville's time, anatomists had identified the retia mirabilia in dolphins. The nineteenth-century naturalists had connected these vessels to the sperm whale's ability to stay underwater for more than an hour. Melville would've read about this in his *Penny Cyclopædia* and even seen an illustration of them there (see fig. 32). Here Melville also read explanations by the English naturalists Hunter and Owen that whales and dolphins have an "almost total absence of valves," which they believed explained how harpoons can kill and bleed out a huge whale. In "The Pequod meets the Virgin," Ishmael uses this same explanation to describe the pools of blood and gore surrounding a recently killed old sperm whale. But the truth is, even if whales have fewer valves, this doesn't have much to do with their ability to stop bleeding. Marine biologists today know that sperm whales simply have a far greater volume of blood in relation to their body weight than humans, which is part of their ability to dive and stay

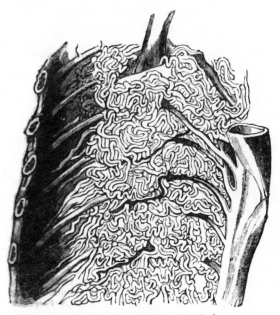

Arterial plexus in the Dolphins. (Breschet.)

FIG. 32. Illustration from the "Whale" entry in Melville's *Penny Cyclopædia* (1843) of the dolphin's retia mirabilia, or the "labyrinth of vermicelli-like vessels," as Ishmael terms them.

down, so perhaps the naturalists and whalemen were just amazed at the sheer volume of blood that they saw in the water.[14]

WHALE DISSECTION IN THE NINETEENTH CENTURY

While writing *Moby-Dick* Herman Melville was only a decade from the chance to observe toothed whales in a way similar to that of Justin Richard at the aquarium. The first cetacean kept in captivity in the United States seems to have been a beluga whale captured from the St. Lawrence River. In 1860 it was delivered by crate on a locomotive to the newly opened Boston Aquarial Gardens, not long after Louis Agassiz and Harvard opened their natural history museum across the river in Cambridge. The beluga whale survived for more than a year. During the beluga's life in the tank it was occasionally strapped with reins around its head in order to tow a woman on a clam-shaped boat, replicating a sort of Venus scene for visitors. P. T. Barnum loved the idea.

FIG. 33. Artist-whaleman Robert Weir's illustration of scooping out spermaceti on deck from a small sperm whale's head (1857).

Over the next couple years he paid to bring a few "white whales" of his own to New York City. Barnum had more difficulty keeping them alive, however, despite building a pipe under several city blocks that pumped in water from New York Bay. On July 13, 1865, his American Museum burned down in a fire. The two belugas boiled to death.[15]

In addition to the inability to observe whale behavior in a controlled setting behind glass, the naturalists of Melville's day had only barely begun to understand the internal anatomy of large marine mammals. In 1839 Thomas Beale did not provide for his readers an illustration of the internal organs of the sperm whale. No one anywhere at the time had published these types of drawings. This was in part because the body of a whale itself was usually too large to bring on deck or to access, half-sunk, even after all the blubber had been stripped off. The sperm whale's head, especially that of the male, can be over twenty feet long. So only occasionally did they hoist a sperm whale head up on deck. (See fig. 33.) Ishmael says in "The Sphynx" that "the beheading of the Sperm Whale is a scientific anatomical feat," since the whale has no distinguishable neck and this part of the whale is the "thickest part of him." In "The Great Heidelburgh Tun," Ishmael describes the men extracting oil from the head by precariously hauling it up vertically out of the water along-side of the hull, which was the common method.[16]

If before the 1840s any naturalist completed a careful dissection of a sperm whale that had beached, the reports had not made it to European or American professional circles. The time it took to travel and pass information, the challenges of preservation and transport, and the lack of proper tools, often made it impossible to take advantage of a beached whale of any species for scientific purposes. So naturalists such as Beale and Bennett pieced together information of the sperm whale's internal anatomy from a variety of sources—especially dissections of smaller whale species, such as dolphins. While at sea in 1835, Dr. Bennett dissected a fetal sperm whale, found in the body of its mother after it was killed. Bennett believed that the fourteen-foot-long male was only hours from being born. He was able to describe this fetus's digestive and respiratory system, including stretching out 208 feet of intestine. Even professional whalemen, such as James Colnett, occasionally captured young sperm whales to learn anatomy and to teach each other how to properly cut the blubber strips and extract the spermaceti (see earlier fig. 28). Thus, in "A Bower in the Arsacides," Ishmael explains that on another ship on which he sailed the crew once picked up a "cub Sperm Whale" to make sheaths for the harpoon barbs out of the whale's "poke or bag [stomach?]," after which Ishmael used a hatchet and his knife, "breaking the seal and reading all the contents."[17]

It was not until 1845, the year Melville returned to Boston, that an American doctor named J. B. S. Jackson published a paper in the *Boston Journal of Natural History* that claimed to record the first full scientific dissection of a sperm whale. A young 3,000-pound, sixteen-foot-long female had been brought up to Boston on a railroad car after being killed about fifteen miles out from New Bedford. Jackson's paper includes a lovely, detailed illustration of the multichambered stomach, but that was it.[18]

WHALE VOICE, HEARING, AND ECHOLOCATION

As Justin Richard and I are speaking about the sensory life of cetaceans, we watch through the glass the two beluga whales swimming below the waterline. Like narwhals and bowheads, beluga whales do not have dorsal fins. This likely evolved to ease their movement under ice and to avoid a convenient handle for polar bears. The sperm whale, like the fin whale, has a relatively small, almost vestigial dorsal fin far toward its tail at the end of its dorsal ridge. What Melville describes as Moby Dick's "high pyramidal white hump" is the sperm whale's dorsal ridge and its dorsal fin. Presumably this helps them track underwater like a boat's keel.

Surfacing, one of the belugas gives a wet trumpet sound, like one of those vuvuzela horns famous from World Cup soccer matches. The belugas also regularly emit all sorts of clicks and squeaks, as well as deep bellows. Through the glass we hear a variety of clicks and rattles, too, especially when one comes close to investigate us.

Beluga whales are especially loud and varied in their noise-making. Mariners nicknamed them "sea canaries." Beluga whales make sounds above the water out of their blowholes. Audible below the surface, they produce clicks and rattles through internal organs in their head. Like nearly all the toothed whales, belugas have a sound-making valve within their nasal passage, often called the monkey's muzzle, *museau de singe*, or phonic lips, which make noises in combination with internal air sacs and the surrounding musculature. Belugas have an extraordinary range of sound out of their blowholes—think whoopee cushions and bagpipes—along with still more of a language *under* the sea, which is primarily, it seems, sonar beams focused and augmented via their oil-saturated forehead, also known as the melon.[19]

Sperm whales have phonic lips, too, but they do not seem capable of making any noises with these through their blowhole at the surface, other than the gasp itself—which certainly can be loud when under stress. Melville's derision of J. Ross Browne's roaring whale and Ishmael's satisfaction with the metaphor of sperm whales as stoically silent are nearly true on the surface of the ocean—but it is not so below. Unique from any other whales, the sperm whales and their close relatives, the dwarf and pygmy sperm whales, have two exceptionally asymmetrical nasal passages. While both passages travel through the oily head-

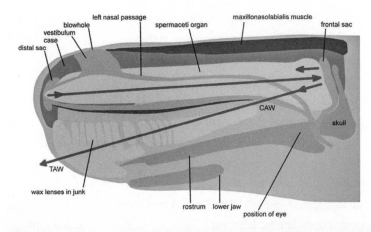

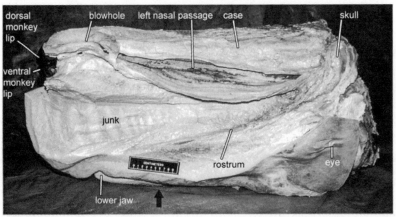

FIG. 34. A modern dissection of the head of a sperm whale calf. The line with arrows represents the theoretical acoustic pathway for sound production for echolocation and communication (TAW = terminal acoustic window; CAW = connecting acoustic window). See Huggenberger, et al., 2016.

matter, the wider diameter air passage leads to the phonic lips, which is situated below the blowhole. The more narrow air passage leads directly to the blowhole and is responsible for all of the air exchanged in breathing. (See fig. 34.) Although beluga whales, dolphins, orcas, and all other toothed whales, also have only one blowhole, unlike sperm whales, their two nostrils, or *nares*, are short and internally symmetrical.[20]

Although the surgeon-naturalists Beale and Bennett did not have a clear understanding of the internal anatomy of the nostrils within the sperm whale's head, they, like Melville and other experienced whalemen, knew well the different oil sacks within the sperm whale's enormous noggin, even if they did not know their utility for the animal's survival. Whalemen knew that the majority of the sperm whale's head is filled with two oil-rich sections. The upper part is what they called "the case," a term still used by biologists today. The case is a roughly cylindrical mass of soft white tissue soaked with a fluid known as spermaceti oil, which does indeed crystallize and turn whitish and waxy when exposed to cold air. Dr. Bennett wrote that it has a creamy, bland flavor, like "very fresh butter." Beneath the case and on top of the rostrum is "the junk," which also has compartments of oily tissue similar to the spermaceti organ. The junk is more solid and is divided by denser tissue into segments or lenses. Ishmael accurately explains that the junk is a "honeycomb of oil."[21]

For the whalemen's market back home, oil was generally divided into right whale/bowhead oil; sperm whale oil from its body; and the most valuable stuff, the spermaceti. The oil from right whales and bowheads was known in the nineteenth century more generally as "whale" or "train" oil, for use in lamps and lubricating some of the machinery of the Industrial Revolution. The blubber from the body of sperm whales burned cleaner so it commanded a better price. Sperm oil was also an excellent lubricant for small machinery, such as ship's chronometers and sewing machines. From 1833 to 1840, for example, Charles W. Morgan's company provided the sperm oil to lubricate all of the government's coastal lighthouses. The highest grade spermaceti, from both oily sections within the sperm whale's head, were processed together ashore into high quality sperm oil and into a specialized wax that was manufactured into high quality candles.[22]

As for the oil's natural utility for the whale, Ishmael argues in "The Battering Ram" that the fluid-saturated case and junk in the sperm whale's head might serve animals to assist with buoyancy and, as Ishmael says, act like a shock absorber, a ship's fender, to cushion during "inconsiderable braining feats." We'll take up this topic *head-on* later.[23]

Melville, along with all the naturalists and whalemen of his day, would have been *dumbfounded* to learn of the spermaceti's more significant role in underwater communication, navigation, and finding food.

Surgeon Beale knew that sperm whales communicated underwater for several miles, by some sort of "signals," but how exactly they did so "remains a curious secret."[24]

The majority of the sperm whale's life is spent in complete darkness. Even in the clearest ocean, sunlight does not extend beyond some 650 feet down. Research into animal sonar had begun in the 1790s, notably by an Italian bishop named Lazzaro Spallanzani who experimented with gruesome ways of systematically removing each sense from bats in order to confirm that they somehow navigated in complete darkness only by sound. Yet Spallanzani still could not figure out their secret. Echolocation wasn't identified for another one hundred fifty years. Naturalists in Melville's day knew that sperm whales had to somehow capture food in darkness and that even whales that were blind or with deformed jaws survived somehow in the wild. Surgeon Beale liked the theory that the sperm whale floated quietly in the deep sea with its mouth open, alligator style, which attracted squid into his jaws with his mouth, tongue, and glistening white teeth. At least a few American whalemen bought into this theory, too.[25]

Scientists did not begin to recognize the vast array of sperm whale clicks, creaks, clangs, and grunts below the water—and its potential role in finding food and navigating in darkness—until over a century after *Moby-Dick*. This largely began with scientists Bill Schevill and Barbara Lawrence, a husband and wife team from the Woods Hole Oceanographic Institution who submerged microphones to listen to the sounds of belugas for the first time in 1949. This led to their theories that toothed whales could echolocate in the same way as bats. In the 1970s scientists began to take a close look at the chemistry of the two spermaceti organs, which each have a different makeup than, say, the oil within a beluga's melon or nearly any other toothed whale. The consistent lipid composition in the sperm whale's head, along with, perhaps, its internal temperature gradients and the skull shaped like a satellite dish, seem to provide an exceptionally useful method to focus outbound sound waves. In recent decades biologists have identified complex intraspecies social communication and foraging behavior in the sounds of sperm whales. A

2003 study that used an array of hydrophones concluded that the sperm whale's "monopulsed clicks," also known as codas, are "by far the loudest of sounds recorded from any biological source."[26]

Justin Richard explains to me that sperm whales produce underwater sounds through methods similar to belugas. The sperm whale's phonic lips and air sacs, however, are far in the front of their heads. The spermaceti organs seem to amplify and focus back out these echolocation sound waves that likely bounce off the bowl-shaped skull. (See fig. 34.)

There is a slim possibility that baleen whales might be capable of some echolocation, too, but their moaning sounds underwater seem to be more for intraspecies communication, rather than for deep-sea navigation and finding prey. Studies that first began in the 1950s and '60s with Cold War technology searching for Russian submarines, led by Bill Schevill and another colleague, Bill Watkins—the same pair that recorded baleen rattle—led scientists to understand that a blue whale's low, subsonic vibrations can communicate to other blue whales swimming over a thousand miles away, perhaps across entire ocean basins.[27]

The depth-sounding technologies on modern oceanographic and commercial fishing ships use the same premise as that of toothed whales, sending down a series of sonar clicks to the bottom. One type of this equipment is called the CHIRP, an acronym for Compressed High Intensity Radiated Pulse. You can hear the clicks in the lower holds of a steel ship when the CHIRP is operating. During my own years at sea I've also heard through the hull the actual clicks and whistles of dolphins, through both wood and steel. I'm sure that early whalers in the forecastles of the *Acushnet* and the *Charles W. Morgan* must have heard this, too, at least from the smaller whales. It's a believable yarn that the whalemen called these sounds "carpenter fish." The clicks and knocks from echolocating and communicating toothed whales might have sounded to mariners like hammers tapping underwater.[28]

How exactly sperm whales and other whales sense *each other*, are aware of other individuals, remains unclear. At the aquarium, Richard explains that beluga whales seem to have good perception across a wide range of frequencies, but this is largely known from experiments on smaller whales in captivity. Although whales have tiny ear slits behind the eye that lead to internal structures through an auditory canal, the toothed whales—from dolphins to sperm whales—actually first receive sound through the acoustic fat in their lower jaws. Beale, Bennett, and their colleagues in the 1840s had identified the hearing apparatus within the sperm whale's head, so they assumed sperm whales had at least some sense of hearing through the earholes. Ishmael explains that into the tiny ear opening of the sperm whale "you can hardly insert a quill," and that the right whale also has a small canal with a membrane over the ear canal. This is mostly true; there is a plug that's only visible in dissection. Where did Melville get this detail?[29]

American whalemen believed sperm whales could hear—and see— well enough that they had to row up quietly to not startle the animal. At an earlier point in the voyage, Ishmael says, "The sudden exclamations of the crew must have alarmed the whale." At one point during the final chase, Ahab approaches Moby Dick head-on so as to be less visible to the animal, which was a popular conception of nineteenth-century mariners—that whales are less aware of what's in front of them. Observers today recognize that sperm whales can get nervous if a boat is in front, especially if they're not used to them. But the whales rarely if ever turn their heads sideways to look with their eyes at the approaching vessel, as would a seabird. More often they move their forehead toward a vessel to use the echolocation clicks more directly. It could be, too, that the whalemen didn't need to be quiet at all. Sounds made on the surface might not register significantly with sperm whales. The subsistence hunters in Lamalera, Indonesia, for example, chant and bang the side of their boats while hunting sperm whales.[30]

Well aware of the serious ethical conflicts of keeping highly intelligent animals in captivity, Justin Richard explains that the belugas and other marine mammals at Mystic Aquarium have provided research opportunities that would be impossible for him and his colleagues otherwise. Even just small anecdotal moments teach volumes. For example, when he worked as a trainer and scuba diver here, he learned to correlate the underwater beluga sounds to general behaviors, such as play or hunger. Once when he and his fellow trainers slid open a plastic Ziploc bag underwater, Juno and Kela rushed over to a corner together, trying to hide. "It was bonkers. We wondered if we'd replicated something in the wild, like the sound of a killer whale." When Richard travels up to the Arctic he's struck by how hard it is to learn about the whales in their own environment, even about belugas, which are especially coastal compared to most other cetaceans. Still further, he's amazed by how, even with so many technological advancements, so much remains unknown about marine mammals.

Juno comes up to the surface, breathes, and rolls over to look at us with his left eye. Richard explains that the scientific community still remains ignorant of some fairly basic anatomical and sensory information, especially about the largest of whales. Because how could you, for example, design an experiment in the wild that tests the ability of a sperm whale to smell or hear? More specifically, if sperm whales make their echolocation clicks within their heads using air in their nasal passages, how do they do so at great depth with all that pressure—when supposedly they have pushed all the air out?[31]

"Dissect him how I may, then," Ishmael says in his reverie on the tail, "I but go skin deep; I know him not, and never will."[32]

PLATE 1. The whaleship *Charles W. Morgan* sailing on Stellwagen Bank Marine Sanctuary in 2014 beside a humpback whale.

PLATE 2. One of the whaleboats of the *Charles W. Morgan* beside two humpback whales on Stellwagen Bank (2014).

PLATE 3. A "grand armada" of hundreds, perhaps even thousands of whales, photographed in the Indian Ocean by Tony Wu (2014). These are mostly females, calves, and juveniles, but there were a couple large males in the pod, too. Note the sloughing of skin and the defecation at the upper right.

PLATE 4. North Atlantic right whale skim feeding just beneath the surface.

PLATE 5. A white sperm whale calf photographed in 1995 by Flip Nicklin, who believes the calf was a female.

PLATE 6. Almost certainly the same whale as in plate 5, now grown, photographed off the Azores in 2016.

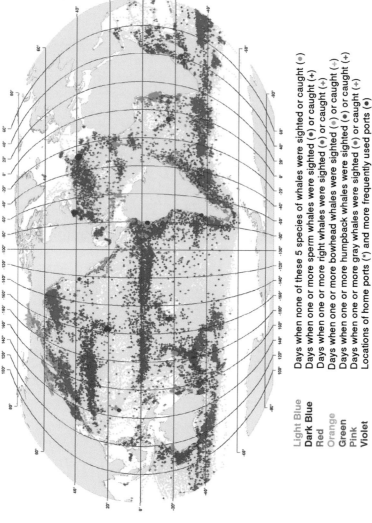

Light Blue   Days when none of these 5 species of whales were sighted or caught (•)
Dark Blue   Days when one or more sperm whales were sighted (•) or caught (+)
Red   Days when one or more right whales were sighted (•) or caught (+)
Orange   Days when one or more bowhead whales were sighted (•) or caught (+)
Green   Days when one or more humpback whales were sighted (•) or caught (+)
Pink   Days when one or more gray whales were sighted (•) or caught (+)
Violet   Locations of home ports (*) and more frequently used ports (•)

PLATE 7. Historic whaling under sail from 1780 to 1920, with data from logbooks representing about ten percent of American whaling voyages during this time as published in Smith, Reeves, Josephson, and Lund (2012).

PLATE 8. Image of defaunation and environmental impact on terrestrial versus marine ecosystems as published in McCauley, et al. (2015). Human impact on ocean systems has been slower, but with increased technological advances and global warming (middle bar, IPCC data) we will likely, rapidly, increase our impact, both in the number of marine extinctions and in heavily altered ecosystems.

PLATE 9. A freshly dead giant squid that washed up on the southern coastline of Wellington, NZ, in 2018.

PLATE 10. A great white shark tears into the fluke of a Bryde's whale carcass off South Africa.

PLATE 11. The enormous *Porites* coral "Big Momma," with a circumference of about 135 feet (2014). The biologists of the National Marine Sanctuary of America Samoa estimate this "colossal orb," as Ishmael would call it, at more than five hundred years old.

PLATE 12. Magnificent frigatebird (*Fregata magnificens*) with deflated red gular pouch.

Equator

Java Sea

Straits of
Sunda

AUSTRALIA

INDIAN
OCEAN

of
Hope

NEW
ZEALAND

Ch. 17

WHALE AND HUMAN
INTELLIGENCE

> To scan the lines of his face, or feel the bumps on the head of this Leviathan;
> this is a thing which no Physiognomist or Phrenologist has as yet undertaken.
>   Ishmael, "The Prairie"

I recently sailed with a watch officer named Adrienne Wilber, known
by her shipmates as "Heartbreak." When she is not working on school
ships, she fishes for salmon and groundfish off Sitka. Heartbreak told
me that sperm whales have become a significant problem for fishermen
off the coast of Alaska. The whales have learned to eat black cod (also
known as sablefish, *Anoplopoma fimbria*) caught on the deep gangions
of the fishermen's longlines.

"We watch the sperm whales sound," she said, "and you know they're
going down to eat our fish. I've seen video footage of it. The sperm
whales have learned to essentially floss the fish off the longlines. And
they seem to be teaching it to their young." Heartbreak explained that
the sperm whales have learned to recognize the sound of the fishermen's
hydraulic winches, like a dinner bell, swimming in to feed when the
fishermen begin to haul back.[1]

SPERM WHALE INTELLIGENCE

Sperm whale experts today see that sort of whale behavior in Alaska as not only evidence of higher intelligence, but indicative of culture, by which researchers such as Hal Whitehead and Luke Rendell mean the flow of information between animals. More specifically, they accept the definition of culture as "behavior patterns shared by members of a community that rely on socially learned and transmitted information." For example, the ability to communicate is genetically inherited, but click dialects in clans of sperm whales, like a language, are more than that: sperm whale click dialects are part of their nonhuman culture.[2]

Beginning in earnest in the 1960s, behavioral ecologists have studied bottlenose dolphins in particular for examinations of marine mammal intelligence and culture. Scientists such as Kathleen Dudzinski have identified in these small whales extraordinary evidence of both. Dolphins have what is referred to as a "theory of the mind," meaning that an individual recognizes a self, with thoughts and ideas, and that this self sees another individual and recognizes that he or she must also have his or her own individual thoughts and ideas. Experiments in captivity and in the wild over the past half-century have also shown that dolphins, like chimpanzees and elephants, recognize themselves in mirrors. They can understand the permanence of an object, meaning if you take it away from their view, they know it still exists. Dolphins will seem to play with inanimate objects, both alone and in groups. In addition, dolphins form social alliances at multiple levels. They have long-term social memory of these alliances, exemplified in a study in which a female bottlenose dolphin recognized the sounds of her ally, the signature whistle of another female, after twenty years apart.[3]

Though far harder to study and assess, it seems probable that sperm whales are capable of similar, if not equivalent or even more advanced, cognitive perceptions and abilities.

In the mid-nineteenth century naturalists such as Surgeon Beale believed that sperm whales were more intelligent than other large whales.

Some whalemen thought so, too. This perspective serves, of course, one of Ishmael's narrative purposes, which is to build up Ahab's foe as far more than a "dumb brute," as his pious first mate believes. In "The Chart" Ishmael implies that sperm whale migrations are not only instinctive, but rather a "secret intelligence from the Deity." On the first day of the chase, Ishmael refers to the "malicious intelligence ascribed to" Moby Dick. Fortunately for us, because our narrator would be simply intolerable otherwise, Ishmael was not armed with the information that the sperm whale has the largest brain, at seventeen pounds, of any living organism on Earth. Although he'd read from Surgeon Beale the size of the brain cavity, Ishmael says in "The Nut" that the sperm whale's brain is only a "mere handful," leaving him to comically contort an elaborate, tongue-in-cheek rationalization that the sperm whale's high intelligence must be measured to include the entire spinal column.[4]

Beginning in the 1960s, the sperm whale's brain was the source of much conjecture, led by John Lilly, as to its possible superhuman intelligence. Scientists quite recently have confirmed that the toothed whales, with the bigger brains, do seem capable of higher human-like cognitive and cultural activities than, say, the smaller-brained baleen whales. A southern right whale's brain, for example, is a third the size of the sperm whale's. Yet today, when considering brain mass in relation to total body size, the sperm whale's brain is not entirely extraordinary. At this point we just can't reach too many conclusions about intelligence simply by the sheer size of a brain.[5]

PASSING FABLES ON HUMAN INTELLIGENCE

While the two whale heads are lashed on either side of the *Pequod* in the Indian Ocean, Ishmael continues to elevate the sperm whale and take humans down a notch by satirizing some of the scientific thought of the 1850s. Ishmael calls upon the methods of two of the fad-sciences of his day. He fingers around the sperm whale's face using the ideas of physiognomy in "The Prairie," and then he feels around the sperm whale's head using the concepts of phrenology in "The Nut."

When Melville was in London for the winter of 1849, in addition to buying the Gothic novels *Frankenstein* and *Castle of Otranto* (1764), he purchased an illustrated English translation of Johann Kaspar Lavater's *Essays on Physiognomy* (1798), the popular study of how the features of one's face reveal a range of characteristics. Both Lavater's physiognomy and the pseudoscience of phrenology, the latter of which mapped the regions of the brain and diagnosed personality by feeling the texture of the skull, were all the rage in Europe and the United States at the time. They were not always taken entirely seriously, though. Captain Robert Fitzroy, for example, was a devotee, while his cabinmate, young Darwin, was skeptical. By the 1850s physiognomy and phrenology mostly provided good entertainment in the way so many of us today enjoy taking quizzes to determine our IQ, personality type, or ideal job.[6]

Lavater, however, earnestly took his physiognomy studies into the animal world. He analyzed the head and facial features of organisms from lions to insects. He did not discuss whales, but he believed the elephant to be the wisest of animals. And based on the slope of face, the size of its mouth in proportion, and its lack of eyelids, Lavater declared fish to be "the most stupid of creatures."[7] (See fig. 35.)

Lavater is exceptionally attentive and precise regarding human noses, hence Ishmael's comical rationalization that the *lack of a nose* instead increases the impression, if not the reality, of the sperm whale's sublime genius.

Regarding phrenology, in order to analyze "the bumps on the head of this Leviathan," Ishmael correctly identifies the Austrian physician Franz Joseph Gall as the founder of this other cephalo-centric pseudoscience. Gall's phrenology derived, somewhat reasonably, from the concept that the brain is a muscle and that it has certain areas devoted to personality traits. This actually carries some truth with modern neuroscience, although the mapping emphasis is now more on brain function and tasks. Gall and the phrenologists believed that certain parts of the muscle would enlarge or atrophy based on use, so the rises and depressions on the skull could reveal the person's characteristics. At some point Melville likely picked up Johann Gaspar Spurzheim's *Phrenology*,

FIG. 35. To accompany his plate on fish in *Essays on Physiognomy* (1848), Johann Kaspar Lavater wrote of the hammerhead shark in the middle (2) and the fish at the upper left (3): "A monster, 2. How infinitely distant from all that can be called graceful, lovely, or agreeable! The arched mouth, with the pointed teeth, how senseless, intractable, and void of passion or feeling; devouring without pleasure or satisfaction! How inexpressibly stupid is the mouth of 3, especially in its relative proportion to the eye!"

or the Doctrine of the Mental Phenomena (1832), or a similar summary in his *Penny Cyclopædia* or elsewhere, since Ishmael's use of the phrases "self-esteem," "veneration," and organ of "firmness" match the sections labeled on Spurzheim's diagram of the head—and comically correspond to the regions of the battering ram of the sperm whale (see fig. 36). The organ of firmness, Spurzheim explained, is about obstinacy and constancy of principles, which are suited to command. "It is true," wrote Spurzheim, "that persons endowed with this feeling in a high degree, constantly say, *I will.*" The "Organ of Firmness" reads like a diagnosis of Ahab, even as Ishmael puns on the sperm whale's hump.[8]

Now, in addition to providing some humor as we follow Ishmael's forecastle logic, "The Nut" and "The Prairie" serve the novel's blue ecocritical themes in three significant ways.[9]

First, this is another place to stick it to the scientific community who has never experienced life on the ocean. "Physiognomy, like every other human science, is but a passing fable," Ishmael says. He appreci-

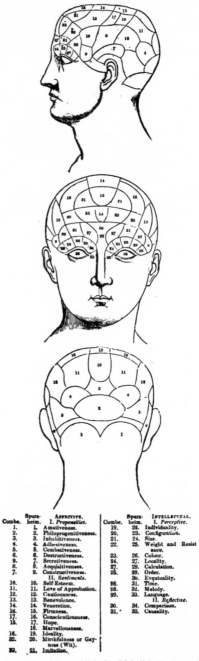

FIG. 36. Spurzheim's phrenological diagram as published in Melville's *The Penny Cyclopædia* (1840). The three sections on the head, which correspond to the battering ram of the sperm whale's head, are Numbers 10, 14, and 15, the exact words Ishmael uses in "The Nut."

ates careful inquiry and learning, but Ishmael wants to make clear, as in "Cetology," that any human construction to explain the world in neat confident systems is futile. Nobody can read the whale: "I but put that brow before you. Read it if you can."[10]

Second, especially for the twenty-first-century reader, intelligence and sentience are the surest ways to establish a connection, to build a sympathy, even an empathy for the whale. As we try to decide, to rank, which animals should be eaten or not, which should be sacrificed for consumer products or not, and which should be allowed to be kept in captivity for our enjoyment/education or not, the debate often hinges on intelligence, self-awareness, culture, and theory of mind.

Third, Ishmael's dabbling in these pseudosciences to try to describe the sperm whale's head continues the proto-Darwinian decentering of the human and the elevation of the whale. Through his anthropomorphic humor, Ishmael shows that intelligence and wisdom are slippery concepts, and the tools used by Anglo-Western societies are faulty, if not useless, regardless of whether these measuring sticks consider achievements in literature, language, or the analysis of the bumps on a human's head.

Ishmael finds physiognomy and phrenology ironic and funny. Queequeg, Shakespeare, George Washington, and the sperm whale all have venerable foreheads. Ishmael would be quite happy, it seems, to have sperm whales steal his fish from his hooks off the coast of Alaska. But for Ahab, the failure of humans to be the smartest and wisest, the failure to understand beyond what we can, and the failure to feel and see beyond the metaphorical surface, sets the top of his skull, his organ of firmness, all a-steam. "Of all divers, thou hast dived the deepest," Ahab says to the sperm whale's head. Like Hamlet holding the human skull, Ahab in "The Sphinx" looks down at the severed carcass: "O head! thou hast seen enough to split the planets and make an infidel of Abraham, and not one syllable is thine!"[11]

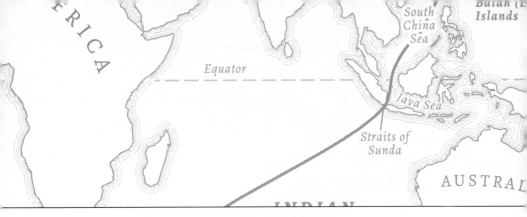

Ch. 18

AMBERGRIS

I think it may contain something worth a good deal more than oil; yes, ambergris.

        Stubb, "The Pequod meets the Rose Bud"

The *Pequod* sails the entire length of the Indian Ocean, squeezes up
through the islands of Indonesia, and now cruises northbound in the
South China Sea. Here the crew *smells* another whaleship in the region.
The lookouts aloft confirm a ship on the horizon and then identify the
French flag. The men observe two whales alongside the other ship under
"a cloud of vulture sea fowl." The French whaleship, the *Rose-bud*, has
two dead whales that had died on their own, although Stubb thinks
he might have stabbed a cutting-spade into one of them at an earlier
point. One is a "blasted whale," partially decomposed and rancid, and
the other is a whale that had seemed to die and then dry out, smelling
still worse, having expired perhaps due to "a sort of prodigious dyspep-
sia, or indigestion," leaving no significant blubber layer on its carcass.[1]

    Stubb rows over to the *Rose-bud*. In a comic, bawdy scene, he tricks
the French captain into abandoning the two rotting whales. When the
Rose-bud sails off, Stubb carves into the more rancid whale, the dry one.
He slices in underneath the ribs, then deeper until he retrieves about six

handfuls of goopy, cheese-like ambergris. Ishmael explains that this is "worth a gold guinea an ounce to any druggist."[2]

"The Pequod meets the Rose Bud" is the first part in one of Melville's tidy chapter pairs on the subject of ambergris, which is a product of the ocean as mysterious as the giant squid and intimately connected to that same animal. Knowledge of ambergris was a rare and chance privilege usually reserved only for those deepest-sailing mariners. In the second chapter of the pair, "Ambergris," Ishmael explains more about the substance with his natural historian hat on, attempting to answer the basic question of what ambergris actually is. He begins with the true story of how in 1791 a Captain Coffin, a Nantucket-born whaleman, enlightened the English House of Commons about the true origin of the stuff. Nantucket whalemen, likely the first to commercially hunt sperm whales beginning around 1712, were indeed able to *confirm* the provenance, which had been guessed as early as 1574 by a French botanist who deduced that it came from whales because of the encased squid beaks.[3]

Much but not all of Ishmael's descriptions in this pair of chapters is true to how we understand the substance today. Ishmael suggests the following: ambergris could be found in "the inglorious bowels of a sick whale"; ambergris smells faintly of perfume; ambergris is "unctuous and savory withal"; ambergris "is of a hue between yellow and ash color;" and inside the ambergris, Stubb finds "hard, round, bony plates" that turn out to be "pieces of small squid bones embalmed." Ishmael goes about four for five on ambergris.[4]

THE PHYSETER'S FECOLITHS

In 1947, aboard the whaleship *Southern Harvester*, an English biologist named Robert Clarke watched the men kill a fifty-two-foot sperm whale. They winched it up on deck from the stern ramp and extracted the meat, blubber, teeth, and all the other products that were still in demand during the mid-twentieth century. At the very last moment, just before the men shoved the viscera overboard, Clarke shouted. I like to imagine that he hollered "Avast!" Clarke had noticed a swelling in the intestines. He sloshed the gigantic slithery entrails out of the way, upon which he commenced cutting open the rectum. Clarke extracted a 342-pound boulder of ambergris, which looked more like a car-crushing meteorite. As he carved down into the lower layers, he found the ambergris to be softer and more yellow.[5]

Clarke would become the world's expert on ambergris. In his own Stubb-like poke at France, he preferred the pronunciation to be the anglicized *ambergreez* rather than *amber-gree*, even though the word derives from the French for amber and gray. Clarke explained that ambergris forms only in the rectum of sperm whales and potentially in pygmy sperm whales. He estimated that ambergris is found in about one of every one hundred sperm whales, regardless of sex, which was actually the same ratio proposed in 1724 by a Dr. Zabdiel Boylston of Boston, after he discussed the matter with his Nantucket whalemen friends.[6]

A sperm whale can eat a ton of squid a day. Normally the whale regurgitates out its mouth the indigestible squid beaks or other hard matter, as does an owl or a cormorant, but sometimes the material passes all the way through the sperm whale's multichambered stomach. Clarke proved convincingly that the sperm whale's body, to reduce irritation in the intestine, builds a protective, cholesterol-rich compound around these indigestible squid beaks to form a compacted ball, a concretion, that builds and transforms over the years. This was not entirely new information. In 1783 the German physician Franz Schwediawer put forth nearly the same explanation in a paper read before the Royal Society by none other than Joseph Banks. In whales, sometimes the blockage of ambergris passes uneventfully out of the anus. Contrary to the general thinking of Ishmael, the whalemen, and the reference works of Melville's day, Clarke and biologists now do not believe that ambergris is necessarily a symptom of illness, although it could eventually lead to the animal's death if it fully clogs the lower intestines.[7]

As Ishmael explains, and Melville had written of earlier in *Mardi*, there has been much confusion over the centuries as to the exact nature of ambergris, whether it be derived from animal, vegetable, or mineral. The confusion is apparent today when newspapers and magazines name ambergris as whale poop or whale vomit.

Down in the tank room at the Natural History Museum in London, Jon Ablett had shown me a container with ambergris. The ambergris was a black, waxy lump about the size of a plum. I have an awful sense of smell, but the dark smudgy clump smelled to me of dirt. Ablett said it smelled to him like coffee. When ambergris is fresh, like Stubb's handfuls, it apparently actually smells, well, like it came out of a whale's ass. This is the one aspect of ambergris that Ishmael seems to misrepresent on purpose. Fresh ambergris does not have "a faint stream of perfume." Melville knew this, but it fit his fiction better for the fresh stuff to have a pleasant odor. He had put a little check mark in his copy of Beale's *The Natural History of the Sperm Whale* where it described the "exceedingly strong and offensive smell," so he knew the truth. As another example, the author of an 1844 article in the *American Journal of Pharmacy* wrote that fresh ambergris is "unctuous when pressed between the fingers; smell somewhat resembling old cow-dung." Nor is the most valuable ambergris fresh from the sperm whale, that goop that Stubb holds. The highest quality ambergris is actually the substance after it has cured and hardened over years floating on the surface at sea. The thing about ambergris, which is why fragrance crafters use it, international epicures put it in hot drinks, and others enthusiasts swear it's an aphrodisiac, is that the substance develops a sweet, woody, musky seaweed scent, especially after it has weathered on the ocean or on the beach for a decade or more. Ambergris is most notable for its property of retaining and fixing odors from *other* sources. For this reason, Ablett believed that this particular piece of ambergris that he showed me at the Natural History Museum, which I kept snuffling like a buried truffle, was once stored in a coffee can.[8]

Ishmael and Stubb do not overestimate the value of fresh ambergris back ashore, however. In 1858 the whaling schooner *Watchman* brought back to Nantucket four barrels of ambergris worth a value at the time of

$10,000. This was more than the value of their entire, full cargo of whale oil. During one of the last voyages of the *Charles W. Morgan* in 1913, the first mate wrote after catching two sperm whales the previous day:

> Cut one Whale and looked him over as we thought well for ambergris Mr Christian saw small lump floting by Railed it up found it was what we had been looking for Cut [No: 2?] Whale he had none after Breakfast Cauncluded to have look at first Carcas Capt Church and Mr Christian lowered waist boat and got along side Carcas and opened it up [more?] found 13:3/4 Pounds of Ambergris.[9]

Ambergris is still in use today. Despite the demand, modern chemistry has yet to entirely replicate the scent and fixing properties to the satisfaction of the experts. Genuine sperm whale ambergris found on the lonely coasts of New Zealand, for example, sells at about 35 US dollars per gram as of 2018—that's over $15,800 a pound—depending a great deal on quality and the amount ordered.[10] In 2016 the *Times of Oman* reported that a fisherman named Khalid Al Sinani and his two fellow fishermen lassoed an approximately 175-pound chunk of ambergris floating at sea, which they expected to auction for the equivalent of about 2.6 million US dollars.[11]

In *Moby-Dick*, after Stubb tricks his way into a small fortune—which he tragically will never be able to collect—Ishmael declares: "Now that the incorruption of this most fragrant ambergris should be found in the heart of such decay; is this nothing?" Like the industry of whaling itself, Melville reveled in the chance to show how the ugly, dirty deep revealed both man's hypocrisies and nature's treasures.[12]

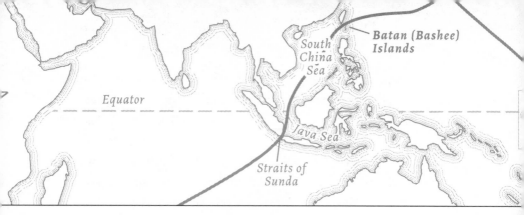

South
China
Sea

Batan (Bashee)
Islands

Equator

Java Sea

Straits of
Sunda

Ch. **19**

CORAL INSECTS

While all between float milky-ways of coral isles, and low-lying, endless, unknown Archipelagoes.
    Ishmael, "The Pacific"

In *The Tempest* (1611), William Shakespeare wrote of coral within some
of the most famous lines of sea literature in the English language:

Full fathom five thy father lies;
Of his bones are coral made;
Those are pearls that were his eyes:
Nothing of him that doth fade,
But doth suffer a sea-change
Into something rich and strange.
Sea-nymphs hourly ring his knell:
        [*Burden*, ding-dong.]
Hark! now I hear them,—ding-dong, bell.[1]

A spirit named Ariel sings these words as he deceives a prince into
imagining his father drowned after a supposed shipwreck. Ariel sings
that the dead king's bones have been formed into coral in a strange,
magical transformation: the image of a dead human now a part of the

living deep.[2] The year before he began to compose *Moby-Dick*, Melville engulfed all of Shakespeare's plays for the first time while he lounged on a couch at his in-laws. He later used coral in his novel to also evoke death, deep, and magic.[3]

In "The Castaway," only a few days after Stubb tricks his way into a handful of ambergris, Ishmael spins off Ariel's song with his story of the tragedy of Pip, the young African-American steward. After Pip leaps out of a whaleboat a second time during a hunt, Stubb leaves him floating in the water. (Melville had once witnessed a nearly exact event with a sailor aboard the *Acushnet*.) "Man is a money-making animal," Ishmael says to explain the officer's cruelty. Stubb assumes the other boats behind him will rescue Pip, but those men end up sailing in a different direction chasing other whales. So the least powerful person of the *Pequod* floats entirely alone in the sea, foreshadowing Ishmael's fate.[4]

I read Pip's submersion in the South China Sea as an imaginary sinking into his own loneliness of self and into the deepest of contemplation experienced within the outermost wilderness of the ocean, a view normally reserved for the sperm whale alone. This is the view Ahab thinks he wants to see. Pip is suddenly surrounded above, below, and in all directions by an ocean that is beyond human reach or influence. Ishmael describes:

> The sea had jeeringly kept his finite body up, but drowned the infinite of his soul. Not drowned entirely, though. Rather carried down alive to wondrous depths, where strange shapes of the unwarped primal world glided to and fro before his passive eyes; and the miser-merman, Wisdom, revealed his hoarded heaps; and among the joyous, heartless, ever-juvenile eternities, Pip saw the multitudinous, God-omnipresent, coral insects, that out of the firmament of waters heaved the colossal orbs. He saw God's foot upon the treadle of the loom, and spoke it; and therefore his shipmates called him mad.[5]

Pip survives. By chance, he is rescued by the lookouts of the *Pequod*. Yet he is rendered insane from the trauma. Just as in Ariel's song, Pip begins to sound the death knell, "Ding, dong, ding!" He now performs as a Shakespearean wise fool, forecasting the ship's destiny to sink and strike the rocks at the bottom, to go, as Fleece puts it, "fast to sleep on de coral."[6]

As a castaway imagining his sinking into the deep, Pip does not hallucinate fish or whales or sharks, but instead "coral insects" and the colossal orbs that they spin, as if spiders, under God's supervision. Melville's choice of words and images here went beyond the connection to *The Tempest*. As early as the 1750s, the scientific community and then the general public understood that small animal-like polyps excreted skeletons to form the various shapes—fans, horns, and globes—that rose from the ocean bottom and grew on rock to form entire reefs. Naturalists also knew in Melville's time that the brilliantly colored corals and the most enormous of these reefs were found almost exclusively in warm, relatively shallow tropical waters. Though known then to not actually be true insects, but colonial shell-building anemones, the common name of "coral-worms" or "coral insects" was in regular use, including by Dr. Bennett and Lieutenant Maury.[7]

Charles Darwin also used the common name of "coral insects" in his *The Voyage of the Beagle*. In 1836, shortly after the circumnavigation, Darwin wrote the paper that became the prevailing theory on the formation of different types of reefs. Most of Darwin's ideas on coral reefs would prove true. He explained that atolls and other reef systems were formed by the subsidence of volcanic peaks and calderas: coral began to grow on top and on the edges of the land that was slowly sinking over millions of years. Aware of and interested in this public debate, Melville in his earlier novels alluded to the competing theories of reef formation, partly as satire, but also with seemingly genuine fascination about the "coral insect ... this wonderful little creature" and their role in vast, geological systems that not only shaped reefs, but had implications for the formations of the Earth's continents and notions of what we now call *deep time*.[8]

In *Moby-Dick* Melville crafted for Pip a vision of the deep with a coral bottom that swirled with the profound of the Divine. Two centuries after *The Tempest*, several works published during Melville's time do much to reveal how mid-nineteenth-century ideas on faith and natural theology became associated with coral, and I want to tell you about three here in particular.

First, literary scholar Martina Pfeiler recently discovered the story "The Drowned Harpooner," which appeared in the *Nantucket Inquirer* and was then reprinted in *Graham's Magazine* in 1827. In this story a whaleman in the South Pacific, whom the author calls Jonah Coffin, is sucked under the surface by a line attached to a sperm whale and then dragged down into a sort of vacuum in the whale's underwater wake. Jonah can breathe and see into the "bottomless profundities," which includes "an extensive forest of coral inhabited by shapes indescribable." Jonah is eventually rendered unconscious, then rescued alive after the sperm whale turns to come back up to the surface to die.[9]

Second, around the same time, a magazine article titled "Works of the Coral Insect" was a lesson in *The First Class Reader*, a popular book published for American schools. The anonymous author was astounded by the concept of the sheer vastness of the numbers of tiny coral organisms and the amount of time that would be necessary to create some of the islands in the Pacific and the Great Barrier Reef of Australia. This author believed, just as Emerson had lectured in "The Uses of Natural History" in Boston, that the reefs just below the surface were continuing to build *upward*, to provide still more land for humankind in the coming centuries. "These are among the wonders of His mighty hand," wrote the author, "such are among the means which He uses to forward His ends of benevolence. Yet man, vain man, pretends to look down on the myriads of beings equally insignificant in appearance, because he has not yet discovered the great offices which they hold, the duties which they fulfil, in the great order of nature."[10]

The third publication is a fascinating illustrated booklet published in 1847 in New York that is a compilation of the writing of contemporary scientists, titled *Relics from the Wreck of a Former World*, which sails along the same currents as Louis Agassiz and Lieutenant Maury, emphasizing even in its lengthy subtitle no difficulty aligning an Earth that is millions of years old with the Biblical teachings of the Christian faith, allowing for a vast range of extinct animals of "fantastic shapes" and "the most elegant colors." Reverend Thomas Milner, a student of astronomy and a popular scientific author, was quoted at length. Milner wrote that the formations of coral reefs were perfect examples of "the power of Nature to effect her vast designs through apparently feeble and insufficient agents." On the same page of his discussion of the "coral insects," is the editor's consideration of the concept that a grain of sand might hold its own ecosystem and that the distances of space and light might be unimaginable: "And when we reflect that if it were possible for us to attain to those distant spheres [stars], we should look, not on the limits, the blank wall of Creation, but only into *fresh* fields of Creation, Power, and Wisdom, we feel that our earth and all that it inherits is a mere speck in space, an atom amid the vast Universe."[11]

Scholars don't know whether Melville read any of these works, but if they were not a direct inspiration they reveal an aspect of the public perception of coral in an age that was still a century before underwater color photography and scuba. These publications and *The Tempest* show in part why Ishmael connects God and mortality with coral when pondering Pip's philosophical plummet to the ocean's depths. They also add to our reading of Ahab's existential rages against that blank wall of the White Whale. For Melville and his contemporaries in the mid-nineteenth century, the coral reef was as sublime as the iceberg: stunning, unknowable, terrifically beautiful and yet fatally dangerous to the mariner. And perhaps most of all, coral was a symbol of the creative powers and brilliance of God.

Nor did Melville need any of these books or articles to feel spiritually moved by coral reefs. While in the South Pacific, he rowed and sailed over clear tropical waters that held these wonders. In French Polynesia, even right in Papeete harbor, he gazed, as he wrote in *Omoo*, "down in these waters, as transparent as air," to see "coral plants of every hue and shape imaginable." Melville had swum out to fringing reefs in the South Pacific. He described spearfishing around them by the light of torches. He might have even seen for himself specimens of the enormous *Porites* corals. Colonies of these corals off the South Pacific islands where he sailed and rowed can be truly "colossal orbs." Scientists believe *Porites* can be some of longest living organisms on Earth. In the waters of American Samoa, for example, is "Big Momma," also known as "Fale Bommie." This colony is a forty-two-foot-wide globe and estimated at five hundred years old. The top of its dome is at a depth of thirty feet beneath the surface: full fathom five.[12] (See plate 11.)

In 2017 researchers in the journal *Nature* declared that due to the effects of global warming, the entire Great Barrier Reef has died at an unprecedented rate: more than sixty percent of the coral, especially in the north, experienced extreme bleaching from the most recent high temperature event. This is more than six hundred miles of lifeless white coral that likely will not recover during my lifetime.[13] My God.

Melville and his contemporaries watched and understood that invasive plants and animals were altering the Polynesian Islands, and that Americans and Europeans were corrupting and demolishing native cultures of "noble savages" in the South Pacific. Less understood then was that the Pacific Islanders themselves in their relatively recent migrations had vastly altered these islands, too, deforesting, spreading invasives, and hunting to extinction such species as the giant flightless moa (*Dinornis* spp.) in New Zealand. Yet I suspect that none of the nineteenth-century cultures, indigenous or colonial, could fathom that humans could ever influence the growth of coral reefs. In *Omoo* Melville retells part of a song from the elderly Tahitians:

> "A harree ta fow,
> A toro ta farraro,
> A mow ta tararta."
> The palm-tree shall grow,
> The coral shall spread,
> But man shall cease.[14]

Read today, Pip floats over coral that now represents an entirely different type of bleached death. Pip is African American, but he could be an icon of a Native American or a Pacific Islander or any individual human reminder that social justice and environmental justice are inseparable. As the least powerful on the ship, a castaway, Pip is a symbol of those most vulnerable to anthropogenic climate change, depleted water resources, and the coastal pollution that is left in the wake of the hubris, the harpoons of petro-capitalism. We're in the middle, of course, of a sea change of the very worst kind. It's enough to make you lose your mind.

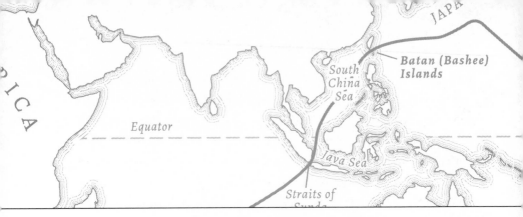

South
China
Sea

Batan (Bashee)
Islands

Equator

Java Sea

Straits of
Sunda

JAPA

PICA

Ch. 20

GRANDISSIMUS

But what is this on the chest? I took it up, and held it close to the light, and felt it, and smelt it, and tried every way possible to arrive at some satisfactory conclusion concerning it.
    Ishmael, "The Spouter-Inn"

In "The Cassock" the crew of the *Pequod* continues its slow northeasterly track over the South China Sea as they finish butchering the sperm whale that Stubb killed after abandoning Pip. "Look at the sailor," Ishmael narrates, "called the mincer, who now comes along, and assisted by two allies, heavily backs the grandissimus, as the mariners call it, and with bowed shoulders, staggers off with it as if he were a grenadier carrying a dead comrade from the field."[1]

Ishmael's mincer cuts off its "dark pelt," turns it inside out, stretches it "so as almost to double its diameter" and then hangs it in the rigging to dry. Later, the mincer cuts holes in the penis pelt for his head and arms and makes a coat of it. The description of this garment matches Queequeg's "poncho," which Ishmael had tried on at the beginning of the novel in their room in New Bedford. Ishmael describes the garment as heavy, "shaggy and thick, and I thought a little damp."[2]

The grandissimus currently on display at the New Bedford Whaling Museum fits Ishmael's dimensions and description almost exactly. It is too narrow for me to fit inside, though. The skin must not have been stretched out. If I thought there was even a chance I would fit, I was going to ask for a special favor from Michael Dyer, my colleague there.

"The vague label is not so the school kids can't figure it out," Dyer tells me. "It's for the volunteer staff. The staff didn't want it there at all. But the president of the museum said, *I'm sorry, we are putting it there.* He let them label it *grandissimus*, though, the whaleman's word, instead of saying directly on the sign that it's a sperm whale's penis."[3]

So the four-and-a-half-foot-long grandissimus from the late 1700s stands in a vertical glass case in a prominent hallway. It is black, but worn tan and brown in spots, like an old leather couch. It tapers to a point at the tip, like an elf's hat. I'm certain that thousands of visitors breeze past this penis each year, unaware of what it is. This is perhaps also the case for thousands of readers of *Moby-Dick* who skim past Queequeg's "door mat" in "The Spouter-Inn" chapter, then later doubt their own impression when considering Ishmael's description in "The Cassock."

Melville surely saw a sperm whale's penis himself. Yet as accurate as he was to the best of his knowledge regarding marine biology, oceanography, and whaling practices, it just might be that he fabricated this penis poncho panorama purely out of his own perverse panache. Dyer knows of no American whalemen who ever actually skinned and stretched a whale's penis so that it could fit over a grown man for the purpose of a work smock. None of Melville's known fish documents mention a garment like this, either. Dyer thinks it is quite reasonable, though, since whalemen used penis leather for a variety of work on ships. They used penis leather to muffle oarlocks, for example, and maybe even for knife sheaths.

Sometime later I take up the question with the founder and director of the Iceland Phallological Museum, a collection devoted entirely to animal phalluses.

"We don't have any other references other than the novel *Moby-Dick* to such uses," Hjörtur Sigurdsson tells me. "However, we think it's quite plausible that the penis skin of a sperm whale could be used as a cassock. The full specimen we have stands 1.75 meter high [5.7 feet], weighs 75 kilos [165 pounds], and the skin of that specimen, properly cut, could fit a slim man, I'm sure."[4]

The specimen on display at his phallological museum stands vertically in a glass case of spirits. It was collected from a fifty-two-foot long sperm whale that beached in the summer of 2000 on the north coast of Iceland. It had apparently died of an intestinal blockage. (There's no record of ambergris.) Sigurdsson confirmed that the way that Ishmael described skinning the penis is just how it would've been done. It can be stretched like leather, but not to twice the size. In addition to that massive grandissimus in fluid that the museum displays, you can see two other sperm whale penises that have been dried and hollowed out. One hangs on the wall and holds flowers.

In *Moby-Dick* Melville placed "The Cassock" directly after "A Squeeze of the Hand," crafting a subtle chapter pair to shift the tone after the drowning of Pip's soul and sanity. In "A Squeeze of the Hand" Ishmael has the job of squeezing out clumps from the spermaceti to soften it for boiling. The homoerotic suggestions of the imagery are undeniable, regardless of the time period: "Squeeze! squeeze! squeeze! all the morning long; I squeezed that sperm till I myself almost melted into it; I squeezed that sperm till a strange sort of insanity came over me; and I found myself unwittingly squeezing my co-laborers' hands in it, mistaking their hands for the gentle globules."[5]

Anatomically similar to that of an elephant or a bull, the penis of the sperm whale and other toothed whales retracts into a single coil inside the urogenital slit. (See fig. 37.) A female sperm whale has three slits close to each other: two mammaries and the vagina. Other than relative size in species that are sexually dimorphic, these ventral lines are the only way to tell the sex of a whale.

In a later scene in "The Grand Armada," Ishmael describes a mo-

FIG. 37. Engraving of a frenzy of harvest, spectacle, and science around a beached sperm whale on the North Sea coast of the Netherlands, by Jacob Matham from the engraving of Hendrick Goltzius (1598). Note the measuring of the penis.

ment when sailors look over the rail of their small boats and see "young Leviathan amours in the deep." A bull sperm whale cannot grasp a female underwater, so the flexible penis might be prehensile, meaning it goes into the cow whale's vagina and helps hold them together during their brief copulation. Few if any mariners or naturalists in the nineteenth century could describe whale sex from observations at sea, yet Ishmael says, "Some of the subtlest secrets of the seas seemed divulged to us in this enchanted pond." In 1840 Dr. Bennett was one of the only people who described the act: "Like other cetaceans, they couple *more hominum* [like humans]: in one instance, which came under my notice, the position of the parties was vertical; their heads being raised above the surface of the sea." So Melville ended his "whale amours" note with, "When overflowing with mutual esteem, the whales salute *more hominum*."[6]

Twenty-first-century biologists still do not know much about sperm whale sex. Our understanding today is mostly deduced from observations of smaller whale species in captivity. Rare observations of sperm whale sex in the wild, such as Bennett's, have recorded male and female sperm whales touching their ventral sides together while briefly suspended, vertically or horizontally, which biologists assume to be an act of copulation. Experts believe that sperm whales, based on their behavior and social groups, only mate in pairs per season, or a male might fertilize a few different females. Again, there's no evidence of any kind of harem, as Ishmael claims in "Schools & Schoolmasters." By contrast, other whale species, such as dusky dolphins (*Lagenorhynchus obscurus*) likely reproduce through sperm competition: a female receives sperm from multiple males during a close period of time. Male right whales, with their over-eight-foot penis, go for volume of sperm in order to compete to pass on their genes. Male right whales probably have the largest testes of any animal on Earth: each ball is 6.5 feet long and weighs more than 1,100 pounds. In 1725, the American naturalist Paul Dudley wrote that the right whale's "Stones would fill half a Barrel."[7]

Generations of literary scholars have examined the phallic and homoerotic jokes, references, and deeper meanings in *Moby-Dick*: from the name of the White Whale, to Ishmael and Queequeg's relationship, and on to Ishmael's little quips about the narwhal's horn. (For the record, the slang connotation of "dick" wouldn't be used for penis until decades after the publication of the novel.) It's almost as if the more time the crew is out to sea, the more innuendo Ishmael laces into the story. After seeing the whales make love in the "Grand Armada," several descriptions in the chapters that follow, "Schools & Schoolmasters," "Fast Fish and Loose Fish," and "Heads or Tails," all include anthropomorphic or direct jokes about human sexual relationships—many of which are misogynistic and, at most charitably, stereotypical and dated.[8]

Literary scholars have taken Melville's phallic references quite seriously. At the start of the twentieth-century revival of critical appreciation for *Moby-Dick*, D. H. Lawrence wrote that the cassock scene in *Moby-Dick* is "the oddest piece of phallicism in all the world's literature," and that the white whale himself is "in one of his aspects, a purely phallic symbol: for the deep sexual passion is the passion within the waters, the sensual depth." Scholars have interpreted the loss of Ahab's leg as a Freudian symbol of castration, and others as a true castration after the captain's fall in Nantucket before the voyage. My favorite paper on the subject, "The Serious Functions of Melville's Phallic Jokes," was published by Robert Shulman in 1961 in the austere journal *American Literature*. Shulman's essay reveals, just as this book in front of you, the political and social wave of his generation. Shulman carefully examines each sexual reference in *Moby-Dick*, reading these not just as satire, but as hostile, social defiance of religious and political norms. Writing as a liberal professor in the early 1960s, Shulman read Melville's phallic jokes as the author's erect middle finger to The Man. Shulman raised an argument around the final meeting of Ahab and Moby Dick, suggesting it is loaded with phallic imagery.[9]

It's also possible that Ishmael's phallic references are primarily for the sake of Shakespearean pit humor, to break up the levity of his drama and to address the reality of how men at sea dealt with distant, close communities at sea where they felt helpless to wind, weather, and dictatorships on board. Certainly, too, all of Ishmael's sexual innuendoes were a way to continue to poke and question religious authority and the Victorian prudishness that sexuality is un-Christian. Melville, after all, had experienced firsthand the more openly sexual cultures in the South Pacific, and twenty-first-century maritime historians are still learning about shipboard attitudes toward homosexuality. Melville's communities at sea might have been far less repressed and far less heterosexual than historians might have previously thought.[10]

In "The Cassock" Melville's mincer, dressed like a monk, stands slicing the blubber into pieces as thin as possible, called "Bible leaves" by American whalemen. Melville ends the chapter with what historian Nathaniel Philbrick wrote "may be the most elaborate, not to mention obscene, puns in all of literature." Ishmael's whaleship mincer, with his black sperm whale grandissimus slipped over his chest, is now "a candidate for the archbishoprick." Even in the 1800s prick was crude slang for penis or a person who is a jerk. The archbishopric is the rank of the archbishop, a lofty post within Christian churches.[11]

All that said, until someone uncovers a reference to an American whaleman actually making one of these garments, it's fair to assume Melville entirely invented the idea of a penis poncho, just to pull the reader's . . . leg.

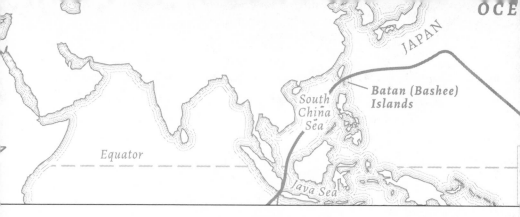

Batan (Bashee)
Islands

South
China
Sea

Equator

Java Sea

Ch. 21

WHALE SKELETONS
AND FOSSILS

But to a large and thorough sweeping comprehension of him, it behoves me
now to unbutton him still further, and untagging the points of his hose, un-
buckling his garters, and casting loose the hooks and the eyes of the joints of
his innermost bones, set him before you in his ultimatum; that is to say, in his
unconditional skeleton.
    Ishmael, "A Bower in the Arsacides"

Having already described him in most of his present habitatory and anatomical
peculiarities, it now remains to magnify him in an archæological, fossiliferous,
and antediluvian point of view.
    Ishmael, "The Fossil Whale"

Biding his time until the season-on-the-line as he percolates on the
White Whale, Ahab steers the *Pequod* up through the South China
Sea. Ishmael, meanwhile, continues to dissect the whale through the
processing and trying out of the blubber. The remaining and deepest
duties of this anatomization are to examine the animal's skeletal system, which leads inevitably to some whale-sized themes that he believes
to be worthy of a condor's quill and a crater full of ink. In these three
chapters, "A Bower in the Arsacides," "Measurement of the Whale
Skeleton," and "The Fossil Whale," Ishmael engages with comparative
anatomy and catastrophism, with the natural theology of the day, and

with the pre-*Origin of Species* understanding of bone, fossil, and man's place in what nineteenth-century natural philosophers had been learning to be a wide and evermore ancient universe.

WHALE SKELETONS

For nearly fifty years Ewan Fordyce has studied whale bones and fossils. He has spent a good part of his career on his knees brushing dirt off fossilized fragments or with a chainsaw freeing a block of lime from a quarry before the miners obliterate the treasures inside. Primarily from this post in New Zealand, but also with other paleontologists from all over the world, he's found dozens of extinct species of ancient whales, giant penguins, and a thirty-foot-long proto-white shark. His office, his lab, and the basement storage under the museum that he manages are piled high with specimens, from tiny cases of grain-sized foraminifera to eight-foot-long blocks of stone and dirt on palettes. In his office, lab, museum, and basement are shells, rocks, beaks, and teeth. For comparison purposes, he collects bones of extant species of whales, too, some of which are propped up against storage shelves as casually as you'd lean a broom against a closet.

Every moment of his research over a blink of a half-century has reminded Fordyce of how *Homo sapiens* represents the briefest of ticks in deep time. I don't know whether he's always been this way, but this perspective seems to have honed a nearly hypnotic, kindly calm to him. Occasionally what he says is actually quite biting and even depressing, yet his delivery is so gentle and even that I barely realize its darkness until later.

It starts easily enough. Fordyce thumbs across the pages of his own dusty copy of the novel. "I remember parts of *Moby-Dick* being really quite exact and revealing," he says. "But when I reread Melville on the skeleton, I think he didn't do a very good job. I was waiting to read about the skull of the sperm whale shaped like a Roman chariot, for example, which is a widely used analogy. But instead we're left groping in the dark trying to understand what he was looking at. He seems to be struggling, really. With anatomy."[1]

Fordyce's point is fair. Ishmael had given some description of the skull and jaw in "The Nut," but in "Bower in the Arsacides" and "Measurement of the Whale Skeleton," although Ishmael suggests a quantitative rigor, he is thin on what the skeleton actually looks like. To open his description in this chapter pair, Ishmael explains that he once visited the island of Tranque (a small island off Chile) in the Arsacides (the Solomon Islands, the entirely other side of the Pacific). Here Ishmael visits a skeleton that the Pacific Islanders had reconstructed from a beached sperm whale. This sperm whale skeleton on Tranque, overgrown with ivy, is seventy-two-feet long, he says, twenty feet of which is the skull and jaw. Fordyce corroborates Ishmael's claim that a sperm whale's skeleton represents about four-fifths of the animal in life. Ishmael believes then that this whale would have been ninety feet in his skin. Ishmael says the animal would've weighed about ninety tons, equivalent to a village of 1,100 people. This is also reasonable if a sperm whale were indeed ever that long.[2]

When he wrote *Moby-Dick*, Melville did not have the opportunity to stand on the balcony of Great Mammal Hall at Harvard with Rob Nawojchik, nose to nose with three whale skeletons of three different species. Nor could he examine a range of cetacean fossil bones with the likes of Ewan Fordyce in a museum in New Zealand. At Harvard, it was not until 1865 that Agassiz's son Alexander organized the acquisition of the museum's first whale skeleton, a North Atlantic right whale from Cape Cod. The articulation of Harvard's full sperm whale skeleton, from an animal killed in the Azores, wasn't completed for display until 1891, the year Melville died. I'm actually fairly certain that before writing *Moby-Dick* Melville never saw an actual full whale skeleton of any species, in any of his travels, even during his short visit to Paris where he could've seen the baleen whale skeleton at Baron Cuvier's natural history museum, the one that Emerson described in his Boston lecture on natural history in the 1830s.[3] (See fig. 38.)

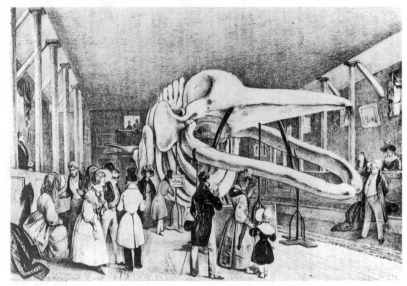

FIG. 38. Baleen whale skeleton at the Museum of Natural History, Paris, by Victor Vincent Adam (c.1830).

In "Bower in the Arsacides" Ishmael explains that he must be truthful in his measurements, because he knows of whale skeletons that exist in museums and curators could call him out on any exaggerations. Ishmael mentions the baleen whale skeletons in Hull and the sperm whale skeleton on the Yorkshire coast, at Burton Constable, which all truly existed and still do. But Melville never traveled to these places. The "Greenland or River Whale" skeleton that Ishmael mentions in Manchester, New Hampshire, he likely plucked verbatim from Thoreau's *A Week on the Concord and Merrimack Rivers* (1849). I suspect this was only a beluga or a narwhal skeleton, anyway. So it seems the most of a large whale skeleton that Melville ever saw *in person* was the skull or lower jaw. He certainly saw illustrations of sperm whale skulls printed in his encyclopedia and perhaps in a scientific journal or two. (See earlier fig. 7.) He also saw line art of full baleen whale skeletons in a couple of his fish documents, but he could not have seen a complete illustration of a full *sperm* whale skeleton, since no one had published anything like this anywhere before 1851.[4]

Melville seems to have gotten most of his skeletal information from Surgeon Beale's detailed descriptions of the skeleton of a sperm whale that beached near the North Sea village of Burton Constable in 1825. Beale described, for example, forty-four vertebrae; Ishmael says "forty and odd" for his skeleton. Beale wrote that the last one is "nearly round, and is about 1½ inches in diameter"; Ishmael calls his final vertebra two inches in width, "something like a white billiard-ball," with two still smaller ones lost because the "little cannibal urchins, the priest's children" stole them to play marbles.[5]

In 1818 Baron Cuvier purchased in London a large but damaged sperm whale skeleton. This seems to have been unknown even to Beale, who thought the only full sperm whale skeleton in collections was the one in Burton Constable. Ishmael's suggestion that Sir Constable himself wanted to make money off the sight was probably a poke at P. T. Barnum and others whose natural history displays in America in the 1800s ran parallel and often intersected with the more august collections, such as those being rapidly assembled by Louis Agassiz at Harvard and by the Academy of Natural Sciences in Philadelphia. The latter was the first one in America, opening its doors in 1812. The tens of thousands of specimens that had returned from the Wilkes Expedition, and which would later form that early core of the Smithsonian Institution, were largely unattended even by 1851 — and the scientifics certainly did not bring back full whale skeletons.[6]

Part of the problem was what remains true even for Fordyce today: whale skeletons were hard to find, to transport, to put together, and then to manage because of the weeping oil that seeps regularly from the bones for years, making a stink and a mess. Over their entire careers, Agassiz and Barnum were eager for a large whale skeleton of any species. But they never got one. When Dr. Jackson dissected the baby sperm whale that had been brought up to Boston on a railway car in 1842, for the first fully documented dissection of a sperm whale on shore, he had to bring the cartload of organs by horse over to the Medical College at Harvard. This was after he stitched the whale back together, presumably for continued public display by an entrepreneur of some kind.[7]

It was not until 1851 that a curator and artist named William S. Wall at the Australian Museum in Sydney published a booklet that Melville

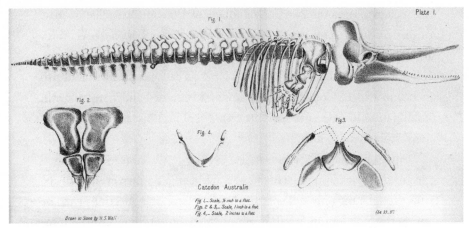

Plate I.

Fig. I.

Fig. 2.

Fig. 3.

Fig. 4.

Catodon Australis

Fig. I.— Scale, ⅛ inch to a foot.
Figs. 2 & 3,— Scale, 1 inch to a foot.
Fig. 4,— Scale, 2 inches to a foot.

Drawn on Stone by W. S. Wall

(Sc. 33..87)

FIG. 39. An approximately thirty-foot long skeleton of a sperm whale curated and illustrated by William S. Wall at the Australian Museum in Sydney (1851). Wall believed this to be a separate species, which he called *Catodon Australis*. Note the vestigial hind limb bones (Fig. 4), which he cut out of another sperm whale.

would've loved to have seen as he was finishing up the novel. Wall described what seems to be the first full sperm whale skeleton assembled, which he then illustrated himself (see fig. 39). Wall had read in the newspaper about a roughly thirty-seven-foot whale towed in by a merchant schooner. After convincing the captain that he needed to have the entire whale, including the lower jaw—which the captain wanted as a souvenir—Wall found four former whalemen from Portugal to help him clean up the carcass and tow it to a beach of an island in the harbor. Here he got special permission to store and cover it in a lime mixture. Wall tracked down the tail, which had been cut off with the blubber and brought to market. Two months later, as the skeleton was nearly clean, he found the missing right flipper at a beach on the other side of the harbor, thanks to a report by a couple boys who described a "strange fish" on the rocks. Wall said the smell of the whole process and the preparation of the skeleton was "most offensive to the senses."[8]

The only parts missing from Wall's skeleton were the vestigial pelvis bones. "The pelvis is hard to find," Fordyce tells me. "You need to know where to look. They're long and thin, floating there. About this size." He found a bone on a shelf and held it up. "They're a missing link."

Fortunately, soon after finding his whale, Wall found *another* sperm whale, this one beached. From this he intrepidly dug out a set of these bones, despite, he says, "some danger from the heavy surf which broke over it." Although Wall understood at the time that cetaceans had vestigial bones that matched the developed pelvis and "hinder legs of ordinary mammals," he did not discuss them in his booklet as any kind of meaningful missing link. For his part, Beale simply described the "rudimentary pelvis" without comment as to any evolutionary significance. Agassiz figured that vestigial parts were part of God's patterning, an affinity with other mammals that He just reduced when making a whale. Ishmael doesn't say anything about the whale's vestigial pelvis. Seemingly useless parts, however, was one of the evidences for the transmutation, the fluidity, of species. Darwin would argue a few years later in *Origin of Species* for evolution by natural selection by, in part, discussing the "rudimentary, atrophied, or aborted organs," such as the vestigial wings on some insects, "the mammae of male mammals," and the teeth in fetal baleen whales that then disappear as they grow.[9]

Back in Sydney, Wall argued that the sperm whale that he articulated was an entirely different species. He created multiple tables with various measurements in comparison to published details of Cuvier's skeletons and the Burton Constable whale, exacting data that Melville would've loved to chide. Wall compiled a table of the length and the circumference of each of the forty-four vertebrae, including the last "globular" one with a length of but three-quarters of an inch.[10]

Ishmael's measurements in *Moby-Dick* are more meaningful to the novel beyond simply satirizing detail-oriented scientists. Ishmael mentions a few times that the skeleton only reveals so much about the whale in life, emphasizing the unknowability of the animal, and thus the sea more broadly, especially to those who spend their lives ashore. This scene of the skeleton in the grove has deep biblical connections, too, including from the Book of Job. God says to Job, when considering the leviathan with smoke coming out of its nostrils: "Canst thou draw out leviathan with a hook? or his tongue with a cord *which* thou lettest down?" Job's leviathan cannot be measured in *Moby-Dick*. The slapstick chiefs, in turn, whacking each other with yardsticks, knock the foolishness all the way across the globe—not just to Pacific Islanders, but to the scientists and priests in America and Europe who argue over the odd inches of truth. Ishmael ends "The Fossil Whale" with American sailors worshipping a whale skeleton in the Mediterranean, being no less reverent or foolish. The "Bower in the Arsacides" scene might even parody the description of a visit by the Wilkes Expedition to a Pacific island that they had named Bowditch Island, after the famous American navigator-mathematician. The expedition's officers convinced an old priest to allow them into a temple, in and around which the Americans conducted a sort of quantitative anthropology, measuring the height of the idols and the dimensions of the tables upon which their gods sat.[11]

Ishmael's measurement of the skeleton, with its comic and spiritual elements, is a prime example of scholar Jennifer Baker's emphasis on Ishmael's "honest wonders." Ishmael aims to impress the reader with the size of the whale. This often requires quantifiable, accurate measurements to convey how extraordinary are the facts on their own, without embellishment. Baker found that in one of Emerson's lectures in Boston, he explained a sentiment that would be Melville's approach to depicting nearly all of the natural history in the novel: "The poet loses himself in imaginations and for want of accuracy is a mere fabulist; his instincts unmake themselves and are tedious words. The savant on the other hand losing sight of the end of his inquiries in the perfection of his manipulations becomes an apothecary, a pedant. I fully believe in both, in the poetry and in the dissection."[12]

In other words, Ishmael's epic yarn *needs* the numbers. He wants to be fact-checked by the curators of the museums. Ishmael wants you to know, however, that these measurements are but one facet. Ishmael, the polymath, is all about your interdisciplinary studies at a time before the bifurcation of the humanities and the sciences or any categories of right or left brain. Ishmael was all about empiricism *and* Romanticism. Idealism and materialism. Locke and Kant. A tattoo of a poem on his arm beside a tattoo of the measurements of a sperm whale skeleton.

FOSSIL HISTORY OF THE SPERM WHALE

After showing off a floor-to-ceiling cast of the fossilized bones of an extinct baleen whale that is about twenty-six million years old, Fordyce (*sneezes*) walks me over to one of his prized finds. This is behind glass: a "shark-toothed dolphin" skull that is over twenty-four million years old. It has triangular, serrated teeth rather than the more conical teeth of toothed whales today. This shark-toothed dolphin might have echo-located and perhaps fed on the giant penguins of its epoch.

This shark-toothed dolphin skull also looks to me like the head of a massive crocodile. Or even the skull of the basilosaurus that Ishmael describes in "The Fossil Whale." Fordyce says that Ishmael's was a true story, one that actually helps explain why this shark-toothed dolphin skull in front of us is not a reptile. In 1839 an American physician named Richard Harlan brought to London some fossilized fragments from a specimen in Alabama that he believed to be a giant reptile. He had named it basilosaurus, which means "king of the reptiles." Fordyce says, "Harlan was widely interpreted as a bit useless, but he would have been struggling with what resources he had at the time." Harlan packed the teeth, perhaps rolled in flannel, and brought them across the Atlantic to London where at the Royal College of Surgeons Richard Owen ("who was really quite brilliant," Fordyce says) examined the remains of this basilosaurus specimen. Owen noticed that the teeth were worn down. Reptile and shark teeth have evolved to fall out, while these fossilized teeth had two distinct roots to anchor it into sockets in the jaw, like mammals. Owen proposed to change the name to Zeuglodon, which means "yoked tooth." This name did not stick, and scientists today recognize the first naming, however inappropriate. A few years after Owen's identification that it was actually an early whale—one of the first fully aquatic cetaceans, it turns out—a fossiliferous collector named Albert Koch collected enough basilosaurus fossils to build, with perhaps other bones and a few rocks, his own extinct "sea serpent." Spurring international debate as to its veracity, involving the likes of

Richard Owen and Charles Lyell, Koch took his skeleton creation to New York City in 1845, then to Boston and abroad, exhibiting it as "Hydrarchos" and playing off the Leviathan in Job that lived before the time of Noah's flood.[13] (See fig. 40.)

In Melville's day, fossils and the recognition of extinction did much to shake the Western world. Baron Cuvier presided as the authority over these sciences in the 1820s. Richard Owen did so in London in the 1830s and '40s. Louis Agassiz carried this torch in America. In 1842 Owen famously coined the very name "dinosaur." Discoveries of this kind stretched the notion of Genesis and God's creation of a perfect world. Fordyce explains that it was geology and paleontology that forced the shifting of paradigms. These were the most public sciences of the day and the most staggering to natural theologians, which included Owen and Agassiz. The three-volume work by Scottish geologist Charles Lyell, *Principles of Geology* (1830–33), for example, not only was profound scientifically, but also was a huge popular hit read by a wide range of audiences. One of Lyell's major points was that the Earth was far older than previously recognized. The Earth was constantly, very slowly rising, subsiding, changing in the same ways that could be observed today—sedimentation, erosion, a small earthquake here, a little eruption there—slowly, incrementally, changing the Earth over millions and millions of years. Melville was among those readers of Lyell, or at least other texts that summarized current geology and paleontology. In his novels before *Moby-Dick*, Melville engaged, with a layman's understanding and often with humor, not only with the theories of coral reef and atoll formation, but also with interpretations of the layers of earth deposits and fossil discoveries. For part of the basilosaurus story, for example, Melville stole outright a bit of poetry from Owen's scientific paper on the topic (which Melville likely read in his encyclopedia). Ishmael's wild, lovely proto-Darwinian closing line about the mutations of the globe getting blotted out of existence is almost verbatim from the closing words of Owen's paper.[14]

Though Ishmael ends *Moby-Dick* with a biblical reference to the

FIG. 40. Broadside for Albert Koch's New York City exhibit of the (inaccurately) articulated fossil bones of *Basilosaurus* sp. (1845).

Earth at 5,000 years at the Flood, in other parts of the novel, such as in "The Fountain," he recognizes our planet as on the order of "millions of ages" old before Noah built the ark. Louis Agassiz and others in the 1840s organized the fossilized ages into primary, secondary, and tertiary periods. Lyell divided these still further. In "The Fossil Whale" Ishmael relays what he might have learned through Agassiz, that extinct whales, even basilosaurus, appeared only in the tertiary fossil layers. This has proven true. In the nineteenth century naturalists dated the rocks by relative age, looking at the layers of rock and the fossils preserved within, which they then compared to the sedimentation rates that they could observe in the present day. Agassiz, from his natural-theological desk, organized the historical layers of the Earth while retaining the supremacy of man as king, as God's chosen species. (See fig. 41.)

While sailing aboard the *Beagle*, Darwin grew particularly inspired by Lyell's work on geologic processes and fossils. He became far more interested in how atolls formed and why he found shells in the hills of the Andes than he was about finches or the causes of bioluminescence. Darwin was regularly seasick during his passages aboard the *Beagle*. So he spent any time he could ashore "geologizing." In "The Fossil Whale" Ishmael allows himself to also be swept away with the cosmic significance of the new understanding of the breadth and age of the Earth and the universe, while still swirling this into his masthead natural theology:

> When I stand among these mighty Leviathan skeletons, skulls, tusks, jaws, ribs, and vertebræ, all characterized by partial resemblances to the existing breeds of sea-monsters; but at the same time bearing on the other hand similar affinities to the annihilated antechronical Leviathans, their incalculable seniors; I am, by a flood, borne back to that wondrous period, ere time itself can be said to have begun; for time began with man. Here Saturn's grey chaos rolls over me, and I obtain dim, shuddering glimpses into those Polar eternities; when wedged bastions of ice pressed hard upon what are now the Tropics; and in all the 25,000 miles of this world's circumference, not an inhabitable hand's breath of land was visible. Then the whole world was the whale's; and, king of creation, he left his wake along the present lines of the Andes and the Himmalehs. Who can show a pedigree like Leviathan? Ahab's harpoon had shed older blood than the Pharaohs'.[15]

Ishmael stretches the geographic extent of the ice age, and he describes more of an Agassizian catastrophism. Yet Fordyce says: "I thought Melville was good there. The Himalayas are uplifted marine rocks, and those contain rocks from Eocene times and the first transitional whales, such as the pakicetids. So Melville was spot on. Where there were now mountain ranges, there were once whales. It doesn't work for the Andes.

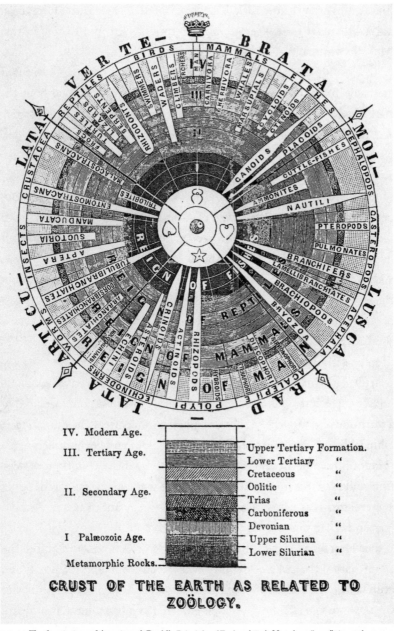

IV. Modern Age.
III. Tertiary Age.

II. Secondary Age.

I Palæozoic Age.

Metamorphic Rocks.

Upper Tertiary Formation.
Lower Tertiary "
Cretaceous "
Oölitic "
Trias "
Carboniferous "
Devonian "
Upper Silurian "
Lower Silurian "

CRUST OF THE EARTH AS RELATED TO ZOÖLOGY.

FIG. 41. The frontispiece of Agassiz and Gould's *Principles of Zoology* (1851). Note how "man" sits at the top with a crown, within the Modern Age, the "Reign of Man."

The Andes are volcanic rock. They're erupted. Whereas the Himalayas are sedimentary rock, they're uplifted from the sea. But I thought that Melville was really good there."[16]

In closing, I ask Fordyce what he thinks of the new name so many are using for this epoch, the Anthropocene. Fordyce says without the faintest shift in tone or calm: "Shocking. True. I'm a paleontologist, so I deal with the impotence of past life, and almost everything I deal with is extinct—so to me extinction is a natural process. It's the fate of all species. Most of them are going to go extinct without leaving descendants, and a few are going to leave descendants. Their lineage alone persists. So it is a natural cycle of doom. Then the next question is: *should people be accelerating this*? Well, I don't think so at all. And I see no hope either. I think it's truly hopeless. You just have to read Darwin and Wallace's paper of 1858 and you realize that they're talking about the struggle of nature, the balance of nature. There is always an excess of young produced. And the numbers in populations never rise because the Earth is saturated. And our species is saturating it furiously. I'm deeply pessimistic."

In *Moby-Dick* these three scenes on whale skeletons and fossils do not seem to plunge Ishmael into a state of helplessness, but instead into a state of wonder in the smallness and insignificance of the human race. Ishmael lives on an ocean that seemed ferocious and beyond the fingerprint of humanity. Yet he was clearly rattled by the new notions of deep time and how this might displace God as the ultimate singular designer and creator. These anxieties and ponderings play out in Ishmael's musings and Ahab's rages.

Put another way, a century later in 1945, which was still before the confirmation that whales definitively evolved from land mammals and even before the recognition of plate tectonics, scholar Elizabeth Foster wrote about the effect of developments in geology on our author's psyche: "At some time between *Typee* and *Moby-Dick*, Melville's universe changed: the benevolent hand of a Father disappeared from the tiller of the world."[17]

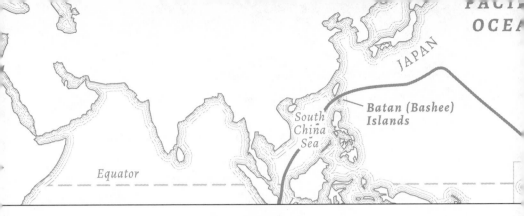

JAPAN

PACI[
OCEA

Batan (Bashee)
Islands

South
China
Sea

Equator

Ch. 22

DOES THE
WHALE DIMINISH?

The moot point is, whether Leviathan can long endure so wide a chase, and
so remorseless a havoc; whether he must not at last be exterminated from the
waters, and the last whale, like the last man, smoke his last pipe, and then himself evaporate in the final puff.
    Ishmael, "Does the Whale Diminish?"

As the *Pequod* continues northeasterly toward the open Pacific, ever
searching for the first glimpse of the White Whale, Ishmael spins his
final essay on the natural history of ocean life. "Does the Whale Diminish?" is a punctuation mark to finish the dissection and anatomization
of the whale as an animal. By the end of this chapter Ishmael puts to
rest all his factual elements about the process of whaling and its significance, and all the information about the natural history of the sperm
whale and its fellow ocean inhabitants, all that he feels he needs as background and scaffold before entering into his final act of the drama.

In "Does the Whale's Magnitude Diminish?—Will He Perish?," the
full name at the start of the chapter, Ishmael confronts human effects
on whaling populations. He answers these two questions by concluding:
no, the magnitude of whales has not diminished over the centuries—
whales are actually at their largest in his present day; and *no*, whales will

not go extinct—although hunting has cut their numbers and influenced their behavior.

DOES THE WHALE'S MAGNITUDE DIMINISH?

In "Does the Whale Diminish?" Ishmael lists the Zeuglodon fossil skeleton found in Alabama as measuring less than seventy feet. He believes this to be the longest known whalelike fossil found on record at the time. This seems true. Ishmael explains that the "ancient naturalists" of Greece, such as Pliny and Aldrovandus, were comically incorrect in their claims that whales were acres wide and eight hundred feet long. Ishmael continues that even in his present day the accounts of sober naturalists greatly exaggerate due to misinformation and lack of experience firsthand. For example, he says, French naturalist Lacépède lists the right whale as three hundred twenty-eight feet long.[1]

Ishmael is correct that most of the historical measurements of whale length and girth were far overestimated. Similar to his treatment of stories about the giant squid, Ishmael tolerates no absurd historical exaggerations, yet he permits himself a smaller stretch in the tradition of a sailor's yarn. So "on whalemen's authority" Ishmael declares that sperm whales can reach one hundred feet long from head to flukes. He never states the actual length of the White Whale himself, only writing of its "unwonted magnitude," which leaves the rest to the reader's imagination—within, er, reasonable parameters. None of Melville's fish documents claim the sighting of a sperm whale of one hundred feet. Nor have I found any similar record, other than the dubious secondhand claim that in 1859 a Swedish whaleman captured a 110-foot-long Mocha Dick. Owen Chase estimated that the whale that sank the *Essex* was eighty-five feet long. Surgeon Beale, by presumably measuring the half-sunk dead whale alongside against the length of the ship's deck—if not sending a Queequeg-like harpooner down with a tape measure— recorded an eighty-four-foot sperm whale. The Nantucket Whaling Museum has an eighteen-foot-long sperm whale jaw, which experts believe might, by proportion, match that of a whale of eighty feet. In

FIG. 42. Francis Allyn Olmsted's illustration, "Pulling Teeth," from a large sperm whale's jaw (1841).

1874 whaleman-author William Davis claimed that he once measured a seventy-nine-foot sperm whale, and Davis heard from another captain that they had caught a sperm whale off New Zealand that was ninety feet long, with a jaw that was eighteen feet. In 1895 the whaleship *Desdemona* captured a whale that they measured at over ninety feet. The New Bedford Whaling Museum has two of this whale's teeth, the largest ever brought home, which are at 11¾″ long. The men would've wrenched these teeth out of the whale's jaw by lashing it on deck.[2] (See fig. 42.)

Today, sperm whale experts are skeptical of even Beale's eighty-four-footer (as was Dr. Bennett at the time). Marine biologist Randall Reeves estimates that the maximum size of male sperm whales is about sixty feet, which still weighs some 120,000 pounds. Sperm whales are the most sexually dimorphic of any whale. Female sperm whales average less than thirty-six feet and weigh less than half of the largest males.[3]

Ishmael's logic is suspect when he compares the size of current animals to those of historical populations by using old illustrations and other ancient references. I read Ishmael as tongue-in-cheek here. Though he believed the gist of the matter, I don't think he really bought into his own pseudoscientific comparisons any more than he peddled in phrenology. Making Moby Dick the largest sperm whale and thus the largest predator that has ever lived serves his yarn and elevates still more his white monster.

In these comic comparisons, Ishmael also continually elevates sailor knowledge and experience above the naturalists ashore, making fun of scientific discourse. He places whales, cattle, and humans all on the same plane, thinking about them all as animal species responding to environmental factors. Literary scholar Elizabeth Schultz argued that this equalization was more revolutionary than it might seem to the twenty-first-century reader. Melville created in the novel a "cetacean-human kinship" that was ahead of his time, encouraging a sympathy for the whale.[4]

Although twenty-first-century biologists agree with Ishmael that whales over previous centuries were not three times larger than they were in the 1850s, a team of researchers from universities in Zurich and Tasmania recently looked at historical data that recorded whale length and found some controversial trends that potentially turn Ishmael's question about the diminishment of the whale's magnitude on its head, and perhaps temper our skepticism of enormous sperm whales reported by nineteenth-century whalemen. The researchers, led by Christopher Clements, looked at the data collected by the IWC of the reported kills from 1900 until 1986, which was when the international moratorium on the hunting of large whales went into effect. During this time whale populations showed a reduction in their average length when stressed by hunting. The case of the sperm whales was most dramatic. Their average length decreased from the start and just kept decreasing until whales killed in the 1980s measured on average 13.1 feet shorter than those killed in 1905, an over twenty percent reduction in average length! Perhaps this was simply due to hunters selectively killing the largest whales available and moving down, but the researchers claim that with sperm whales they can look back at the data and see warning signs that the sperm whale population was nearing a tipping point of overexploitation, even forty years before, four generations of whales.[5]

Could it be then, unknown to Melville and his contemporaries, that the average sperm whale's magnitude had already been diminishing by the peak of American open boat whaling in the 1840s, a hunt that had begun impacting sperm whales since the 1700s?

WILL HE PERISH?

Ishmael recognizes that the topic of the potential extinction of whales at the harpoons of human hunting has been of some debate in the 1840s, especially in consideration of the number of mast-headers searching the globe "into the remotest secret drawers and lockers of the world."[6]

Ishmael begins the discussion by comparing the whale's fate to the pending extinction of the buffalo. He recognizes that "the cause of [the buffalo's] wondrous extermination was the spear of man," but compares the situation to the inefficiency of American whaling. In comparison to men killing forty thousand buffalo in four years, he says, American whaleships on a four-year voyage were content with some forty whales. Ishmael chooses his numbers for poetry, but they are accurate, at least in terms of whaling: a typical successful whaling voyage in the 1840s would capture approximately thirty to sixty whales—and kill or injure several more that they were unable to collect and bring back to the ship. The totals of whales killed for a given voyage were greatly dependent on how many barrels of oil each whale yielded. Catching one or two whales a month on average was considered a successful trip. Ishmael does admit, however, that it was getting harder to find whales—the ships have had to sail farther and farther—and that he believes the hunting by whale-ships around the world has actually altered sperm whale behavior.[7]

The right whales, in particular, he says accurately, were not as commonly found in their previously known regions. Regardless, Ishmael does not believe the right whale population is in danger. He says of right whales (which included bowheads as he knew them):

For they are only being driven from promontory to cape; and if one coast is no longer enlivened with their jets, then, be sure, some other and remoter strand has been very recently startled by the unfamiliar spectacle.

Furthermore: concerning these last mentioned Leviathans, they have

two firm fortresses, which, in all human probability, will for ever remain impregnable. And as upon the invasion of their valleys, the frosty
Swiss have retreated to their mountains; so, hunted from the savannas
and glades of the middle seas, the whale-bone whales can at last resort
to their Polar citadels, and diving under the ultimate glassy barriers and
walls there, come up among icy fields and floes; and in a charmed circle
of everlasting December, bid defiance to all pursuit from man.[8]

Though Ishmael says that American hunters have taken far more right
whales—fifty to one compared to sperm whales—the right whales, he
says, are much like elephants in that they can survive the loss of enormous numbers, especially because the whales have such vast "pasture"
in which to dwell. Whales, like elephants, Ishmael says, have such long
lifespans that it extends their species' survival. Thus Ishmael concludes
confidently: "In Noah's flood he despised Noah's ark; and if ever the
world is to be again flooded, like the Netherlands, to kill off its rats,
then the eternal whale will still survive, and rearing upon the topmost
crest of the equatorial flood, spout his frothed defiance to the skies."[9]

From a maritime history perspective, Ishmael's position on the sustainability of whale populations is unremarkable, even reasonable. The
nineteenth-century whalemen had observed that they were depleting
whales in various regions as they sailed farther and farther from New
England to catch whales. This often happened quickly. For example,
within the two decades of the 1830s and '40s, the whalemen hunted so
many thousands of right whales off the coast off New Zealand that they
crashed the local population by an estimated 90 percent. The whalemen
then moved up into the North Pacific. Within the decade of the 1840s,
which Ishmael recounts as an example of abundance without yet seeing the other end, the men killed tens of thousands of right whales so
quickly that this region too was abandoned to go still farther north for
bowheads. A whaleman in 1845, whose opinion Melville read in his fish
documents, recognized the suddenly poor catches and wrote that "the
poor whale is doomed to utter extermination, or at least, so near to it
that too few will remain to tempt the cupidity of man." This whaleman,

Mr. M. E. Bowles, predicted the global whale industry, for any species, would be abandoned within a century. Right whales, like Steller's sea cows, were coastal and slow enough to be more easily caught by open boat whalemen. Right whales were, essentially, sequentially extirpated in a clear, traceable pattern from the North Atlantic around South America and South Africa into the Pacific. Melville and his contemporaries had not yet seen the widespread developments of a long-range explosive harpoon or the twentieth-century technologies that enabled whaling of all species in polar seas with massive diesel-powered engines and wire winches that reeled full-sized whales onto the deck of a steel-hulled ship so the carcasses could be flensed and processed by a few men as it motored off to kill more whales.[10]

Although human hunting might have reduced the global sperm whale population by about thirty percent in the nineteenth century, sperm whale global health seems to have been less affected than that of right whales. It might be simply that since the sperm whale's range is so much larger and farther offshore that their contact with humans was reduced. Melville's idea that sperm and right whales were swimming away and joining together in larger schools, to avoid the ships and harpoons, was also the perception of at least some American whalemen at the time. It made sense. The hunters depended on this idea to justify the reduction of the whales that they saw and could catch. But it wasn't simply that men were scaring the whales away. They were systematically depleting local populations of sperm whales, too.[11]

Much of Ishmael's logic on this question of extinction, despite his attempts at scientific reasoning, are suspect when viewed in hindsight, due, I think, more to misunderstanding than humor. Today, when biologists talk about the survivability of species, the focus is not just on predation, but about reproductive fitness—ideas that Darwin advanced a few years after *Moby-Dick*, although Ishmael does explain that the gestation period of a mother sperm whale was about nine months (it's actually probably closer to between fourteen and sixteen months). Ishmael does not talk, as a conservation biologist would today, about fecundity or even food availability. Melville and his contemporaries did not know that while a female sperm whale might indeed live to eighty years or older, she gives birth to only one calf every four to six years, and this rate decreases as she gets older, until she probably no longer gives birth at all.[12]

In "Does the Whale Diminish?" Ishmael believes that whales probably attain "the age of a century and more." This fits poetically, building Schultz's cetacean-human kinship, or a "fraternal congenerity" as scholar Robert Zoellner put it. This longevity is actually not outrageous or beyond the thinking then or today. Right whales, of either gender, live at least to sixty or seventy. Earlier, in "The Pequod meets the Virgin," Ishmael tells of an extraordinary find in an old sperm whale that they harpooned: "A lance-head of stone being found in him, not far from the buried iron, the flesh perfectly firm about it. Who had darted that stone lance? And when? It might have been darted by some Nor' West Indian long before America was discovered." In recent decades the scientists and local Inupiat in sub-Arctic Alaska have found a few old harpoon tips of stone and iron buried in the blubber of captured bowheads. Researchers who have dated these harpoons, along with studying the amino acids in the animal's eyes, believe that bowheads indeed live to well over one hundred, and even, just maybe, over *two* hundred years. Melville was surely aware that if there is any trait in animals, or even trees, that creates respect and reverence for our fellow organisms on Earth, it is age.[13]

Now, the most significant point that Ishmael avoids in regard to extinction and depletion, which is more understood today, is that sperm whales had nearly no predation threats for millions of years before humans were able to enter their habitat. Sperm whales evolved long life spans, social groups, long-term raising of single offspring, big brains, large size, and deep-diving abilities to thrive in a very specific niche in the ocean. Orcas, as a pack, can occasionally take a large whale, but biologists do not believe they have any global effect on their populations. If you remove human hunting and return to the era when Ahab first sailed into the Pacific on his first voyage, the era of the late 1790s and the voyage of James Colnett and the publication of "The Ancient Mariner," sperm whale populations in the Pacific were limited almost entirely, if not completely, by fluctuations in their food source and competition within their species.

ISHMAEL AND AMERICAN ECO-GUILT

The notion that an animal or plant could go extinct was in the mainstream consciousness in Melville's time. Species extinction, however, was usually folded by naturalists such as Louis Agassiz into terms that taught that God chose to eradicate certain animals from the Earth with floods and ice ages as part of the progress toward the creation of Adam and Eve. Extinction was part of God's design. This perception potentially absolved humans of responsibility, because even much of the general public understood by the 1850s that human action, such as hunting, destruction of wild habitat, and general human occupation had threatened local animal and plant-life.[14]

As early as 1833 an author in the *American Journal of Science and Arts* concluded his essay on the global fur trade in this way:

> The advanced state of geographical science shows that no new countries remain to be explored. In North America, the animals are slowly decreasing from the persevering efforts and the indiscriminate slaughter practised by the hunters, and by the appropriation to the uses of man of those forests and rivers which have afforded them food and protection. They recede with the aborigines, before the tide of civilization, but a diminished supply will remain in the mountains, and uncultivated tracts of this and other countries, if the avidity of the hunter can be restrained within proper limitations.[15]

This passage was reprinted and paraphrased in multiple other publications, in addition to being read in the halls of Congress. In 1856 Thoreau lamented the loss of large mammals in Concord, Massachusetts—the

bear, deer, wolf, lynx, and beaver—due to hunting for food and to protect livestock. In "The Cabin Table" in *Moby-Dick*, Ahab "lived in the
world, as the last of the Grisly Bears lived in settled Missouri."[16]

Later in "Does the Whale Diminish?" Ishmael singles out the pre-
carious thinning of buffalo herds because this was on the mind of his
contemporaries in America. In his *The California and Oregon Trail*,
Parkman observed that the herds of buffalo had moved westward, and
though he witnessed them in astounding abundance, he knew the expansion of new Americans on their way to Oregon and California would
exterminate the buffalo as well as the nomadic, indigenous people that
depended on them. In 1849 Parkman does not use *if*, but "When the
buffalo are extinct."[17]

Though Melville saw the fragility of animal populations on land, he
thought those of the sea were different. He represented through Ishmael the views of the experts he wanted to trust on this point, such
as Surgeon Beale, who at sea in the 1830s sailed in the new and prolific sperm whale fishery off Japan. Beale believed in a sustainability
of whaling that Ishmael echoed almost exactly. "Such is the boundless space of ocean throughout which it exists," Beale wrote, "that the
whales scarcely appear to be reduced in number. But they are much
more difficult to get near than they were some years back, on account of
the frequent harassing they have met with from boats and ships." Beale
believed the sperm whales were simply becoming more cautious, with
an "instinctive cunning" to avoid humans.[18]

Others observers, such as the merchant Charles W. Morgan in a
speech to the New Bedford Lyceum in 1830, explained that one would
think the sperm whales would be in danger and "this destruction must
eventuate in their extermination," particularly in considering the whale
only has one offspring and they hadn't ever had enemies before man.
But Morgan contradicted this, declaring the greater danger was actually
"an oversupply." Then, in a revision of the speech only seven years later,
Morgan acknowledged that the voyages were taking longer and "a full
ship being now the same exception to a general rule."

In 1841 Charles Wilkes reported after returning from his circumnavigation some concern about whale populations. In his "Currents and Whaling" chapter, the one that Melville incorporated carefully into "The Chart," Wilkes wrote that there is "ample room for a vast fleet" in the Pacific, with space for more ships, but "an opinion has indeed gained ground within a few years that the whales are diminishing in numbers; but this surmise, as far as I have learned from the numerous inquiries, does not appear well founded." Ishmael says this idea has been put forth by "recondite Nantucketers," to give it gravitas relevant to the novel, but he closely follows the path of Wilkes's rhetorical arguments.[19]

Thus, Ishmael's belief that whales can survive the pressure of American hunting in the 1850s reflected the mainstream knowledge of his day. At this point in the novel, as Melville builds up to Ahab's battle, he couldn't possibly show any vulnerability or potential extinction of whales or show his whalemen heroes as careless to their profession's future harvest. Ishmael's arguments, especially from our perspective today, are "wishful, if not desperate, thinking," according to scholar Elizabeth Schultz. With his recognition of the impending loss of buffalo, with the novel ending the way it does, with his faith in man shaken, Melville's depiction of whales as even potentially vulnerable reveals his concerns about the sustainability of humanity's industrial drive.[20]

Consider this: what did Ishmael have to feel suicidal about in "Loomings" at the start of the novel? Why was he walking and brooding along the streets and docks of southern Manhattan wanting to turn a pistol and ball to his head? What was it really about the sea for Ishmael that was cure? He asks, "Are the green fields gone?" Later in "The Grand Armada" he says that even the islands of Indonesia are unsafe from "the all-grasping western world." Even in the farthest reaches of the Pacific islands, Melville and his protagonist were aware of man's heavy hand in habitat destruction and overhunting. Ishmael chooses not to travel west from New York City to cure his urban ills. He goes to sea.[21]

Melville and his contemporaries did not know of too many full extinctions at sea, other than, for example, the Steller's sea cow and the spectacled cormorant up in the far Aleutians or the loss of the great auk in the Canadian Maritimes. But American mariners certainly knew that their coast looked different in 1851 than it had in 1651. They knew this was a result of overfishing and overhunting. The Caribbean monk seal was a rare sighting by 1851—they would be extinct by the mid-twentieth century. Atlantic salmon were in steep decline, locally eradicated in places, due to the damning of rivers and overfishing with weirs. Hunters extirpated North Atlantic right whales from the waters of New England by the mid-1700s, due to shore-based hunting and a long history of whaling in the Gulf of Maine, which might've begun as early as the 1500s. The North Atlantic population of the gray whale was already mysteriously extinct or nearly so before seemingly any European or even Native American hunting. In 1839 an entry in the popular *The Naturalist's Library* lamented the wasteful, selfish destruction of the entire population of fur seals in the South Shetland Islands in only two years, over 320,000 animals.[22]

In 2015 a research team led by Douglas McCauley looked at global extinctions at sea in comparison to those on land. They found that our effect on species in the marine environment has been thousands of years slower than in terrestrial environments. Although the rate of extinction in the ocean is far less—at least as far as we can tell—we have had our effect over the past couple hundred years. And a tipping point might be ahead. McCauley's group found that humans have "profoundly" decreased the abundance of large whales and smaller marine fauna (see plate 8). Considering threats of climate change, mismanagement, and habitat loss in the coming decades, McCauley and his colleagues concluded: "Today's low rates of marine extinction may be the prelude to a major extinction pulse, similar to that observed on land during the industrial revolution, as the footprint of human ocean use widens." They looked at factors such as global warming, increased shipping traffic, increased dead zones, and the increasing mining of the ocean floor. Animals at sea certainly have more wild space to roam and potentially can respond more naturally to global warming and ocean acidification, but they too can be overwhelmed. Ishmael's "polar citadels," which he believed would harbor whales, are now melting and opening up lanes for shipping traffic, new areas for fishing, and further and farther all-grasping.[23]

Perhaps we've underestimated the psychological and sociological effects of the Industrial Revolution on those that were in its earlier gears. Just as so many of us do today, Melville, Thoreau, and others of their generation also felt socially and individually guilty about human impact on the natural world—even if this helplessness might have been tempered or rationalized in a larger perception of God's role in the natural order and what Americans had begun to term manifest destiny. Is the foul, choking, smoking scene of "The Try-Works," in which the whalemen are "burning a corpse," an indictment of Ishmael's industrial age? Tortured Ahab sees his mind and fate on railroad tracks. Melville is fully aware, and seemingly more sympathetic than most, of the soil on which he stands—on which still waged the systematic genocide of Native Americans. Melville saw firsthand the foul impacts of Western man on Polynesian cultures in the South Pacific. He read of it happening in the American West. Did Ishmael hope that he could escape back out to sea, to get away from all of these visions, back to the true "preadamite" wild, a pure ocean beyond the touch of man?[24]

If Ishmael begins to recognize that whales are endangered by hunting, if these endless reefs are no longer pristine, this might be just too all-crippling, just too Pip-like maddening—right?

WHALE POPULATIONS TODAY

Although it's nearly impossible to calculate how many whales early Native Americans, Indonesians, Japanese, and Europeans killed when they first starting pushing off their coasts in order to harpoon or net coastal marine mammals, recent scientists, such as Tim Smith and his colleagues of the World Whaling History Project, have worked with a variety of source material, including Maury's logbooks, as well as a range of sampling methods and statistical models to get a sense of how many whales there were before Melville's day.

In comparison to nineteenth-century whaling under sail from small boats, modern twentieth-century industrial whaling was far more devastating to whales, especially the large rorquals. The open boat killing of sperm whales peaked in the late 1830s at about five thousand sperm whales a year. The discovery of petroleum in 1859 and the Civil War greatly diminished the open boat whaling fleet around the world, but then sperm whale kills began to rise again in the mid-1920s, unrelated to, although poetically beside, the last voyage of the *Charles W. Morgan* and the wreck outside New Bedford of the last working American whaleship, the *Wanderer*, in 1922. Twentieth-century whalemen, capable of capturing the large and fast rorquals, found new markets for whale products. Several large nations, including the US, killed and used whales for farm animal feed and for other products such as margarine, soap, bone meal, and cosmetics. Baleen was nearly useless, replaced by light metals and then plastics, but spermaceti oil continued to have a market as a high-grade lubricant as well as in a range of products such as inks, detergents, cosmetics, and hydraulic fluids. The Japanese and Norwegians also caught whales for human food. The modern hunt of sperm whales peaked in the mid-1960s, with a global catch by all nations of close to 22,000 sperm whales killed per year. And many of the modern industrial hunters turned to sperm whales only after the blue, sei, and fin whales proved harder to find.[25]

Hal Whitehead has overlaid the numbers of sperm whales killed by commercial hunting against his best possible estimates and modeling of sperm whale global populations (see fig. 43). He cautiously suggested that about 1.1 million sperm whales swam the global ocean before the start of the 1700s, before Nantucket whalemen began capturing them regularly off New England. By the publication of *Moby-Dick*, there were perhaps some 800,000 sperm whales left, and then, at the nadir, down to about 300,000 sperm whales anywhere in all oceans. Whitehead believes that the moratorium in the 1980s came just in time—before a full crash of the global sperm whale population. No one has been able to conduct a

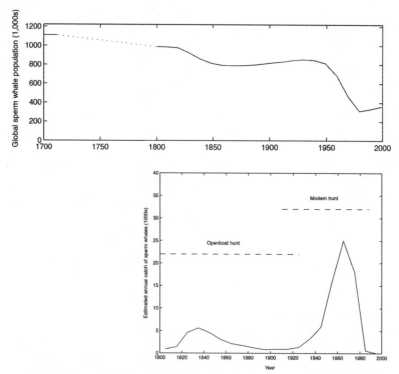

FIG. 43. Estimated global sperm whale population, 1700–1999 (above), and estimated sperm whales captured and killed by hunting, 1800–1999 (Whitehead, 2003). The lower figure does not account for the significant numbers of sperm whales killed or injured that whalemen were unable to bring back to the ship to process. The upper figure, however, does incorporate estimates of those "stuck but lost."

systematic, reliable global population count of sperm whales since then, but as of 2018 the IUCN lists sperm whales tentatively as "vulnerable."[26]

Ishmael suggests in "Does the Whale Diminish?" that sperm whales are aggregating into large armies for protection. This might not have been as absurd as it sounds. Modern biologists have found that sperm whales in the Atlantic rarely join their matriarchal *units* together, swimming as ten to twelve females and offspring, while those in the Pacific almost always join units, forming *clans* of about twenty or thirty individuals. Observers from Captain James Cook to Frederick Bennett to Hal Whitehead have observed scenes of thousands of whales together in transit. It's not out of the realm of possibility that these clans would ally together, especially if the mothers of these units are killed, but the

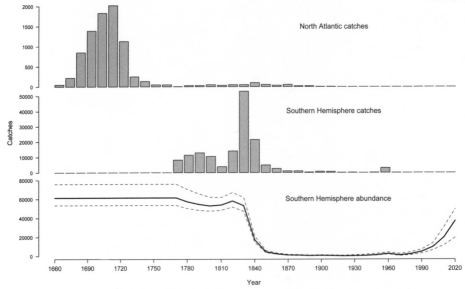

FIG. 44. Based on historical logbooks and other data, the capture of North Atlantic right whales peaked in the early 1700s (top, Laist, 2017), while the captures of southern right whales peaked in the 1830s (middle, IWC, 2001). Note the difference in scale: how many more southern right whales were killed throughout the Southern Hemisphere, but catches for both areas are greatly underestimated due to loss of records. Below is the estimated trajectory of abundance of southern right whales based on mtDNA research from current whales (Jackson, et al., 2008). Researchers have not yet been able to do this abundance modeling with the right whales of the Northern Hemisphere, but both species never recovered and are in imminent danger of extinction.

evidence is not strong enough that human hunting under sail adjusted sperm whale behavior in this way. Whitehead and Rendell believe the large Pacific clans of sperm whales might aggregate to help each other to defend against killer whales.[27]

For the right whales, estimating global populations and historic hunting pressure has overlapping but different sets of complexities. Right whales seem less equipped to survive human predation. The number of North Atlantic right whales killed probably reached its peak a century earlier at over 2,000 animals for the decade between 1710 and 1720. By the mid-1700s whalemen barely found and harpooned any right whales at all along the North American coast.[28] (See fig. 44.)

Over the course of the *Charles W. Morgan*'s thirty-seven whaling voyages from 1841-1921, the longest run of any whaleship in history, the men killed very few North Atlantic right whales, if any—ever. This species is today the most critically endangered of the large whales. In *Moby-Dick* Melville did not have the men of the *Pequod* lower their boats to try to catch whales until the South Atlantic. This was historically accurate. In 1849 Captain McKenzie wrote to Maury that for his whole career he "never did—nor ever expected to find right whales on the out ward passage till I reached Lattd 30.° South." On the first voyage of the *Charles W. Morgan*, the men did not see a whale until forty days into the voyage when they attempted to harpoon a pilot whale just north of the equator, off what is now Liberia. A month later they spotted a couple southern right whales off El Rio de la Plata. Then, on the other side of Cape Horn they finally caught a small female sperm whale. The *Commodore Morris*, leaving Woods Hole in the summer of 1845, did much better, spotting near the Azores two pods of sperm whales less than three weeks into the trip, capturing one large one. By the time they were at Cape Horn, they'd seen sperm whales, pilot whales, humpbacks, and finbacks, but, no kills and no sightings of right whales of either species. They weren't wasting time, anyway, as they were now bound for the Pacific where the chances were far better.[29]

Right whale expert David Laist believes that pre-hunting populations of North Atlantic right whales might have been ten to twenty thousand individuals. Likely because of long, sustained early hunting by man and now primarily because of ship strikes, entanglements with fishing gear, stress from ocean noise, and shifts in copepod abundance with climate change, the North Atlantic right whale is now the lone critically endangered species of large whale. A recent analysis in 2015 found that only 458 individuals of North Atlantic right whales remain on Earth. And any slow recovery seems to have been reversed, judging from recent deaths and lack of births. For the southern right whales, the pre-hunting historical abundance was likely anywhere from 55,000 to over 100,000 individuals, with open-boat whaling beginning in earnest in the late 1700s when whalemen began to travel down to the South Atlantic and then into the Indian Ocean and the Pacific. The southern right whale population across the entire hemisphere is now probably over 25,000 individuals, and is on the rise from a low in the 1920s of only three hundred or so, at the brink of extinction. The North Pacific right whales, which Melville never hunted but which are the animals Ishmael mentions as being killed by the thousands in the "north-west," the Gulf of Alaska, are now as precariously endangered as the North Atlantic right whales, although very little is known as to their true numbers.[30]

THOREAU'S OCEAN: THE SAME AS AHAB'S, ISHMAEL'S, AND NOAH'S

In October 1849 Thoreau began a series of trips to Cape Cod, that thin strip of land that embraces Massachusetts and Cape Cod Bays and points out to the Gulf of Maine. Today this is, and likely has been for millennia, the spring habitat of the North Atlantic right whale. Although there remain less than five hundred individuals on Earth, I've seen them eating in April right off the beach, skim-feeding for copepods, for brit. The North Atlantic right whale spends the summers feeding on zooplankton concentrations in coastal New England and the Canadian Maritimes and then swims south in the winter to mate and give birth off the coast of Georgia and Florida. They raise one calf every three to five years and have a twelve-month gestation cycle that has evolved to match that migration.[31]

Thoreau almost certainly never saw a North Atlantic right whale. Melville probably didn't, either. The seemingly imminent loss of the North Atlantic right whale, which likely would've already happened decades ago if it weren't for the inspiring and committed work led by a handful of biologists and environmentalists working out of Provincetown and elsewhere, is not simply about a loss of diversity or even a blow to human cultural values. The loss of whale abundance in the waters off New England has far-reaching ecological implications that are only beginning to be understood. The reduction of the relative abundance of large whales since human hunting has affected entire ocean ecosystems. By using genetic data, Joe Roman and Steve Palumbi estimated that the population of humpbacks, fins, and minke whales in the Gulf of Maine before whaling was between six and twenty times more than today. In other words, where researchers now estimate 10,000 humpback whales, Roman and Palumbi believe there was once near a quarter million.[32]

The impact of this loss of whale abundance is not simply that there would now be more squid and fish in the water. Whales energize ocean ecosystems; they are crucial in "lively grounds" from the top down. Even today in their diminished populations, baleen whales are enormous consumers and predators, skimming and gulping, for example, between eight to eighteen percent of all net primary productivity off the northeast continental shelf of North America.[33] Roman, along with another colleague, James McCarthy, estimated that all that marine mammal excrement in the Gulf of Maine accounts for more input of nitrogen — a nutrient essential for phytoplankton — than all of the rivers that feed into this shoulder of the Atlantic.[34] And the individual and aggregate biomass of large whales when they die, sink, and decay, creates entire localized ecosystems as they degrade. Thus, just as predators and grazers are essential to healthy ecosystems on land, when more whales swam in the Gulf of Maine, marine life of all kinds was more abundant — more swordfish, more herring, more copepods — far more than what Melville would've seen sailing away from New England in 1841.

As Melville was composing *Moby-Dick*, Thoreau was walking along the beaches of Cape Cod, judging the locals eating pilot whale sandwiches. He was stunned by the open expanse of beach and indifferent sea. Thoreau wrote in *Cape Cod*:

> Though once there were more whales cast up here, I think that it was never more wild than now. We do not associate the idea of antiquity with the ocean, nor wonder how it looked a thousand years ago, as we do of the land, for it was equally wild and unfathomable always. The Indians have left no traces on its surface, but it is the same to the civilized man and the savage. The aspect of the shore only has changed. The ocean is a wilderness reaching round the globe, wilder than a Bengal jungle, and fuller of monsters, washing the very wharves of our cities and the gardens of our sea-side residences. Serpents, bears, hyenas, tigers, rapidly vanish as civilization advances, but the most populous and civilized city cannot

scare a shark far from its wharves. It is no further advanced than Singapore, with its tigers, in this respect. The Boston papers had never told me that there were seals in the harbor. I had always associated these with the Esquimaux and other outlandish people. Yet from the parlor windows all along the coast you may see families of them sporting on the flats. They were as strange to me as the merman would be. Ladies who never walk in the woods, sail over the sea. To go to sea! Why, it is to have the experience of Noah, — to realize the deluge. Every vessel is an ark.[35]

Thoreau's perspective on the ocean was that it was a cold, indifferent, wild otherworld. His view, in print only a few years after *Moby-Dick*, has poignant convergent similarities to Ahab's final thoughts at the masthead, the day he was to die in the hunt for the White Whale, as well as Ishmael's apostrophe to the merciless ocean in "Brit." Both Melville and Thoreau compared the American view of the shore to that of the sea. They both saw a cold reality to the shipwreck and the storm.

For all its ferocity and indifference and otherness—its *outlandishness*—the ocean in the American mid-nineteenth century still held the healing, reassuring view of a watery wilderness beyond human reach.

ASIA

PACIFIC
OCEAN

JAPAN

Batan (Bashee)
Islands

South
China

Ch. 23

MOTHER CAREY'S CHICKENS

Omen? omen?—the dictionary! If the gods think to speak outright to man, they will honorably speak outright; not shake heads, and give an old wives' darkling hint.—Begone!
    Ahab, "The Chase—First Day"

One night soon after the ship enters the Pacific, the ocean where he will meet the White Whale, Ahab clops over on his ivory leg to visit with Perth, the blacksmith.

"Withdrawing his iron from the fire," Ishmael says, "[Perth] began hammering it upon the anvil—the red mass sending off the sparks in thick hovering flights, some of which flew close to Ahab."

"Are these thy Mother Carey's chickens, Perth," asks Ahab, referring to the sparks, "they are always flying in thy wake; birds of good omen, too, but not to all;—look here, they burn; but thou—thou liv'st among them without a scorch."[1]

Although this bird is mentioned only once directly by name in *Moby-Dick*, it's a compelling animal on which to hover for a moment before the *Pequod* sails into the tempest. Again, Melville leans on sailors' experience with ocean species while also drawing upon sailors' lore and his numerous fish documents on his farmhouse shelves.

A Mother Carey's chicken is a storm petrel. Most ornithologists today agree to about twenty-two separate species of these birds in two families, the Hydrobatidae and the Oceanitidae. Nearly all storm petrels have mostly charcoal-brown feathers, blue-black webbed feet, and a black beak with a short tubenose to excrete salt—like tiny albatrosses. Generally only a bit heavier and longer than your average swallow, with similarly shaped wings, storm petrels are the smallest of the ocean-going seabirds. Yet storm petrels can survive several months or longer on the cold open ocean, in part due to a thick layer of fat under the feathers. Aboard the *Charles W. Morgan*, James Osborn read exulting prose about storm petrels in his copy of Good's *Book of Nature*, which included, inaccurately, that these birds survived on oil alone, especially from dead whales and fish. The word petrel might come from "Petrello," Italian for little Peter, since these birds pitter-patter over the surface, more often nibbling on plankton or fish eggs while still in the air, a behavior that has evoked for some the biblical story of St. Peter stepping over the waves. Ahab's comparison to these birds as sparks flying aft is appropriate, because when not hovering just over the surface, the storm petrel's flight is quick and sharp. They're known to follow ships, perhaps to benefit from the food stirred up in a vessel's wake.[2]

Sailors in Melville's time, like Ahab, did indeed call storm petrels "Mother Carey's chickens," which might derive from the Latin *Mater Cara*, meaning Virgin Mary, the protector of sailors. The common name was so well known that in the 1830s John James Audubon felt compelled to add it to the label of his painting of two Wilson's storm petrels (see fig. 45).[3]

Before writing *Moby-Dick* Melville surely heard them called Mother Carey's chickens at sea. As one example of how sailors like Melville used the common name, consider an entry on August 19, 1849, by harpooner Isaac Jessup of the Long Island whaleship *Sheffield*. Two days outward bound, he wrote: "To day is the 1st Sabbath I spend on sea & I must own it is far from a pleasant one. Seasickness is taking away all my strength & rendering me unfit for labor of any kind. Of home or church or anything else I have scarcely thought. Our course is changed to easterly. Mother Carey's chickens are abundant." I love that little

WE II

Wilson's Petrel. Mother Carey's chicken.
1 Male. 2 Female.

FIG 45. John James Audubon's *Wilson's Petrel—Mother Carey's chicken* (*Oceanites oceanicus*) in his *Birds of America* (1827–1838).

entry not just as an example of the sailor's name for storm petrels, but also for its reminder of the strains of life at sea and the ever-present lens of the Christian faith for so many American mariners—the Starbucks out at sea.[4]

And Melville clearly had a soft spot for storm petrels. In his "The Encantadas" (1854), published only a couple years after *Moby-Dick*, Melville wrote in more detail about storm petrels, describing them as "this mysterious humming-bird of ocean, which had it but brilliancy of hue might from its evanescent liveliness be almost called its butterfly, yet whose chirrup under the stern is ominous to mariners as to the peasant the death-tick sounding from behind the chimney jam."[5]

Just as Coleridge drew from sailors' superstitions with his albatross in "The Ancient Mariner" and Melville did earlier in *Moby-Dick* with his "sea-ravens" in "The Spirit-Spout," Melville used Mother Carey's chickens for poetic effect in "The Forge." Melville knew that sailors often associated storm petrels with dangerous weather. In that passage from "The Encantadas," the storm petrels' chicken-like clucking frightens the mariners. Though some superstitious voyagers believed the birds were divinely sent to warn them of danger, to aid the men, other mariners thought these petrels *caused* the storms and even that the birds were a morph of witches. Mad Ahab welcomes the Mother Carey's chickens as a *good* omen, a sign of the coming typhoon.[6]

Storm petrels do not cause storms, of course, but their visibility to humans may increase during times of heavy weather, as Melville's fish documents explained, because the little birds seek refuge on or near ships, or, perhaps, more likely, this is simply a time when ships are sharing the same element with these tough seabirds. I once held one that had landed and got trapped in a small boat of our schooner during a gale in the Caribbean Sea. It felt as light as a child's balled sock. When I worked on a lobsterboat in Long Island Sound, we saw storm petrels only during nasty weather. Presumably they were blown in by a gale from the North Atlantic, or that's simply when they prefer to come closer to shore. And once when sailing quietly on the Grand Banks on a small boat, I saw storm petrels trailing astern. It was the first time I had ever heard their chirrupy-chirping sound. Storm petrels have a light clucking, chortling call, which I think might also have contributed to the "chickens" nickname. They followed astern of my boat for a couple days, appearing around dusk and early evening. There was something spooky and bat-like about their little black following flight.[7]

In "The Forge," after Ahab's dark comment about the blacksmith's sparks flying astern like Mother Carey's chickens, the captain asks Perth to forge a harpoon strong enough to kill the White Whale. In one of the most heretical scenes in the novel, Ahab then commands the harpooners to come temper the iron with their blood, blessing the weapon under the name of the Devil. The image of the sooty gray Mother Carey's chickens as sparks sets this scene, charges the lighting, for the literal and metaphorical storms that loom at the end of this chapter and in the days to follow.

ASIA

PACIFIC
OCEAN

JAPAN

South
China

Batan (Bashee)
Islands

Ch. 24

TYPHOONS AND
CORPUSANTS

Ah, poor Hay-Seed! how bitterly will burst those straps in the first howling
gale, when thou art driven, straps, buttons, and all, down the throat of the
tempest.
    Ishmael, "The Street"

In "The Candles" a storm in the Sea of Japan suddenly engulfs the
Pequod. With images of predation and warfare, this "direst of all storms,
the Typhoon," ambushes like a "tiger of Bengal" and pounds them "like
an exploded bomb upon a dazed and sleepy town." Ishmael provides no
preamble and cuts immediately to the action after the initial blast. His
language is, for him, relatively spare. The three mates try to manage the
chaos as the ship tears along without any sails set, just steering by the
tiller and the force of the wind pushing against the hull and the bare
masts. Starbuck believes that the typhoon coming from the eastward
is an evil omen, since it's roughly the direction they're heading as they
sail toward the White Whale. He, and much of the crew, believe God
follows this warning with a wave that punches a hole in the place where
Ahab normally stands in his whaleboat.[1]

Starbuck realizes that they have not thrown their lightning-rod
chains into the water. As he gives the order, Ahab shouts "Avast!" The

captain commands the crew to leave the chains on deck. He declares that he wants the lightning to "have fair play," a better chance to blast apart the *Pequod*.

Then streaks of lights appear at the tips of all three masts and on each yardarm. The crew and Ishmael call them the "corpusants," a name derived from the Latin *corpus sanctum*, or holy body, also known as St. Elmo's fire. Ishmael describes the glowing as "God's burning finger had been laid on the ship," a biblical writ of pending doom. Under the flame-like streaks, the crew stare aloft, speechless and motionless. Stubb tries to put a positive spin on the vision. Starbuck sees yet another bad omen. Ahab shouts, "The white flame but lights the way to the White Whale!"[2]

Ahab then places his foot on the kneeling Fedallah, grasps the final link of the lightning-rod chain with his left hand, and raises his right arm in defiance. He delivers a blasphemous soliloquy on fire and lightning, questioning God's knowledge and power. As the corposant flames rise in response, Ahab bellows: "Light though thou be, thou leapest out of darkness; but I am darkness leaping out of light, leaping out of thee!"[3]

Starbuck and the crew shrink, terrified by Ahab's heresy—and still more so when the harpoon that he had specially crafted to hunt Moby Dick—the one baptized in "pagan" blood in the name of the Devil—now catches fire at the tip. Ahab seizes the harpoon, pulls it on board, and blows the flame out in front of them like a candle. "I blow out the last fear!"[4]

Aghast, all the sailors scatter. Ahab remains. He stands proud and "so much more a mark for thunderbolts."

Now, what's perhaps most notable about the way that Melville crafted his heavy weather events in this novel is that he did so with so little of the meteorological detail or grand frothy-plumed descriptions that most twenty-first-century readers expect in a sea story. Melville described no grueling, sublime, lengthy scenes of crashing around Cape Horn or heeling through hurricanes, the types of salty adventures that were made famous by narratives such as Dana's *Two Years before the*

Mast, or later taken to the highest level by Joseph Conrad in *Typhoon* (1902). The tiger attacks. The bomb explodes. Ishmael barely yarns of portentous calms or any approaching meteorological signs that foretell this bad weather. Ishmael never describes a specific wave height or wind speed. He does not mention changes in barometric pressure or air temperature. He does not discuss strategies for steering or sail plans. In his storms in *Moby-Dick*, which include Father Mapple's multilayered storm sermon on Jonah and the squall in "Forecastle — Midnight," Ishmael does not describe the characteristics of clouds. He barely describes the sounds of storms, beyond a few words of roaring thunder and cracking lightning.

TYPHOON KYLE, FOR EXAMPLE

During my own first voyage aboard the barquentine *Concordia*, we sailed on the edge of typhoon Kyle in the South China Sea. In 1993, an average year, meteorologists recorded fifteen typhoons. As we sailed along that November we had, fortunately, *not* heard that only a few months earlier in Typhoon Koryn, the 12,522-ton merchant ship *Lian Gang* sank in fifty-foot waves and one-hundred-mile-per-hour winds not that far out from Hong Kong. The British captain and three other crew members drowned. The rest were rescued.

Typhoons, hurricanes, and cyclones are all different regional words for the same thing: massive circular systems that rotate around a low pressure, with winds of over sixty-four knots. Typhoons in the North Pacific, as with hurricanes in the North Atlantic, more commonly occur in the autumn. So hitting bad weather in the South China Sea in November was not unexpected. Melville appropriately slammed his fictional *Pequod* with a typhoon in Japanese waters around December. He wrote, too, through Captain Bildad, that the *Pequod* on a previous voyage had lost all its masts due to a typhoon in the same region, "on Japan."[5]

Weather forecasts, now delivered by email to ships no matter how far at sea, show a weather map displaying and predicting atmospheric conditions in the region and the predicted tracks of high- and low-pressure systems. In the mid-nineteenth century mariners relied on their knowledge of clouds, changes in barometric pressure, and shifts in force and direction of wind and swells to forecast local weather events. Whaleships in Melville's day carried barometers. Captain Valentine Pease aboard the *Acushnet*, for example, recorded the pressure in his abstract log for Maury. At one point in *Moby-Dick*, Starbuck goes below on the *Pequod* to check on the barometer. Perceptive captains and watch officers observed the skies carefully. Without engines they had fewer options, but they could respond to short-term shifts so the ship sustained less damage. For long-range planning regarding best regions and time of year, as well as day to day developments, whalemen made weather plans based on their past experience, their fellow mariners, their library of published narratives, and they could also read advice in reference charts and reference books, such as those published under the guidance of Lieutenant Maury, who had compiled wind and storm patterns in the same way he collected whale sightings.[6]

The most common route for typhoons swirling in the northwest Pacific is to bend northerly *before* the Philippines and bash into the waters of Japan, just as one pounces on the *Pequod*. However, in 1993 our Typhoon Kyle kept plowing westerly over the Philippines, where at least eight people died because of the weather. When our ship was midway across the South China Sea, from what I can piece together, the storm continued but I believe we were to the northwest of the system, which is the safest sector, as the wind is more astern. As I took my tricks at the wheel, pelted in the face by rain, frightened by the shrieking wind, and desperately trying to keep the vessel on course (I was once, to my shame, replaced by the engineer because I failed to do so), our ship was actually moving toward safer seas. Kyle continued to pick up still more speed until it was categorized as a full typhoon and then a category two, notching a maximum of seventy-five knots before it pummeled the coast of Vietnam, killing at least seventy-one more people, and eventually dissipating into land, losing the heat and water that fuels and sustains cyclones.[7]

After two difficult days and one excruciating evening aboard the *Concordia*, we continued motor-sailing north until we were able to shut down the engine and sail under shortened canvas. Just as Ishmael describes as the typhoon eases in *Moby-Dick*, "the ship soon went through the water with some precision again." I remember being off watch and sitting on the after superstructure that morning. I was still velcroed up to the nose in my foul weather gear. I leaned my back against a life raft on the windward side and watched exactly what Ishmael describes in "The Needle": "Next morning the not-yet-subsided sea rolled in long slow billows of mighty bulk, and striving in the Pequod's gurgling track, pushed her on like giants' palms outspread." Sore and bruised, I saw these enormous swells rear up and then slide underneath as the hull heeled over.[8]

I wrote in my journal that during the height of the weather, we recorded a maximum of forty-five knots of wind. I had recently read *Two Years before the Mast* for the first time, and I couldn't help imagining myself like Dana in his little ship rounding Cape Horn. My forty-five knots was nothing to sneeze at, but I've been in stronger winds since then, and in retrospect, without knowing the forecasts at the time, I think the captain did a good job of keeping us safe. But I'm sure the office fielded their own storm of phone calls from worried parents when the students called home from Hong Kong and told of sailing through a "typhoon." I'm not saying we weren't in dangerous conditions, but I don't trust my memory. I know too well now how the perception of wave height and wind speed alters with experience.

ISHMAEL'S STORMY LANGUAGE

In the 1840s, Noah Webster was openly exasperated at the word "gale." In his definition in Ishmael's "ark," *An American Dictionary of the English Language*, Webster writes:

> GALE, *n* A current of air; a strong wind. The sense of this word is very indefinite. The poets use it in the sense of a moderate breeze or current of

air; as, a *gentle gale*. A stronger wind is called a *fresh gale*. In *the language of seamen*, the word *gale*, unaccompanied by an epithet, signifies a vehement wind, a storm, or tempest.[9]

When it came to writing about heavy weather in *Moby-Dick*, Melville played the part of the poet rather than the captain. Without any sense of indignation or righteous concern for delineation, Ishmael uses gale, squall, storm, tempest, typhoon, and hurricane almost interchangeably. He throws in the names of regional winds such as Euroclydon, Levanter, and "Simoom." In "The Whiteness of the Whale" Ishmael explains that the white squall is a snow-filled micro-storm, while in Pip's earlier forecastle scene Ishmael more accurately describes the white squall as a burst of wind whipped and frothed white by the velocity—which Ishmael uses here more for its racial and religious implications.

Although Melville didn't seem troubled by the fluidity of these terms, Webster was not alone in wanting some standardization. Decades before Maury began to collect observed wind speeds and currents from mariners, Sir Francis Beaufort, an English surveyor and later rear admiral of the Royal Navy, developed in 1805 a scale from 0–12 that remains, though tweaked, in international use today. Melville's characters don't mention the Beaufort Scale, but it was in use among some American mariners, although seemingly rarely used, if ever, by the whalemen. Robert Fitzroy was the first to officially record the scale on a regular basis during his voyage with Darwin on the HMS *Beagle* in the 1830s. In those days there was no way to accurately quantify wind speed in any real fashion aboard ship, so if not using Beaufort's numerical scale, the wind was recorded in logbooks in qualitative phrases such as "gale" or "fresh gale" or "furious gale," without a universal agreement as to what these meant precisely. For example, in the abstract log of the *Acushnet*, Melville's Captain Pease used descriptors such as "light," "fresh," and "heavy gale" and added in his remarks such phrases as "rugged" and "squally."[10] (See earlier fig. 16.)

The most definitive source of reference for the working mariner in the United States in Melville's time was *The New American Practical*

Navigator, written by Nathaniel Bowditch. Simply known as *Bowditch*, this text has been the American sailor's bible, continually revised since the late eighteenth century. Bowditch, as a man and a text, was a source of pride for the new United States. When the crew of the *Essex* abandoned ship after being struck by the whale, the officers grabbed two copies for their small boats. Ishmael mentions *Bowditch* a couple times in *Moby-Dick*, synonymously with a practical interest in seamanship and navigation. In the twentieth edition of 1851, *Bowditch* (which after his death in 1838 had been continually revised by his son) gives advice on wind speed to mariners, presumably to standardize materials for Maury. Bowditch follows Beaufort's system, listing from "calm" (0) to "fresh breeze" (5) to "hurricane" (12), here defined primarily by the amount of sail the ship could carry. Bowditch also listed a series of letters for the abstract log, e.g., "l" for lightning, "q" for squally, and, my favorite: "u" for "Ugly, threatening appearance of the weather." Bowditch wrote a short chapter on winds, which included the dangers of the "*ty-foongs*." He explained that a few experiments had been conducted to try to quantify wind speed, especially by "the space passed over by a cloud or any light substance."[11]

Bowditch remains the definitive reference for American mariners even today, continually revised and published now by the National Geospatial-Intelligence Agency of the US government. Every American mariner sitting for hours taking his or her exam to be a licensed captain has access to a large volume with the huge word "Bowditch" in gold on the binding. The Beaufort Scale still goes up to Force 12: the true hurricane, typhoon, or cyclone at sea. When Sir Francis Beaufort first defined a Force 12 in the early 1800s, he described it as wind "which no canvas could withstand."[12]

In the 1970s meteorologists developed for still more precision, separately from the Beaufort Scale, the Saffir-Simpson Scale, which delineates categories of hurricanes from one to five. The thirty-three men who drowned when the American merchant ship *El Faro* sank north of Puerto Rico in 2015, succumbed to hurricane Joaquin, a high-end category four with winds over 155 miles per hour.

AHAB'S LIGHTNING AND CORPOSANTS

The steel masts and hull of the *Concordia* would have dissipated a lightning strike if we'd been hit during Typhoon Kyle or any other storm. Unlike wooden ships such as the *Charles W. Morgan* and the *Pequod*, a steel ship is naturally "grounded" to open an easy path of electricity all the way down to the water. Lightning still can damage steel ships, but usually only in the form of burned-out electronics.

In Melville's mid-nineteenth century, at the historical apex of wooden tall ships and the sheer number of men sailing around the world in various trades, lightning was of utmost concern. Before Maury, Benjamin Franklin was young America's only real investigative scientist who had any international respect, and, famously, lightning was Franklin's priority for research. After his kite and key experiment in 1753, which resulted in his lighting rod invention for buildings on land, other inventors began to fix lightning rods and grounding chains to ships at sea. These are the chains Starbuck realizes he wants to get in the water during the typhoon aboard the *Pequod*. They seem to have been rarely if ever installed on American whaleships, although they appear to have been standard equipment on naval vessels, such as Melville's *United States*.[13] A seaman named Henry Eason, while sailing in the equatorial Atlantic aboard the USS *Marion*, wrote in his journal of a morning in 1858:

> I was at the wheel when very suddenly, a very heavy report was heard, it was so sharp & sudden that many persons, who were below, at the time, thought 'twas a gun fired to break a water spout; the lightning that preceded it struck our lightning conductor, which is an iron one reaching from the main royal truck into the water, this the electric fluid broke into twenty or thirty pieces, & splintered the main chains a little. If it had not been for the Conductor we should probably have lost our mast.[14]

In "The Candles" Ishmael describes accurately the problems with lightning rod and chains. They were awkward and needed to be set up by the sailors *before* lightning struck (see fig. 46), and the chains themselves might not be substantial enough to contain all of the electricity from a strike, hence how Henry Eason described a few of them breaking after the bolt. In "Midnight, on the Forecastle," still working in the throes of the typhoon, Stubb and Flask banter over the merits of lightning rods, turning it into a comic debate on fate and faith as they lash down the palms of an anchor, metaphorically tying the ship's hands to follow Ahab's quest.[15]

Melville did not experience a typhoon himself, since he never sailed in the northwestern Pacific, but he likely experienced storm or hurricane-force winds at some point in his travels at sea. Before *Moby-Dick* he definitely experienced lightning and thunderstorms on the water, although it's unknown if he ever was at sea for an actual strike. On his trans-Atlantic voyage just before beginning the novel, he saw St. Elmo's fire. On Saturday, 13 October 1849, the day after he went aloft "to recall the old emotions" and the very same day that he wrote of Coleridgean natural theology *and* the passenger who jumped overboard to commit suicide, Melville wrote, even as the man still sunk to the bottom:

> By night, it blew a terrific gale, & we hove to. Miserable time! nearly every one sick, & the ship rolling, & pitching in an amazing manner. About midnight, I rose & went on deck. It was blowing horribly—pitch dark, & raining. The Captain was in the cuddy, & directed my attention "to those fellows" as he called them,—meaning several "Corposant balls" on the yardarms & mast heads. They were the first I had ever seen, & resembled large, dim stars in the sky.[16]

Descriptions of St. Elmo's fire were not new. They were recorded and even illustrated at least as early as 1680 and appeared in editions of Maury's popular works (see fig. 47). In 1840 Dr. Bennett wrote of the "peculiar sickly and unnatural light" on the mastheads, a "globular form" during thunderstorms and tempests. Bennett observed them "about the size of a tennis-ball" and having something to do with atmospheric electricity and evaporation since he only witnessed them when

FIG. 46. Illustration in W. Snow Harris's *On the Nature of Thunderstorms; and on the Means of Protecting Buildings and Shipping* (1843), showing the dangers of men trying to set up conducting wires in the middle of a storm.

FIG. 47. An illustration of St. Elmo's Fire in a revised edition of Maury's *Physical Geography* (1891).

it was raining. Wilkes wrote of the "Corpo Santos" during which the officers "felt electric shocks." An entire article on the subject appeared in an 1845 issue of the popular *Penny Magazine*. In *The Voyage of the Beagle* Darwin observed St. Elmo's fire when the ship entered the Río de la Plata in 1832. Darwin's description shows how electricity in the sky and the bioluminescence in the water had for sailors similar effects, and even causes: "We witnessed a splendid scene of natural fireworks; the masthead and yard-arm-ends shone with St. Elmo's light; and the form of the [wind] vane could almost be traced, as if it had been rubbed with phosphorous. The sea was so highly luminous, that the tracks of the penguins were marked by a fiery wake, and the darkness of the sky was momentarily illuminated by the most vivid lightning."[17]

Shortly after the publication of *The Voyage of the Beagle*, Dana wrote of his own "corposants" in *Two Years before the Mast*. His ship was homeward bound in the North Atlantic when the sailors all saw the electricity as fraught with meaning. Dana wrote that the sailors believed that the vertical direction of "the ball of light" predicted fair or foul weather and that "it is held a fatal sign to have the pale light of the corposant thrown upon one's face." The dropping of the light proved accurate, as Dana and his shipmates aboard the *Alert* soon felt raindrops, some thunder, and then a loud blast of lightning. The rain poured down, and then the lightning continued for hours. During the storm the sailors were completely silent, watching the "electric fluid" flow over the anchors and some of the lines.[18]

THE NINETEENTH-CENTURY STORM AS AN ACT OF THE ALMIGHTY

Writers often use a storm to transport a character and the audience to an entirely different world. Consider how Dorothy is plucked up in a tornado to enter the land of Oz. In the opening storm of *The Tempest*, Ariel delivered not only heavy weather, but corposants, and scholars argue that the Ancient Mariner's vision of dancing "death fires" was St. Elmo's fire, too. In *Moby-Dick*, the gale in "The Lee Shore" rips the reader from the rational land. Next, the gale at the Cape of Good Hope says goodbye to the Atlantic, from here the men begin to kill and the story opens up to the supernatural. Then the typhoon in the Sea of Japan bursts a new entry into the final chase toward the equatorial Pacific. By 1850 storms had become well-established in fiction, drama, and poetry as a trigger, an association, an objective correlative, for psychic and emotional turmoil. For *Moby-Dick*, Melville drew in particular from the use of storms and lighting in *King Lear* (1606), as well as the Gothic novels, such as *The Castle of Otranto* and *Frankenstein*, both of which Melville purchased on his most recent trip to London just before beginning *Moby-Dick*.[19]

Scientifically, by 1851, authors such as Bowditch and Maury had described accurately the general movement of hurricanes in the North Atlantic, and less so in other parts of the world. Thanks in part to Benjamin Franklin's observations that a single storm system tended to move north and eastward along the US East Coast, scientists and mariners at the time understood how weather systems tracked, and, for circular storms, they knew about the calms in the center, but were just learning the strategies to avoid this weather or how to sail to the safer sectors around this kind of cyclonic system. Mariners and scientists did not know *why* these storms formed or the true speed of the wind that these strongest storms could accumulate. Much of this would be understood and strategies practiced by the end of the nineteenth century, enough to at least try to proactively get out of harm's way.[20]

This transition from passive fatalism regarding heavy weather was happening right around the publication of *Moby-Dick*. In 1848 British mariner Henry Piddington published a book about "the law of storms," expressly for mariners, which accurately described how circular storms spiraled and moved forward in relation to the equator. Piddington included careful diagrams on how captains should maneuver the ship. For mariners at sea, these storm avoidance strategies were often, and for some still are, from the natural theological perspective: scientific explanations and practical theories to explain the acts of God—although it's not without significance that this entailed a steadily lessening sense of helplessness from the captain. In contrast to an author such as Maury, Piddington does not mention God in a single explanation, yet, on his title page, right under a passage from Francis Bacon extolling the importance of careful, inductive reasoning, he includes a stanza from Falconer's *The Shipwreck*—as if just to be safe—which says "With thee, great Lord! 'whatever is, is just.'"[21]

A lovely example of the connections that Ishmael-like mariners "with the Phædon instead of Bowditch" in their heads, made between weather, religion, and the poetic sublime—the power and wonder of God—is in the journal by artist-whaleman Robert Weir.[22] He wrote after days of heavy weather in which the ship screamed along at an extraordinary fourteen knots off the Cape of Good Hope in 1856:

It was worth ones life to let go a hands graspe from the rigging for a moment—again the gale would come upon us with redoubled fury—

Making the sea look like driven snow—and lifting showers of water from the crests of the waves—which fell upon us like heavy rain—One may imagine a sailors life is a cheerless one in such times, and so it is to some—but not so to me for I love to be on deck and watch the sea & sky—and hear the Almighties voice in the storm—it makes me feel that we are actually in the great presence of the Omnipotent. It serves to remind all that God is still there and watching over them—the sea is His and He made it. How little we think so—And yet we know that His slightest thought could send us all to destruction—could send the whole Universe—I should think sailors above all men should be Christians—because they seem to live in the deapths of dangers,—and God is eternally saving them from destruction.[23]

So Ishmael's storms and how the characters respond in *Moby-Dick* are literary and reflective of American culture at the time. Both sailors and landlubbers read God's will in storms. Melville's Christian tradition was Calvinism, which taught about trying to read and interpret signs of the Divine's intentions.

To learn more I go to literary scholar and Shakespeare expert Dan Brayton, who explains to me while he's sanding his boat that Melville's Calvinism asked, *Are you going to Heaven or are you going to Hell? He [God] has already decided. Now you've got to figure it out.*

Brayton continues: "Ahab has a real beef with that God. King Lear also rages against the winds, at the elements, at the Gods. Throughout Shakespeare, and throughout the Renaissance, and up to Melville, people were always writing about bad weather in terms of what it meant about God's will towards us. It's sort of what we do now with climate change. Every new roll of thunder is a warning."[24]

Today, more and more Americans, in a true crisis of faith, are recognizing that the twenty-first-century storm is, in fact, more in the hands of man.

In 2018 the Geophysical Fluid Dynamics Laboratory, a division of NOAA, revised their report on global warming and hurricanes. The scientists concluded that there is still not enough data to state confidently whether the increase in greenhouse gases, such as carbon dioxide, produced by human activities has a "detectable impact" on the number or intensity of these storms. They just do not have the data going back far enough. What they do confidently declare is that anthropogenic warming is causing more rainfall around tropical cyclones, in the range of 10–15% within a circumference of the storm, and that due to global warming "tropical cyclone intensities globally will likely increase on average. This change would imply an even larger percentage increase in the destructive potential per storm, assuming no reduction in storm size." Since 1851, from when reliable records have been reconstructed, ten of the fifteen most active hurricane seasons that slammed into the Caribbean and the US East Coast have occurred between 1997 and 2017.[25]

[*Ahab takes peg leg, made of hard white petroleum-based plastic, and pushes gas pedal to the floor.*]

ASIA

PACIFIC
OCEAN

JAPAN

Batan (Bashee)
Islands

South
China

Ch. 25

NAVIGATION

Science! Curse thee, thou vain toy.
  Ahab, "The Quadrant"

After the *Pequod* survives the typhoon in the Sea of Japan, Ahab trims
sail for the equator. A century before GPS, radar, and depth sounders,
Ahab had safely steered the *Pequod* to the other side of the world by
navigating with his charts, with celestial navigation, and with deduced
reckoning.

In Ahab's time getting the ship's latitude with a sun sight at the sun's
highest point in the middle of the day was relatively easy, as long as you
had a nautical almanac or a copy of *Bowditch* on board to help with declination. Declination is how far north or south is the sun in relation to
the equator due to the tilt of the Earth and its orbit around the sun.

Getting the ship's *longitude* in the mid-1800s required more finesse,
planning, and math. Whaleship officers, who rarely had time or interest in lunars or other advanced celestial trigonometry, most commonly
calculated their longitude by: (1) the ship's chronometer—an expensive
timepiece but available to American whalemen in the 1840s; (2) keeping a regular record of the compass direction that the ship was steered;
and (3) keeping a regular record of the average speed that the ship sailed
over the water, which was determined by estimation from experience

or by measuring on marked rope how quickly a block of wood, a chip log, was dragged away from the ship. (See fig. 48.) Calculating longitude with this direction and speed of the ship over time, which measured distance from the prime meridian of 0° longitude at Greenwich, England, was known as part of deduced reckoning or, more commonly, as "dead" reckoning. This was loaded with potential error because of currents, variable helmsmanship, and leeway, the sideways sliding of sailing ships.[1]

In "The Quadrant," "The Needle," "The Log and Line," and Stubb's musings in "The Doubloon," Ishmael does not get into all the details, such as the means of calculating longitude, but Melville clearly understood navigation at sea and wrote of everything accurately enough to get us reasonably to the equator in the Pacific.

In "The Quadrant," just before the typhoon, Ahab takes an angle of the sun appropriately at midday. He finds the sun at its highest-most point, its "precise meridian," as Fedallah, the sun-worshipper, looks on. Mariners since before Columbus measured the height of the sun, stars, and planets with a variety of protractor-like instruments, beginning with the kamal and the cross-staff. These were improved to the octant, quadrant, and sextant, which were all variations of the same tool and all available in the 1840s. On the first voyage of the *Charles W. Morgan*, the captain used an octant. The officers of the sinking whaleship *Essex* grabbed two quadrants along with their copies of *Bowditch*.[2] (See fig. 48.)

With a pencil, Ahab writes the angle on the special plate that's been fixed to his ivory leg: "Ahab soon calculated what his latitude must be at that precise instant."[3] Ishmael skips declination, but if you're taking a noon-sight every day, day after day, you have a good sense of what that is without looking at the tables, and even a super accurate timepiece isn't essential. Ahab then looks back at the quadrant, fingering its "numerous cabalistical contrivances," and begins to rage at the device:

The world brags of thee, of thy cunning and might; but what after all canst thou do, but tell the poor, pitiful point, where thou thyself happen-

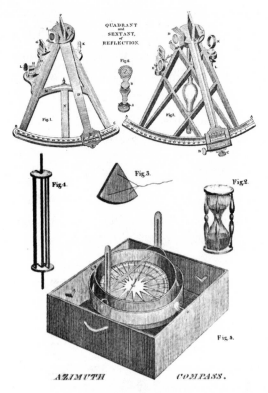

FIG. 48. The primary tools of nineteenth-century ship navigation as pictured in Nathaniel Bowditch's *American Practical Navigator* (1851): above, a quadrant (left) and sextant; middle, the tools for calculating speed over the water, including sand glass, chip log, and spool (minus the line); and below, a compass.

est to be on this wide planet, and the hand that holds thee: no! not one jot more! Thou canst not tell where one drop of water or one grain of sand will be to-morrow noon; and yet with thy impotence thou insultest the sun! Science! Curse thee, thou vain toy; and cursed be all the things that cast man's eyes aloft to that heaven, whose live vividness but scorches him, as these old eyes are even now scorched with thy light, O sun! Level by nature to this earth's horizon are the glances of man's eyes; not shot from the crown of his head, as if God had meant him to gaze on his firmament. Curse thee, thou quadrant![4]

Ahab smashes the instrument with both his living foot and his dead peg leg. The sailors look on, horrified. Ahab orders the ship to start heading toward the equator. "No longer will I guide my earthly way by thee," Ahab seethes at the quadrant, "the level ship's compass, and the level dead-reckoning, by log and by line; *these* shall conduct me, and show me my place on the sea." It's not without meaning that a typhoon blasts the ship in the very next scene.[5]

Ahab, Lucifer-like, lashes out at his inability to rise beyond what he can know. Perhaps part of Ahab's rage here is that he cannot surmount Agassiz's chain of being to be more like a God, as had Bulkington at the start of the voyage and as Ahab had declared he would in "The Quarter-Deck." Ahab does not want to have to look skyward. In rejecting this celestial science, Ahab becomes more animal, choosing to navigate more intuitively with more basic methods, believing man was not meant to gaze up at the stars. You could very well substitute Ahab's quadrant for a GPS, a smart phone, or even the internet in general— really any technological device, advancement, or even abstract scientific theory. What can these actually do for us beyond providing information? They cannot help with what Ahab wants to know: the future, if and how God controls him, what's the meaning in one human life, why is he so personally and ceaselessly raging against the White Whale, and all the unknowables that we humans are unable to unmask. The workings of the celestial bodies—the sun, the stars, the moon—and everything humanity has learned become folly to Ahab.[6]

In this way is Ahab prescient, representative of what was becoming, and still is, an enormous, psychological rending in parts of Western cultures? As reliance on and devotion to science grew, adding skepticism to creation, miracles, and other religious beliefs, is it possible that Ahab's speech expresses Melville's concern about a growing, gnawing gap in the ability of Judeo-Christian scriptural teachings to provide that human, intrinsic need for something that resembles faith: pilotage?[7]

When the energy of the typhoon flips the compass, a phenomenon that had been documented in Melville's fish documents, mad Ahab uses the knowledge gained from his forty-years' experience at sea to flaunt his mastery. Now rational science is on his side. Yet for the sailors, Ahab's repair of the compass is as much black magic as their captain's calculation of latitude by the sun, but far more foreign. The men are dependent on his navigation for their lives, watching his blasphemous rejections of higher knowledge, which they know would keep them safe. The crew fall into a far deeper, superstitious gloom at their fate.[8]

Now Ahab vows to navigate only with dead reckoning. The gloomy implication of the phrase is not lost on Ishmael. Starbuck recognizes it, too, but as a less reliable way to navigate. Then the scene in "The Log and Line" delivers still another omen related to their navigation. The officers had been merely estimating the speed of the ship visually and hadn't bothered with the log and line. The rope had rotted. When Ahab puts it out, the old Manxman warns him that it'll break, but Ahab won't listen. The line snaps.

South
China
Sea

Batan (Bashee)
Islands

PACIFIC
OCEAN

Ch. 26

SEALS

> For not only are whalemen as a body unexempt from that ignorance and super-
> stitiousness hereditary to all sailors, but of all sailors, they are by all odds the
> most directly brought into contact with whatever is appallingly astonishing in
> the sea.
>
> Ishmael, "Moby Dick"

In late June 1794, Captain James Colnett's ship *Rattler* sailed southbound off the coast of Chile, bound for home. They hadn't caught that
many whales, but at least no one had died of scurvy. At around eight
o'clock at night the men on watch heard something over the side. Colnett described the scene later: "An animal rose along-side the ship, and
uttered such shrieks and tones of lamentation so like those produced
by the female human voice, when expressing the deepest distress as to
occasion no small degree of alarm among those who first heard it."[1]

Colnett explained that the cries of the animal were the closest thing
he'd ever heard to "the organs of utterance in the human species." The
wails lasted for three hours and seemed to get louder as the ship sailed
away. Colnett believed that it was a female seal that lost her "cub." Or
maybe a seal pup had lost its mother. For his men, however, the wailing
of the animal "awakened their superstitious apprehensions."

Nothing terrible happened directly after they heard the crying animal, but after rounding Cape Horn an enormous wave crashed over the back of the ship, filled the quarterdeck with water, smashed in some of the stern, and ruined the captain's charts. By the time they neared St. Helena in the South Atlantic, not only was the *Rattler* "almost a wreck," but a couple of their men were ill from scurvy and apparently still traumatized by the wailing sea creature. "We had one man indeed, who was literally panic-struck by the appearance and cries of the seal in the Pacific Ocean," Colnett wrote. As they approached St. Helena, Colnett said that if they had spent another day at sea, the man would have died.[2]

For *Moby-Dick* Melville appears to have scooped up Colnett's scene with the screaming seal, dipped the yarn in a little dark Gothic sauce, and plopped it right into his chapter "The Life-Buoy." Ishmael explains that as they sail past a clump of small rock islands approaching the equatorial Pacific to hunt for the White Whale, the men hear "wild and unearthly ... half-articulated wailings ... crying and sobbing." Some of the crew of the *Pequod* think the cries are from mermaids. The oldest sailor declares that the wails are the souls of drowned men. When Ahab clops up on deck with his peg leg, he, like Colnett, explains to his crew that the sound is only from seals.[3]

Ishmael then delivers an astounding perspective when read today: "Most mariners cherish a very superstitious feeling about seals," Ishmael says, "arising not only from their peculiar tones when in distress, but also from the human look of their round heads and semi-intelligent faces, seen peeringly uprising from the water alongside." The sailors in *Moby-Dick* find their superstitious fears about the cries to be confirmed. It *was* a bad omen—because a few hours later one of their shipmates falls from aloft and drowns. And when the men learn the next day that the crewmen of the *Rachel* are lost, the old Manxman attributes the seals' cries to be those same departed spirits.[4]

From a marine biology perspective, Melville squished his geography to help his fiction. Unless you imagine the final meeting with the White Whale on the far eastern side of the Pacific, near the Galápagos, there are no regular rocky island rookeries or haul-outs for marine mammals anywhere near the implied cruise track of the *Pequod* as it approached the equatorial Pacific. Along the cold-water coastline of Chile, Colnett and his crew perhaps heard the cries of a South American fur seal (*Arctocephalus australis*), but the safer guess is that they heard the larger, more common, South American sea lion (*Otaria flavescens*), another species of eared-seal. Both are found in that part of the world, feeding on fish and squid and hauling out on islands and coasts to mate and give birth. Biologist Claudio Campagna, who was part of a recent study of the vocalizations made by South American sea lions, read Colnett's account for me and explains: "It could have been a sea lion female. Sea lion females call their pups with sounds that really may be described as a human female in distress."[5]

From an environmental studies angle, Ishmael's comments that most sailors have a superstitious sympathy for seals is surprising for the time period, even with Colnett's narrative in mind. Bashing and hunting seals, sea lions, walrus, and sea elephant for oil and furs was big business and a sort of little brother industry to whaling. Hundreds of vessels in the 1800s did both. At some point during his travels at sea, Melville surely saw and heard seals beside his ship, especially near the Galápagos, even though by then large-scale hunting of pinnipeds at any and all accessible rookeries in the Southern Ocean and South Pacific had been exploited for decades. The whalemen would often eat the meat of the pinnipeds they captured. Nineteenth-century mariners' accounts seem more disgusted and afraid of pinnipeds, rather than seeing any of the human identity or intelligence in their big eyes and whiskers that Ishmael describes, or even his comparison in an earlier scene in *Moby-Dick* in which Ahab is like a proud, bull sea lion presiding over the cabin table. For example, author George W. Peck sailed along the same coast of Peru and wrote in 1845 that he believed the South American sea lions to be "unnaturally hideous" and "the compelled agents of some diabolical spell or inevitable doom." Peck described the sea lions regularly swimming out to ships: "their breathing is always like sobbing; their cries are wails." Especially in the Arctic, whalemen were often forced to kill these marine mammals to supplement their oil stores or to feed themselves. A few did sympathize with these pinnipeds, as Ishmael implies. For example, Mary Brewster aboard the whaleship *Tiger* explained that a sailor stopped short of killing a walrus alongside the ship because "it looked so innocent." But the whalemen still regularly harpooned and clubbed them to death. Adeline Heppingstone, the captain's daughter aboard the New Bedford whaleship *Fleetwing*, once wrote of a seal she watched over the rail: "One dear little fellow came close alongside of us and looked up into my face." Heppingstone didn't seem to mind the killing, though, because "the skins make warm clothing for them." In *Moby-Dick* the harpooners Queequeg, Tashtego, and Daggoo barely consider the pinnipeds at all: "the pagan harpooners remained unappalled" by the screaming seals. In the same way they were unmoved by the giant squid, experienced at sea like Ahab, these men do not make any anthropomorphic connections to the sound of these animals.[6]

So Ishmael's story about the superstitions that Yankee mariners had about seals, because of their human-like faces, suggests that there actually has *not* been, as is often theorized, an entire cultural shift in the way Americans thought about smaller marine mammals—the idea that *everyone* hated all marine mammals until the mid-twentieth century. Perhaps there really is something hardwired in some of our affection for animals that look or act like humans, as Ishmael spoke of his huzzah porpoises—or maybe there is something unique to seals and their long history in selkie stories. That is, if we don't need seals and sea lions for our livelihood or have not been taught to eat them. To jump-start advocacy and attention to the oceans and coasts, environmental activists in the 1950s chose animals such as baby seals and penguins and dolphins, which each had identifiable human-like behaviors or expressions. The 1964 film footage of the clubbing of fluffy, white, baby harp seals on the ice of the Gulf of St. Lawrence made international news with profound, immediate impacts. Activists for environmental and animal-rights movements found an ideal symbol in these baby seals. Meanwhile, the movie *Flipper* (1963) and then the television show of the same name had been portraying a clever dolphin version of Lassie, the dog.[7]

Ishmael recognizes a century and a half earlier that you'd have to have a stone heart to dislike dolphin play. The perception jump that had to come, which *Moby-Dick* advanced in 1851, although often toward other ends, was that large whales, without overt human features or human-like behaviors, deserved empathy and attention, too. After the popularization of seals and dolphins, it was not old age or sentience, but the *vocalizations*, ironically, that next built sympathy for the large whales in American popular culture. These were not the human-like moanings of sea lions, but the first underwater, exceptionally otherly, organized "songs" of humpback whales, reproduced by Roger Payne (a big fan of *Moby-Dick*), who distributed the little records into every single copy of the January 1979 issue of *National Geographic* magazine.[8]

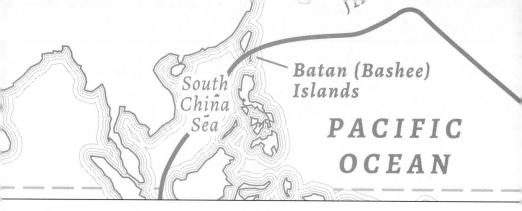

South
China
Sea

Batan (Bashee)
Islands

PACIFIC
OCEAN

Ch. **27**

THE FEMININE AIR

*It really amazed me that you should find any satisfaction in [Moby-Dick]. It is
true that some men have said they were pleased with it, but you are the only
woman—for as a general thing, women have small taste for the sea.*
  Herman Melville, in a letter to Sophia Peabody Hawthorne, 1852[1]

Despite a procession of ill omens, the *Pequod* leaves astern the typhoon,
the wailing seals, and a drowned shipmate as the ship marches under
sail toward the equator. The opening to the scene in "The Symphony,"
the lovely, sunny calm day before the final chase, provides a nineteenth-
century benchmark to examine how American authors portrayed gender in relation to the ocean and its inhabitants, the salty version of "man
vs. nature."

"The Symphony" opens this way:

It was a clear steel-blue day. The firmaments of air and sea were hardly
separable in that all-pervading azure; only, the pensive air was transpar-
ently pure and soft, with a woman's look, and the robust and man-like
sea heaved with long, strong, lingering swells, as Samson's chest in his
sleep.

   Hither, and thither, on high, glided the snow-white wings of small,
unspeckled birds; these were the gentle thoughts of the feminine air; but

to and fro in the deeps, far down in the bottomless blue, rushed mighty leviathans, sword-fish, and sharks; and these were the strong, troubled, murderous thinkings of the masculine sea.[2]

Here Ishmael feminizes the wind and air. Above the ocean, this feminine air is clear, lovely, superficial, snow-white, and soft—yet also duplicitous, waiting, and wicked, as the biblical Delilah who could pounce in the shape of a storm. The sun here at the equator, "aloft like a royal czar and king" is a male god, presiding over the intermingling "throbbing" sex between a newly wedded feminine air and her manly, deep-thinking, violent husband: the sea.

Although described as male in this late scene, Ishmael rarely genders the ocean in any way throughout the story, besides one moment when he describes the sea like a tigress in "Brit" and at the end of "The Dying Whale," when Ahab declaims the waves as his brothers, having been "suckled by the sea." Ishmael's Nature, taken more broadly, is female. In "The Whiteness of the Whale," Nature, capitalized, throws "spells" and enlists "her forces." In "Schools & Schoolmasters" Ishmael equates Nature with the sea, and thus adds to the impression of the White Whale as an agent of God and Nature. Accurately depicting the older males as solitary, Ishmael says: "He will have no one near him but Nature herself; and her he takes to wife in the wilderness of waters, and the best of wives she is, though she keeps so many moody secrets."[3]

When a sailor from the *Pequod* shouts out from the masthead for a spout it is always "There *she* blows!," which seems to be true to the language of the American whaleman at the time. Of the at least eleven whales that the men of the *Pequod* kill during the novel, all are either male or an unspecified gender.

No significant female characters appear in *Moby-Dick*. Women ashore help tend the inns and fit out the ship's stores. As the *Pequod* sails farther out to sea, wives, mothers, and lovers are mentioned in song and banter, usually in reference to sex, then later by Ishmael, Starbuck, Stubb, and Ahab as reminders of the safety of home. Ursula K. LeGuin, when defending her *Searoad* (1991) against criticisms that she did not include any good young male characters, explained: "Well, there aren't any 'good young women' in *Moby-Dick*, but it's still a good book." (*Searoad* is a neglected masterpiece, by the way, especially in its powerful reflections on the ocean, gender, and the literary myths that have crafted our views.)[4]

Disappointing, if not offensive, to the twenty-first-century reader, several of Ishmael's references and puns in the story, such as his jokes about the narwhal's tusk and the ornate metaphors in "Fast Fish and Loose Fish," equate women with whales and products. "The lady then became that subsequent gentleman's property," Ishmael jokes, winking, "along with whatever harpoon might have been found sticking in her."[5]

But it is that opening description in "The Symphony" that was Melville's most significant evocation of gender and the sea in *Moby-Dick*. It's a dated, Victorian image: the ocean is a personified male, hiding the predatory, cannibalistic violence under the surface. The male sea can be calmed at times, temporarily, by the feminine air, female Nature. The image is exemplified in one of the paintings that Ishmael in "Monstrous Pictures of Whales" chides for its cetological inaccuracy—but the gender stereotypes in this image of the knight Perseus rescuing fair Andromeda from the ocean monster are overt (see fig. 49).

Yet why, with over forty years at sea under sail alone and so often in a small boat rowing beside these animals in a time before commercial offshore engines and steel, has Ahab not begun to feminize the sea in the same way as did fictional protagonists in later years, such as the wrinkled veteran fisherman Santiago in Hemingway's *The Old Man and the Sea* (1952)? Scholar Rita Bode has shown maternal imagery in *Moby-Dick* to be as pervasive as the phallic subtleties. So why wouldn't Ahab, if "suckled by the sea" and so intimate with its ecology, perceive the ocean and its creatures as matriarch, or lover, or at least *not* consider the sperm whale and the sea as his combatants?

For one, it might be that Ahab just has far more rage and pent-up madness about that white whale of Nature taking off his leg. Ahab is more driven by anger and revenge. Certainly, too, there's a bit of the general historical gestalt, perhaps stereotyped, as a nineteenth-century

FIG. 49. *Perseus and Andromeda* by Guido Reni (1575–1642), a painting that Ishmael references directly in "Monstrous Pictures of Whales."

man versus nature ethic that encourages a violent masculinity. Perhaps, too, Melville is incapable of seeing beyond the gender stigmas of his time, although it is telling that Ahab's closest moment to a more mutual relationship with the ocean, as well as his reflection on home and his wife—"I widowed that poor girl when I married her"—is here in "The Symphony." Ahab momentarily sees the feminine air calming the murderous sea. He sheds a saltwater tear that mixes with the saltwater ocean. Is this Ahab coming the closest to a more familial, eco-logical, holistic relationship with the sea? Ahab almost abandons his quest. Does he almost bless the ocean creatures unaware? Certainly we expect at this moment a Coleridgean sea snake to come slithering out from under the hull.[6]

But no. Ahab turns. Ahab sees God's plan too fixed on iron rails, his life as a predator of the whale preordained, the sea not a place of kind-ness or equity or empathy, but one of indifferent universal cannibal-ism, which he connects to masculinity: "By heaven, man, we are turned round and round in this world, like yonder windlass, and Fate is the handspike. And all the time, lo! that smiling sky, and this unsounded sea! Look! see yon Albicore! who put it into him to chase and fang that flying-fish? Where do murderers go, man!"[7]

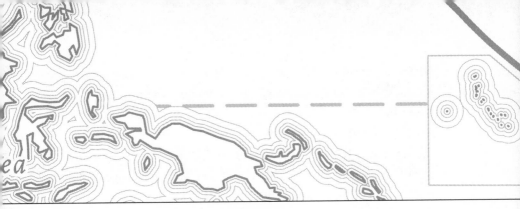

Ch. 28

NOISELESS NAUTILUS

But the moon's bright wake was still revealed: a silver track, tipping every wave-crest in its course, till each seemed a pearly, scroll-prowed nautilus, buoyant with some elfin crew.
　　Melville, *Mardi*[1]

At last! Wild Ahab finally finds the White Whale! In "The Chase—First Day," Ishmael, the crew, and the reader see the famed sperm whale, in person, on the equator, for the very first time.

Ishmael says that Moby Dick is "seemingly unsuspecting" of the battle that is to come. The White Whale glides through "fleecy, greenish, foam." White bubbles dance on the smooth equatorial seas, filling around his dazzling, glistening white hump and forehead. Hundreds of flittering seabirds hover around the White Whale in a cloud, dabbling on the surface with their feet. One of the birds stands at the end of a shattered lance shaft still stuck into the whale's back. You can reasonably imagine this as a tropic bird (*Phaethon* spp.), a bright white seabird that Melville had written about in previous novels, and here in *Moby-Dick* it "silently perched and rocked on this pole, the long tail feathers streaming like pennons."[2]

Melville's description of the first vision of the White Whale is one of those paragraphs you can read aloud again and again, purely for the poetry of the prose. One especially lovely reference here for the natural historian is the line that opens the paragraph. The men of the *Pequod* are in their small, light whaleboats, sails set, and paddling toward the whale: "Like noiseless nautilus shells, their light prows sped through the sea; but only slowly they neared the foe."[3]

While writing *Moby-Dick* Melville marked up the page of Surgeon Beale's *The Natural History of the Sperm Whale* that cited a long passage about the argonaut, or the paper nautilus (Family Argonautidae). In the margins Melville scribbled two ideas on the poetic connection to these tiny ocean octopuses, most of which is illegible, but it was something to do with Sinbad's sea anchors from *The Arabian Nights* (1706). Beale's passage was a quotation from Peter Mark Roget's *The Bridgewater Treatises on the Power, Wisdom, and Goodness of God, as Manifested in the Creation: Animal and Vegetable Physiology Considered* (1834), one of nine popular books of science commissioned by the Earl of Bridgewater upon his death in order to advance natural theology. Roget explained that just like a tiny sailboat this cephalopod has a paper-thin shell, "almost pellucid," from which extends two specialized "tentacula" with membranes that catch the air. The paper nautilus rows and steers with its other arms, Roget said. And if the wind gets too strong, it dives and pulls the shell down under the surface.[4]

Paper nautiluses apparently sailing on the surface like boats had inspired naturalists and writers ever since Aristotle, leading Linnaeus in the early 1700s to name them after the Greek myth of the hero-sailors, the Argonauts, who voyaged aboard Jason's ship *Argo*. In 1807 William Wood wrote in his three-volume book of natural history that some believed the "ancients" got the very idea of the art of sailing from these animals. (See fig. 50.) Wood, like Roget, seems to have used these octopuses and their sails as evidence to prove that God is the ultimate designer and creator, and that any transformation of species, any evolution of form, would be impossible.[5]

Soon after came the French naturalist Jeannette Villepreux-Power, who first studied paper nautiluses in the 1830s. Many credit her with the invention of the aquarium, because in order study these animals she an-

FIG. 50. Aquatint of a paper nautilus by William Daniell in William Wood's *Zoography, or, The Beauties of Nature Displayed* (1807).

chored wood cages in the sea and then pumped sea water through hoses into tanks into her small coastal laboratory in Sicily. Her foundational experiments, which Melville could have read about in a detailed illustrated entry in his *Penny Cyclopædia*, advanced what would later be confirmed by twentieth-century marine biologists: these paper nautiluses are females only—the males are but tiny shell-less functionaries—and the female's egg-case shell is secreted and repaired during its lifetime by the sail-like membranes at the end of the two arms. Villepreux-Power did not discount the "sailing" behavior, but she did not witness it for herself. It turns out that the membranes do not serve to propel the animal by the wind at all. Illustrations, like those in Wood's natural history, seem to have been pure fiction.[6]

For Melville, who tucked this legend into a brief reference to open a crucial scene of the novel, the paper nautiluses were just one of thousands of rare ocean wonders, unknown to those ashore, witnessed almost exclusively by the fishermen and whalemen who traveled quietly along the surface. Apart from shell collecting, public interest in subtidal invertebrates had only just begun in the 1850s. The first public aquarium opened in London as part of the Regent's Park Zoo in 1853; this was followed in the United States by the collections of P. T. Barnum, who set up aquaria at his American Museum in New York City.[7]

At the end of this stunning pivotal paragraph in "The Chase—First Day," Ishmael connects the opening allusion of the men in their quiet whaleboats like tiny paper nautilus to the White Whale himself, now first seen as a glorious Greek ship flagged by a tropic bird: "the painted hull of an argosy."[8]

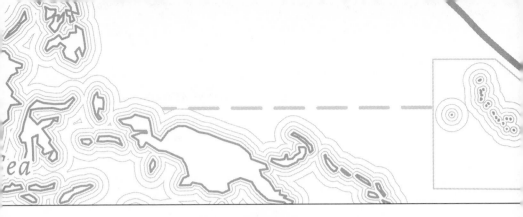

Ch. 29

SPERM WHALE BEHAVIOR

Retribution, swift vengeance, eternal malice were in his whole aspect, and spite
of all that mortal man could do, the solid white buttress of his forehead smote
the ship's starboard bow, till men and timbers reeled.
  Ishmael, "The Chase—Third Day"

The final three-day battle against Moby Dick raises the believability
of three aspects of sperm whale behavior that inspire fear and drama:
humans can locally track a sperm whale over multiple days; sperm
whales are aware of their human hunters; and a sperm whale can be
consciously aggressive to humans in three ways: with its tail, jaws, and
head.

Yet earlier in the novel, in addition to his musings on sperm whale
intelligence, Ishmael shows through the events of the story other aspects of sperm whale behavior that create an alternative view to an aggressive, evil sperm whale. These observations instead evoke a sympathy
for this species, notably that sperm whales suffer when harpooned and
sperm whales behave in human-like units with social structures that are
similar to ours.

TRACKING SPERM WHALES IN THE TWENTY-FIRST CENTURY

Marta Guerra Bobo steers the power boat *Grampus* behind a sperm whale at the surface of Kaikōura Canyon, New Zealand. Shallow, welded of aluminum plate, and powered by a 115-horsepower engine that can hurl the boat at speeds of up to 30 knots, *Grampus* is less than twenty feet long—two-thirds the length of the boats the American whalemen rowed and sailed in the mid-1800s.

Guerra keeps the boat back and well clear so as not to spook the large male. Each time the animal spouts—about once every ten to twenty seconds—she says "blow" into a hands-free microphone set up by the wheel (so she can collect the exact timing later). The whale's dorsal fin, his hump, pokes out of the water. Because it's a calm day, forward of the hump is a portion of the whale's gray back and head that's visible floating over, or more through, the surface. Guerra can see wrinkles in the skin and the large bunched ball of flesh to the upper left of the head that surrounds the s-shaped blowhole. Forward-slanting, the spouts are thick, bushy, white, and clearly discernible, even against the overcast sky.

When Guerra shifts the engine forward to keep up with the whale, she tries to do so during the spout.

"I don't know if it matters," she says. "But it's possible the shift in gear startles them."[1]

Guerra's assistant, Rebecca Bakker, is ready with the camera.

"He's getting ready to go," Guerra says.

The ridge of the whale's spine curls up as he arches. Bakker clicks image after image as the tail rises, facing down as it sheds sea water, then the tail slowly folds over, upwards, aloft, to display its ventral side, spreading wide and tall into the air before sliding under the surface. The sperm whale dives down with barely a splash.

"It's Tiaki," Guerra says. She recognizes a few nicks out of the trailing edge of the left fluke and a scalloping on the edge of the right fluke. Tiaki means "guardian" in Māori. He is one of two dozen or so regular males that are in the region for part of every year, going back to 1988 when whale watchers began keeping records.[2]

Guerra speaks the exact time of the dive into the microphone. She confirms the hydrophone trailing astern is recording Taiki's clicks as he descends. She then steers the boat forward to look for skin floating on the surface or for a cloud of defecation. Guerra motors directly over the oily foot, the calmed circle of water where the whale has just dived. Back ashore, she'll calculate the exact depth beneath the boat from high resolution charts, but she doesn't have the depth sounder on now because the sonar pings, which sound eerily like sperm whale clicks, might also distract or disturb the whales. Off the starboard side of her boat, the tiny rapid eddy created by this fifty-foot sperm whale swirls in the sea. Off the port side, Guerra scoops with a hand-net a little wormlike twirl of skin. She sticks the sample into a bag that she'll freeze when she gets back ashore. Guerra taps the location and other data into her electronic tablet, using software that the designers named (true story) "Ahab." Next, carefully watched by two Salvin's albatrosses (*Thalassarche salvini*) who float nearby in hopes they're bringing up fish, Guerra and Bakker sample the temperature, salinity, dissolved oxygen, and chlorophyll-a of the water column down to 1,800 feet in order to build an oceanographic snapshot of the waters in which Tiaki forages.

After the two researchers haul up by hand their measuring device, known as a CTD (for conductivity, temperature, and depth sensors), Bakker plugs into a different hydrophone. This one doesn't record but is only for tracking. She listens for the next sperm whale by semi-circling the receiver underwater, similar to how you'd turn your head back and forth to locate the direction of the wind between your ears. In the headphones the clicks sound like distant, rhythmic, underwater snaps—like a second-hand on an old clock. In ideal conditions, they can hear a male sperm whale from as far as seven miles.

Headphones still on, pointing with her arm, Bakker says: "I can still hear Tiaki on his way down. But it sounds like there's another one that way. He's faint, but maybe two and a half miles?"

Guerra throttles up and drives toward where Bakker pointed, using the GPS to judge distance and a patch of cloud to navigate her new course. The clicks are louder the closer you are, but there can be other factors, such as the orientation of the animal's head at a given moment, the sea state at the surface, or the topography of the bottom. When the sperm whale is about to capture a squid or fish, the clicks speed up until it's almost a buzz or a creak. When a sperm whale ascends, he or she usually clicks at first then goes completely quiet, unless pausing to feed opportunistically on the way back up to the surface. A couple times, even without the hydrophone, Guerra has heard the sperm whale's click through the metal hull of the boat. These were likely the sperm whale's slower clicks, the *clangs*, which might be part of male-to-male communication; they are so loud that they can be heard with hydrophones from perhaps more than twelve miles away.[3]

After speeding and thumping across the waves, driving *Grampus* as aggressively as would Starbuck if he had his hands on a four-stroke engine, Guerra stops the boat to allow Bakker to put the hydrophone back in the water.

"Definitely louder," Bakker says. "About point-four miles that way."

Guerra throttles up again, and off they go.

"I quite like this tracking," says Bakker, who traveled from Holland for the research opportunity. "It's like good-natured hunting."

Kaikōura Canyon is arguably the most accessible and reliable place in the world to see male sperm whales all year round. American and English whalemen started hunting these waters in the 1830s. It's one of the most biologically productive deep-sea habitats on Earth. The coastal shelf drops off almost immediately into the canyon, which can be over a mile deep. The canyon snakes and empties into a wide trough that's nearly twice that. Guerra's research aims to determine why sperm whales favor this particular region of the world. Kaikōura is occupied almost solely by adult males. She uses stable isotope analysis of sperm whale skin, the oceanographic and topographic data where the individuals choose to forage, and a range of other data to develop a sense of exactly what, where, when, why, and how sperm whales forage in this region of the South Pacific.[4]

Guerra was born in Madrid. Her father is a doctor. Her mother, a naval architect, was the first woman to earn this profession in Spain. Guerra has now been studying the sperm whales on Kaikōura Canyon for the equivalent of over fifteen months over a period of four years. She goes out each day the weather permits. She has now had approximately nine hundred close encounters with male sperm whales, nearly all of them just like this one with Tiaki. (My rough guess is that Melville had a couple dozen or so.) Guerra was also once part of a team that got even closer to the males, riding up in small boats to sample the blow and to temporarily tag six different males.[5]

As Guerra and her assistant track the next whale, a fleet of three large whale watch boats with hundreds of tourists are out on the water, too. No more than three boats of any type are allowed around a whale at any time. The boats keep their distance. They do not get in front of a whale. The Kaikōura whale boats use jet engines to reduce their sound impact underwater. They carry their own hydrophones. The captains share information over the radio and sometimes ask Guerra to guide them.

Tourists after sperm whales have a different experience here than when on a boat watching for baleen whales in other parts of the world. Sperm whales usually dive for a solid thirty to forty-five minutes, or even more, and then only come up for roughly eight to ten minutes in between. When sperm whales are on the surface, they rarely do anything except swim slowly forward and blow, as they recover from the dive. You can barely see anything besides some of their back. After the whale dives, offering a vision of the tail, the show is over. The whale watch boat speeds off to try to find another whale. In all her time on the water, Guerra has seen sperm whales breach only a few times near their boat and several more times at a distance. She's seen a few of the other notable behaviors, such as lobtailing—waving the flukes—and surfacing tailfirst, but she has never seen whales raising their head vertically to look with their eyes, known as spy-hopping. My point here is that Melville really did have to boil the sap and add a great deal of fancy to the behavior of the sperm whale at the surface. The inability for Ahab or Ishmael to get a true vision of the whale, even when alongside being butchered, was not only poetic but true. And appreciating the final scenes of *Moby-Dick* is similar to being able to enjoy a sperm whale watch in the twenty-first century: you need to already be compelled about the *idea* of the sperm whale, to have well-formed a detailed imagination of the organism, because that brief vision of what amounts to a distant, floating, puffing log, or even the graceful sinking of its tail, won't do much for you if you're not already fascinated and if you haven't already built an imaginary world before and after that dive.

Tracking whales with Guerra and her research assistant off New Zealand, I realize that their methods had returned to a type of hunter's knowledge similar to what the open boat whalemen once had. Modern researchers, pioneered by the likes of Hal Whitehead, who himself began and continues his sperm whale research in a sailboat, have a much greater understanding of sperm whale *behavior* than did the twentieth-century researchers. The twentieth-century whalemen learned a great deal more about whale *anatomy* as they dissected so many animals killed by explosive harpoons and reeled up onto their steel decks with power winches. Time at sea with Guerra also makes you wonder, in an age before hydrophones, how the nineteenth-century whalemen ever found a sperm whale at all.

TRACKING SPERM WHALES IN THE NINETEENTH CENTURY

At the start of the voyage of the *Pequod*, Ishmael argues in "The Chart" that Ahab will locate the White Whale in a specific region of the world at a given time of year. In "The Chase — Second Day" Ishmael describes how once Ahab has sighted Moby Dick he's able to track the whale over multiple days. He uses his human lookouts and, by dead reckoning, estimates the animal's speed and direction under the water as it swims:

> And as the mighty iron Leviathan of the modern railway is so familiarly known in its every pace, that, with watches in their hands, men time his rate as doctors that of a baby's pulse; and lightly say of it, the up train or the down train will reach such or such a spot, at such or such an hour;

even so, almost, there are occasions when these Nantucketers time that other Leviathan of the deep, according to the observed humor of his speed; and say to themselves, so many hours hence this whale will have gone two hundred miles, will have about reached this or that degree of latitude or longitude.[6]

This aspect of sperm whale's behavior is helpful for the fictional purposes of the novel, and Ahab even overruns Moby Dick on the final day as he underestimates the wind and his ship's speed.

Ninenteenth-century whalemen certainly knew that there's a difference between whales foraging locally and when they're in transit. Whitehead has observed that when sperm whales move to different areas, they tend to swim at a surprisingly slow rate of about two and a half miles per hour, which is more sluggish than a ship under sail in moderate winds. He found that they travel in straight lines, covering about sixty miles a day, which, if the wind were favorable, would be an easy day of progress in transit for a whaleship like the *Charles W. Morgan* or the *Commodore Morris*.[7]

Guerra has read *Moby-Dick*, which she says is obligatory reading if you're getting a PhD studying sperm whales. "Ahab's timetable tracking sounds a bit stretched to hold as a general rule," she says. "But you know, I wouldn't be surprised if the whalemen got it right every once in a while. Spermies can be so predictable some times, and I imagine that if they are on the move on a steady course they would be doing pretty consistent speed. So maybe!"

COULD A SPERM WHALE RECOGNIZE A HUMAN?

The male sperm whales who come back year after year to Kaikōura Canyon seem comfortable with the whale watch boats and *Grampus*. New Zealand whalemen hunted sperm whales in the Canyon as late as the 1960s for only two seasons, but they were destructive. Could some of the individuals remember?[8]

Guerra says that it is the sperm whales new to the region that are more likely to adjust their behavior in response to her boat and the whale watch vessels—they'll dive slightly more quickly or shift directions while swimming at the surface—which she cautiously believes to be nervous, avoidance behaviors. She has never seen the males approach the boats, turn back toward them, or even seem to have any curious interest. Yet decades before her research, one of the founders of the first whale watch company in New Zealand, naturalist Barbara Todd, observed one sperm whale from 1988 to 1991, which they named Hoon, who often floated close beside their boats. Hoon seemingly enjoyed watching the tourists, and he spent far more time at the surface near their vessel than the other whales, exhibiting a variety of behaviors, such as surfacing with his tail first. Todd felt certain the whale recognized their boat and exhibited human-like play.[9]

Certainly there is ample evidence among the smaller toothed whales, such as the bottlenose dolphins and the beluga whales, of individual animals recognizing individual humans, learning from humans, and engaging with them. There's no reason to assume this could not happen between sperm whales and humans. It just rarely if ever does, because our habitats are so physically distant. Only in the past couple decades have free divers without the bubbles and sounds of tanks been able to get close to sperm whales in the water for more than a few moments, but this is, of course, still for short blips of time and only near the surface.

That said, significantly to an environmental reading of the novel, Ishmael never actually describes or even implies that the White Whale senses, by vision or however else, that the *Pequod* chasing him is any different than another whaleship—or even that he particularly "sees" or chooses Ahab.

SPERM WHALE AGGRESSION: TAIL SLAPS, BITING, AND THE BATTERING RAM

Ishmael never fully anthropomorphizes the White Whale in a Peter Rabbit sort of way, nor does he show the world through the whale's point of view. Ishmael does relate over and again, however, building up to these final scenes, a variety of stories that invest in this particular whale an "unexampled, intelligent malignity" that approaches that of a devious and all-powerful human warrior—or at times even an immortal demigod. In his first full introduction to this whale in the chapter "Moby Dick," Ishmael says: "More than all, his treacherous retreats struck more of dismay than perhaps aught else. For, when swimming before his exulting pursuers, with every apparent symptom of alarm, he had several times been known to turn round suddenly, and, bearing down upon them, either stave their boats to splinters, or drive them back in consternation to their ship." Thus Ishmael relays, *secondhand*, that the White Whale was believed to strategize and even duplicitously feign fear.[10]

As the novel progresses over the voyage, Ishmael tells various stories of men injured and killed in the pursuit of Moby Dick, usually by the animal's tail, jaws, or head. These are all sailors' stories of various degrees of separation told to the men of the *Pequod*. With subtle phrases, including "as if" or "seemingly" or "ascribed to him," Ishmael is clear that there is a layer of the yarn to the behaviors of Moby Dick. Ishmael builds the suspense, the *idea* of this exceptional sperm whale. He fuels a debate for the reader and the characters within his story about whether the White Whale is a "dumb brute" as Starbuck would have it, or if Moby Dick is an evil or Divine agent or principle, as Ahab does. Meanwhile, in chapters such as "Moby Dick" and "The Affidavit," Ishmael defends and supports his facts about the history of this individual sperm whale's dangerous behavior and his capabilities.[11]

In the closing three days of the chase, the crew and the reader finally see the White Whale in the flesh, introduced among imagery of the delicate paper nautilus and the streamers of the white tropic bird. Ishmael bears witness to a set of behaviors that at first seem to tilt more toward an imagined monster. Yet they are all actual known sperm whale behaviors. During the first day, Moby Dick chomps a boat in two with his jaws after gnawing and holding it there in his mouth for a time. The White Whale pokes his head vertically out of the water, spy-hopping, and then circles the men in the water. On the second day, Moby Dick breaches. He smashes two more whaleboats, this time with his tail, and he comes up underneath another boat with his head. Moby Dick *seems* to parry and plan against Ahab and his men, both when the animal surfaces and when he travels. On the third and final day of the chase, Moby Dick turns from the whaleboats and hammers his head into the *Pequod*, thus sinking the ship and drowning the entire ship's company, sparing, by chance, our narrator. Most of these behaviors are the actions of what seems to be only a literary, foaming leviathan that's whipped up for the drama of the fictional climax. Yet, as ever, these animal behaviors are all far less fictionalized when you examine them carefully.

Melville's contemporaries were contradictory about the extent of sperm whale aggression toward hunters. Charles W. Morgan, in his speech on the natural history of the sperm whale in 1830, just before making a passing reference to the *Essex*, spoke of the sperm whale as "in general harmless, but now & then individuals are found exceedingly fierce & bold." Surgeon Beale opened his *The Natural History of the Sperm Whale* explaining that the ferocity of this species has been greatly exaggerated, even in previous scientific literature. Melville underlined and marked up his copy of this section where Beale decried the falsehood of the previous accounts of this animal having "a relish for human flesh" and Baron Cuvier's misguided belief that every fish and shark in the sea is so horribly frightened of the sperm whale that they will not even approach a dead carcass. Beale wrote instead of the sperm whale's timidity. When harpooned the sperm whales seem at first paralyzed with fright, he wrote. If they have not been killed quickly, though, they then can "shew extreme activity in avoiding their foes; but they rarely turn upon their cruel adversaries." If boats and men were injured, it was only in the whale's frantic attempt to escape. Seemingly exasperated in his defense of sperm whales, Beale wrote: "Not one [author] has stepped forward to vindicate its history from the absurd and fabulous accounts."[12]

Defending the sperm whale's passivity, however, was not a torch that

Melville wanted to bear for this little fiction project of his. Melville instead turned to the likes of Dr. Frederick Bennett, for example, who was normally more temperate and independently minded than Beale. But Bennett's observations and writings on sperm whale aggression aligned far more conveniently for the story Melville wanted to tell. In *Narrative of a Whaling Voyage*, Bennett explained that after being attacked, sperm whales will not just respond due to pain but will try to protect another whale that has been injured: "Should the animal be allowed time to rally, it often becomes truly mischievous. Actuated by a feeling of revenge, by anxiety to escape its pursuers, or goaded to desperation by the weapons rankling in its body, it then acts with a deliberate design to do mischief."[13]

Bennett went on to relay from various accounts from specific ships each of the aggressive behaviors that Melville attributes to the White Whale in the final chase.

Regarding the tail, Bennett described three incidents of sperm whales killing men with swipes of their tails, including one thwack in the North Pacific by a whale that was not harpooned himself but defending another one who was.

Regarding the jaws, Bennett included a couple of accounts of sperm whales attacking with their teeth. He personally gammed with the whaleship *Augusta* in the South Pacific in 1836, which had a boat on deck that had recently been "nipped completely asunder by the jaws of a harpooned whale." Bennett wrote that sperm whales sometimes continue to bite a boat into fragments or keep its mouth threateningly open for several minutes—thus laying some groundwork for Melville's moment when his White Whale literally holds Ahab's boat in his mouth long enough for Ahab to fanatically grab at the teeth. In historic illustrations and comic books, angry sperm whales are constantly crunching up boats and whalemen in their mouths, as if they were pretzel sticks. Even beyond Bennett's accounts, this seems to have actually had some historical validity. For example, in 1856 whaleman-artist Robert Weir illustrated an actual event in which a sperm whale chomped and lifted

FIG. 51. Artist-whaleman Robert Weir's illustration of a sperm whale battle off Fort Dauphin, Madagascar (1856).

up a whale boat, just as Ishmael describes in that first day of the chase. (See fig. 51.) And later in Weir's voyage, another boat was pierced by a whale's jaw "making a hole as big as the head of a barrel."[14]

Melville himself, while aboard the *Acushnet*, likely heard on the whaling grounds the story of another whaleship, the *Coral*, during a gam near the Galápagos. Only a couple months earlier, a lone sperm whale had chewed a boat into "many Hundred pieces," according to an account of one of the mates. The bull whale, spouting blood, then reportedly turned on another boat and "Eat [that] Boat up," causing the drowning of one of the men before the mate finally killed the animal.[15]

From a modern biological perspective, sperm whales using their mouths to defend against human hunters makes sense. Most terrestrial mammals are aggressive with their mouths, of course, as are some of the other toothed whales, such as Risso's dolphins, bottlenose dolphins, and Cuvier's beaked whales. Modern biologists have watched female sperm whales use their jaws to defend against killer whales. The rake marks gouged into the heads of older sperm males suggest that they might engage in jaw-to-jaw battle, like bucks locking antlers or hippos sparring with their open mouths. (See earlier fig. 12.) Divers have observed sperm whales using their mouths in play, too.[16]

The White Whale in *Moby-Dick* has a "crooked" and "scrolled" jaw.

Surgeon Beale witnessed two healthy sperm whales with scrolled jaws, commenting on how this showed teeth were not essential to their feeding. Beale never saw competition between sperm whales, but the sailors told him this was a common behavior. Beale reported that he knew of no *female* sperm whales with scrolled jaws. Melville clearly agreed that the scrolled jaw was a trait of strength, an acquired scar from an alpha battle for power and status—just as is Ahab's ivory leg. Modern observers have reported a sperm whale's jaw broken from a fight with killer whales, and another account claimed broken jaws from intraspecific battles, but these don't heal in a curl.[17]

A scrolled jaw, in fact, is more likely a birth defect, rather than healed from any kind of battle. Biologists working from the decks of twentieth-century whaleships estimated scrolled or short jaws, i.e., jaw deformity, occurred in about one in every two thousand whales. Dozens of examples have been documented. (See fig. 52.) It does seems to be more common in males, but several females have had scrolled or broken jaws, too.[18]

As for actually *eating* the wood of the boats, the squid that sperm whales swallow nearly whole are often quite large and a range of surprisingly large animals have been found in sperm whale stomachs, such as seals, sharks, and rays. Inanimate objects have also been found in sperm whale stomachs, such as stones, fishing gear, coconuts, and even once a human corpse a day after his shipmates saw the man gobbled.[19] I don't know of any account of wood shards from stove boats found in whale stomachs, but it seems credible.

During the final chase, the White Whale does more than kill with his tail and chop up whaleboats with his scrolled lower jaw. With his head, his "battering ram," Moby Dick overturns whaleboats and then in the final pages of the story turns and crashes into the hull of the *Pequod* so powerfully that he sinks the ship. There are dozens of reliable stories of sperm whales bashing whale boats, although these are usually, if not always, a defense behavior rather than a blind, flurried response in the moment of being harpooned. (See fig. 53.) Bennett reported that he *heard* that "New Zealand Tom," an enormous bull male with a "white

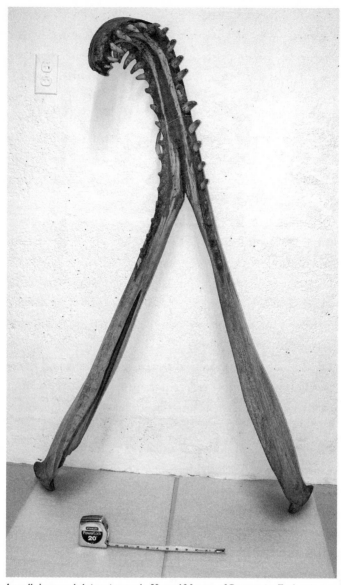

FIG. 52. A scrolled sperm whale jaw given to the Harvard Museum of Comparative Zoology in 1900. The jaw is four and a half feet long, so this was from a juvenile or small female adult.

FIG. 53. Sperm whale smashing a whaleboat with the head as depicted in a painting (c. 1834) by Ambroise Louis Garnery, a work of which Ishmael approves in "Less Erroneous Pictures of Whales."

hump," smashed up all manner of boats. Whaleboats were lightly built with cedar so as to be quick and nimble. This, however, left them fragile to impact. For example, Nelson Cole Haley aboard the *Charles W. Morgan* in 1852 wrote of a calf sperm whale, "taking our boat for his dead mother," accidentally smashing a hole in the bottom of their boat.[20]

SPERM WHALES SINKING SHIPS

What was big news was not a sperm whale smashing a whaleboat, but an entire whale*ship*. The story of the *Essex* is the first and only popular account of a sperm whale ramming its head and *sinking* a full whaleship that Melville would've known when he wrote the novel. Ishmael describes the event in "The Affidavit." Ishmael also writes of the whaleship *Union*, which reportedly sank in the Atlantic in 1807 after accidentally running into a floating sperm whale at night. In 1849 a collision with a whale wrecked the lesser-known Peruvian merchant ship the *Frederic* off the coast of Nicaragua; in 1850 a sperm whale in the midst of the hunt bashed a hole in the hull of the whaleship *Pocahontas*, forcing the ship to run into Rio de Janeiro; and that same year the whaleship *Parker Cook* was rammed *twice* by an enraged whale, although the ship was not sunk. Ishmael does not mention any of these, almost certainly because Melville hadn't learned of them.[21]

In the *Narrative of the Most Extraordinary and Distressing Shipwreck of the Whale-ship Essex,* which was likely ghostwritten in collaboration, the first mate Owen Chase explained that they sighted a pod of whales and lowered the boats. He personally harpooned one bull, who in response put a hole in their boat with his tail. They rowed back to the *Essex* where Chase went to repairing the stove boat on deck by hammering nails around a canvas patch. Meanwhile, the captain and the second mate's boat had made fast to a different whale. Chase soon spotted an exceptionally large bull spouting near his ship, which made for their hull. "He came down upon us with full speed, and struck the ship with his head," Chase wrote. The whale hit the ship's bow, he said, and it felt like they'd struck a rock. The whale continued bumping under the ship and knocked off a portion of the *Essex*'s sacrificial keel. Chase reported that the whale was "in convulsions." He watched the whale "smite his jaws together." The whale then swam across the ship's path, turned, and came directly at the *Essex* a second time: "Coming down apparently with twice his ordinary speed, and to me at that moment, it appeared with tenfold fury and vengeance in his aspect. The surf flew in all directions about him, and his course towards us was marked by a white foam of a rod in width, which he made with the continual violent thrashing of his tail; his head was about half out of water, and in that way he came upon, and again struck the ship." On this second attack, the whale bashed a far larger hole in the hull, sinking the ship for good. The whale swam off to leeward. They never saw it again. They were busy preparing to abandon their rapidly sinking ship, handing down their copies of *Bowditch* and their quadrants.[22]

Although few dispute that the sperm whale hit the *Essex* a second time, there was some public debate about the sperm whale's intentionality, even after the story was published. In an 1834 issue of the *North American Review,* the author declared that "no other instance is known, in which the mischief is supposed to have been malignantly designed by the assailant, and the most experienced whalers believe that even in this case the attack was unintentional." Melville almost certainly read this, since it was republished in Olmsted's narrative. Twenty-first century scholars have also wondered if the banging of Owen Chase's hammer—a perfect twist on the carpenter fish stories—might have even somehow enraged the sperm whales. This is highly improbable. Sperm whales might be agitated by the sound, perhaps intrigued by something heard more distantly, but the idea that the sperm whale would mistake a hammering above the surface, transferred through the hull, for another male, twice, is hard for experts such as Whitehead and Guerra to believe.[23]

News of the sinking of yet another whaleship by the head of a sperm whale, a ship named the *Ann Alexander*, arrived to New York in November 1851, just as the first American edition of *Moby-Dick* was published. An editor friend sent Melville the newspaper clipping reporting the sperm whale smashing into the hull, to which Melville wrote back excitedly: "I make no doubt it is Moby Dick himself, for there is no account of his capture after the sad fate of the Pequod about fourteen years ago.—Ye Gods! What a Commentator is this Ann Alexander whale. What he has to say is short & pithy & very much to the point. I wonder if my evil art has raised this monster." A couple critics at the time even believed Melville whipped up *Moby-Dick* immediately after—to take advantage of the event. Meanwhile the *Utica Daily Gazette* wrote of the absurdity of the sinking of the *Ann Alexander*, which prompted the *New Bedford Whalemen's Shipping List* to fire back in print to not only defend the reality but also satirize how little landsmen knew about what happens out at sea. The following year the captain of the *Rebecca Sims* seemed to have killed the very whale, since two of the *Ann Alexander*'s harpoons were in the animal and its head was injured with shards of wood still stuck in the flesh. Historians have found still other nineteenth- and twentieth-century examples of whales punching holes in wooden hulls and even a couple modern accounts where sperm whales hit steel boats.[24]

So the question then is not really *if* sperm whales attacked boats and ships with their tails, jaws, and head, but, especially during events when they were *not* being harpooned, *why* would they?

Nearly all of the evidence of male sperm whales fighting *each other* by using their mouths or head-butting is from historical, less reliable sources, but they probably do fight occasionally. Bottlenose dolphins will leap out of the water and butt their heads together, seemingly out of male competition. Bottlenose dolphins will use their heads to ram other ocean porpoises and dolphins of different species, too. Killer whales will do the same. If sperm whales do bang their heads together, it has not been commonly observed in areas like Kaikōura where solitary males are foraging or even in the equatorial regions or the breeding grounds when the males are joining groups of females and juveniles. Whitehead and his colleagues, who research primarily in the eastern Pacific and the Caribbean, have observed male sperm whales swimming near groups of females without any observable aggression, and the one time they did see a fight between males, it was over in fifteen seconds and involved their jaws and tails, which seems to match historic descriptions. Bennett presumed that the aggressive behavior of the whale against the *Essex* was because this whale was the "guardian of the school," which, again, has been proven an unlikely Victorian anthropomorphism since males seem to visit the units of females and juveniles only briefly.[25]

Guerra was surprised by Melville's depiction of sperm whale aggression. Her experience on the water, like Beale's, is that male sperm whales are skittish and passive. She had read the story of the *Essex* and other violent clashes with humans. Although she has never seen a whale attacked or in great pain, she very much doubts any intentionality or malice in these animals. Only once on Kaikōura Canyon has Guerra seen what appeared to be potentially aggressive behavior between two males, which she believes was over foraging territory. One sperm whale, one of the regulars, began swimming rapidly toward another until they both met underwater. The other, a transient arrival, was later sighted several miles away while the regular remained foraging in the area.[26]

Today's whale biologists, such as Guerra and Whitehead, cautiously believe that males most commonly identify their competitors' size by clicks and codas. That passively settles disputes most of the time, rather than resorting to jaw-locking or head-ramming.

A couple recent studies have taken up the wreck of the *Essex* by approaching the event not from behavioral observations in the wild, but from a physiological and evolutionary standpoint in the lab, beginning with a series of questions: Could a sperm whale actually sustain a ramming into another whale or a ship? If the animal's bulbous head and the oil sacks within have evolved for sensitive echolocation, why would a whale risk damaging these organs? And could this really be a behavior that would elevate a male's reproductive fitness?

In 2002 experimental biologist David Carrier and his colleagues found that in comparison to the evolution of all other toothed cetaceans, the male sperm whale's bulbous head is far larger than the female's: thus, the size of the head is a secondary sexual characteristic. This, however, might have evolved to produce more effective echolocation sounds for mating, related to the ability to catch larger fish in deeper waters, and/or evolved for louder intimidation sounds. But perhaps this might also involve the slamming of heads? The spermaceti organ, the hard skin at the front of the head, and the fused vertebrae seem physiologically indicative of this kind of aggressive behavior. Certainly other animals put their vital organs, even their lives, at risk for competition within their species. Male elephant seals will tear at each other's heads and necks to establish territory. Cocks fight. Rams ram. Competition in a huge variety of mammals can and does sometimes result in injury or death. And head-butting is common in a range of other cetaceans.[27]

Then in 2016, Carrier and another set of colleagues, led by Olga Panagiotopoulou in Australia, tried to model how and whether the sperm whale might sustain this sort of cranial impact. They consulted with a bioengineer in Japan, and found, just as Ishmael claimed in "The Battering Ram," that the fluid and physiological structures within the head act as a shock absorber in the way a boat fender does between boats and the dock. They examined in particular the lower oil sack, the junk, which is far more suited to impact, since it has evolved transverse tissue partitions. (See earlier fig. 34.) The junk's potential evolution for the purpose of battle is supported by the fact that scratch patterns on the heads of sperm whales tend to be more often on the skin around the junk, not on the top or sides of the heads where the spermaceti sack is less protected.[28]

A fascinating support of this battering-ram-by-junk theory occurred as early as 1845, if we are to read this logbook keeper's words literally. A sperm whale rushed at the hull of the whaleship *Joseph Maxwell* after it had been harpooned. The whale "hove his junk out struck the ship on the starboard bow." The whale was spouting blood afterward and apparently escaped, having done no damage to the ship.[29]

SPERM WHALE SUFFERING AND
"GIVING THY VOICE TO THE DUMB"

Ishmael's depiction of the White Whale's aggressive behavior is instructive, accurate, documented, and possible for an individual male sperm whale. What is perhaps still more surprising are the subtle methods with which Ishmael builds sympathy for the sperm whale, as a species, as he leads up to the closing scenes. In addition to the potential vulnerability to extinction in "Does the Whale Diminish?," it's significant for the 1800s that individual sperm whales suffer in the novel. Ishmael says that sperm whales have formed large schools for safety in response to human hunting. In the novel, true to life, sperm whales regularly live in social families that evoke our own. At one point in "The Dying Whale," Ishmael even suggests a religion of sun worship as they die—a behavior observed by some of the whaleman of Melville's time. In these ways, Ishmael blurs the line between human and nonhuman. He crafts a story in which the depiction of whales is far more complex and nuanced than simply one madman hunting a single otherly, lesser creature. Especially for most twenty-first-century Anglophone readers, who have been taught from the earliest age to see majesty and intelligence in whales, by the time we get to the final chase and see the White Whale at last, the sperm whale as a species is very much not a malignant monster, but a social, awe-inspiring animal with human-like emotions.[30]

When the *Pequod* begins killing whales in the Indian Ocean, Ishmael toys with the reader's emotional response to the hunt. After Ishmael muses on the size of the brain and the implied wisdom of the sperm whale, Ishmael then shifts to tell a story in "The Pequod meets the Virgin" in which the men harpoon and torture an old, sick whale. The large bull is laboriously lagging behind a school of whales, breathing poorly with a stump of a fin and covered with "unusual yellowish incrustations." Ishmael chooses to explain the following even *before* the animal is harpooned, speaking for the whale's suffering because the animal cannot:

> So have I seen a bird with a clipped wing, making affrighted broken circles in the air, vainly striving to escape the piratical hawks. But the bird has a voice, and with plaintive cries will make known her fear; but the fear of this vast dumb brute of the sea, was chained up and enchanted in him; he had no voice, save that choking respiration through his spiracle, and this made the sight of him unspeakably pitiable; while still, in his amazing bulk, portcullis jaw, and omnipotent tail, there was enough to appal the stoutest man who so pitied.[31]

Ishmael pulls back at the end of this passage just short of complete sympathy: the bulk of the beast was, however, enough of a reason to withdraw any protection. The men of the *Pequod* catch up to the whale and all three boats harpoon the animal, who dives, and lays prone and suspended under the water, held "up" by the lines. For the first and only moment in the novel, Ishmael hazards to tuck the reader's imagination inside the head of the whale for a brief moment, using the same language, including the word "phantom," that he has previously had humans use to describe the whales: "Who can tell how appalling to the wounded whale must have been such huge phantoms flitting over his head!" The old whale surfaces. He bleeds profusely as they stab him over and again with lances. The men now see that the animal is blind from growths over his eyes. Yet even then, Ishmael makes sure the pious reader does not dare to hypocritically judge the whale hunters without judging him or herself for creating the economic demand in the first place: "[The old whale] must die the death and be murdered, in order to light the gay bridals and other merry-makings of men, and also to illuminate the solemn churches that preach unconditional inoffensiveness by all to all."[32]

Despite Starbuck's "humane" order to stop, Flask stabs the dying whale in his infected sore, which sends the whale reeling, spraying blood, gore, and presumably pus on all the men's faces and clothes. In a final flurry, the old whale capsizes the boat, wrecks its bow, and then finally dies. When they bring the dead whale alongside the ship they discover the old stone harpoon in his head, elevating the whale still further with his venerable age and the distance he had traveled during his life. To make the "murder" still more shameful, the corpse then sinks before the *Pequod* is even able to get any economic gain from its death: the entire enterprise was a cruel waste. Ishmael doesn't let the reader feel sympathy for the whale too long, though, because in the very next chapter he delivers "The Honor and Glory of Whaling."

Other writers and whalemen in Melville's time had also considered the suffering of whales and even the morality of the entire practice. Reverend Henry Cheever's *The Whale and His Captors*, published only months earlier than *Moby-Dick*, has moments of sympathy and advocacy for the whales themselves. Scholar Mark Bousquet argued that Cheever's is perhaps the first major work that revealed a trickle toward the future reverence most Americans now feel for whales. Cheever was a missionary, a man of the cloth not making his living off the animals. Yet even his sympathies were temporary, favoring the men by the end of his story. In addition, a few working whalemen in their journals and later published manuscripts also expressed some misgivings about taking the lives of these animals. For example, Enoch Cloud, a whaleman on a voyage from 1851 to 1855, wrote of the sorrow of taking of one of God's creatures as it was "bleeding, quivering, dying a victim to the cunning of man."[33]

As Ishmael alludes in "The Whale as a Dish," Anglophone movements on both sides of the Atlantic, led by those in England, had begun to consider animal welfare and the potential for nonhuman suffering and intelligence. In 1749 some Englishmen complained of animal sports such as cockfighting. In 1776 Reverend Humphry Primatt published a book *The Duty of Mercy and Sin of Cruelty to Brute Animals*, which leads with Scripture, "Open thy mouth for the dumb." Primatt wrote that animals feel pain. A decade later, more famously, Jeremy Bentham advanced the argument by proposing: "The question is not, Can they *reason*? nor, Can they *talk*? but, Can they *suffer*?" In France, as early as 1804, Bernard Germaine de Lacépède, who Ishmael declares to be "a great naturalist" but whose illustrations of whales he believed to be awful, wrote a lament about the human hunting of whales that could easily be written today: "They flee before him, but it is no use; man's resourcefulness transports him to the ends of the earth. Death is their only refuge now." In 1822 Richard Martin, known at the time as (true story) "Humanity Dick," pushed through a law in England protecting working animals such as horse and sheep from cruel treatment. The Society for the Prevention of Cruelty to Animals followed in England in 1824. Students at Cambridge in 1829 debated whether Coleridge's "Rime of the Ancient Mariner" might sway public opinion on the treatment of animals. The United States lagged behind by a few decades, but legislation was led by Melville's two home states: New York was the first in 1824 to enact laws against cruelty toward domestic animals, followed by Massachusetts in 1835.[34]

Before the *Pequod* sets sail in *Moby-Dick*, even before the reader has seen a whale death as graphic as the old, sick, injured whale, Ishmael digs at the hypocrisy of the Nantucket Quaker merchants. Ishmael compares violence against whales with violence against our fellow humans. He says of Captain Bildad: "Though refusing, from conscientious scruples, to bear arms against land invaders, yet himself had illimitably invaded the Atlantic and Pacific; and though a sworn foe to human bloodshed, yet had he in his straight-bodied coat, spilled tuns upon tuns of leviathan gore."[35]

HUMAN-LIKE WHALE FAMILIES

In "The Grand Armada" and then "Does the Whale Diminish?," Ishmael anthropomorphizes the whales by suggesting they had begun to aggregate in much larger groups, "immense caravans" of "what sometimes seems thousands on thousands," because the animals are "influenced by some views to safety," and "it would almost seem as if numerous nations of them had sworn solemn league and covenant for mutual assistance and protection." As discussed earlier regarding sperm whale intelligence, does Ishmael imply here, want his reader to embrace, a higher cultural intelligence in whales with this global alteration of behavior? Regardless, it certainly cultivates a sympathy in that they are on the run and victims.[36]

In "The Grand Armada," just before the *Pequod* enters the Pacific Ocean, the crew rows into a vast herd of whales in which they look down and see mothers, calves nursing, and sperm whales making love. Queequeg and Starbuck do not kill any calves, for fear of upsetting the school of whales and jeopardizing the safety of the men in the boats, but naturalists and whaleman at the time had debated how much human-like loyalty the mother sperm whales actually had for their killed calves. (See fig. 54.) It was a common practice to kill a right or sperm whale's calf first in order to ensure the mother would stay close and be easier to kill next. But it is surely not without meaning that Melville has the hunting that does take place in "The Grand Armada" end with several whales killed wastefully, without any whales brought alongside for oil. Consider the effect of this scene on the twenty-first-century reader, how it is interpreted now, or this beside the story in the summer of 2018 of a mother orca in the Pacific Northwest who pushed around her dead calf for a thousand miles and at least seventeen days, the first calf that had been born to their endangered population in three years. Was that mourning?[37]

FIG. 54. Painting of a sperm whale holding its calf after it was harpooned, created by an unknown artist (c. 1830s). Modern marine biologists have observed both female and male sperm whales "gently" holding calves in their mouths.

MOBY-DICK SHOWED A WIDER RANGE
OF HUMAN PERCEPTIONS OF WHALES

No author in the nineteenth century, or arguably ever, came even close to the level of nuance and complexity regarding the human perception and relationship with whales, perhaps with any marine animals, as did Melville in *Moby-Dick*. Honestly, I don't know of a single work of fiction or nonfiction that does so much with ocean organisms until perhaps John Steinbeck and Ed Rickett's *The Log from the Sea of Cortez* (1941) and Hemingway's *The Old Man and the Sea*—and then maybe Farley Mowat's depiction of a fin whale in *A Whale for the Killing* in 1972. Rachel Carson's *Under the Sea-Wind* (1941) was revolutionary in its scientifically tempered consideration of the lives of marine animals, but she chose to barely engage with human perceptions of her fish and seabirds. Similar to how he shows all the men of the *Pequod* considering the doubloon, Melville in *Moby-Dick*, often presciently, shows seemingly every possible way one might consider a whale from a human point of view. Ishmael's multi-nuanced perspective, Elizabeth Schultz wrote, "moves us toward the elimination of nature as abstract other—whether as divinity, as voraciousness, or as endlessly exploitable object." Ishmael combines the rational, objective, scientific, honest Darwinian perspective on the sperm whale, while also celebrating and exploring the whale through the emotional, subjective, poetic wonder of the Emersonian perspective. The narrator splices these together, Schultz argued, to form a third way of seeing this species: by the end of the novel, Ismael has expanded his view to perceive all whales and all of the natural world with a brotherly, proto-ecological, proto-environmentalist eye for interdependency that was far ahead of his time.[38]

Again, Ishmael was not the first to be touched by the sperm whale's wasteful suffering at the hands of man, nor even the first to find parallels to humans in the sight of a whale mother and calf under the water. Where Melville was especially ahead of his time was in his eagerness to reveal all facets of humanity's relationship with the whale, to describe a whale in such depth as an individual, to connect and equalize humans among nonhumans, and, perhaps most significantly, in even having man *fail* before the nonhuman animal, both as an individual and as a species, both physically and morally. Remember, Ishmael is a survivor of an event in which a sperm whale sank the ship that drowned all of his shipmates. Yet he does not rage like Ahab. All of the characters in his story, aside from perhaps sharkish Queequeg, are less than the whale. They are weaker, less just, more hypocritical, more frail and more flawed and irrational than the animals they hunt. In "The Grand Armada" Ishmael compares the gallied whales to herding, sheeplike humans that when alarmed in a crowded theater panic and trample each other to death: "For there is no folly of the beasts of the earth which is not infinitely outdone by the madness of men."[39]

Roger Payne, the marine biologist who revolutionized whale conservation with his recording of whale song in the 1970s, wrote in 2018: "I feel that Melville came closer than anyone has ever come to demonstrating that whales can help humanity save itself—help us make the transition from *Save the Whales* to *Saved by the Whales*."[40]

Back in 1851, Melville's older brother Allan wrote to the publisher about Melville's shift of the title from just *The Whale* to *Moby-Dick; or, The Whale*. He thought the expanded title appropriate, "being the name given to a particular whale who if I may so express myself is the hero of the volume."[41]

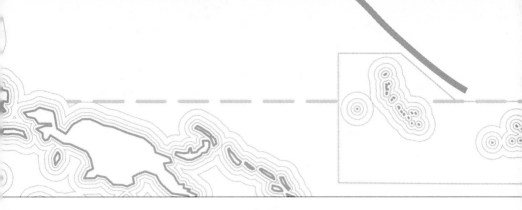

Ch. 30

SKY-HAWK

So the bird of heaven ... went down with his ship.
Ishmael, "The Chase—Third Day"

At the end of the final chapter of *Moby-Dick*, "The Chase—Third Day,"
the White Whale strikes the hull of the *Pequod* with his "solid" forehead. Clinging to the main topmast, Tashtego, the lone Native American on a sinking ship named after a tribe of Native Americans that Ishmael thought to be extinct, nails a red flag to the masthead. The ship
is sinking. From the highest point of the ship, Tashtego hammers until
the water rises to his chest. Ahab had ordered Tashtego aloft there to
replace a long thin flag, a wind vane for the helmsman, which was also
used as a signal to their men in the boats and other ships. In addition,
hammering a ship's flag to the mast was a symbol in the navy that a ship
would not surrender.[1]

As the water rises over Tashtego's head, Ishmael says:

A sky-hawk that tauntingly had followed the main-truck downwards
from its natural home among the stars, pecking at the flag, and in-
commoding Tashtego there; this bird now chanced to intercept its broad
fluttering wing between the hammer and the wood; and simultaneously
feeling that etherial thrill, the submerged savage beneath, in his death-

grasp, kept his hammer frozen there; and so the bird of heaven, with archangelic shrieks, and his imperial beak thrust upwards, and his whole captive form folded in the flag of Ahab, went down with his ship, which, like Satan, would not sink to hell till she had dragged a living part of heaven along with her, and helmeted herself with it.[2]

Melville chose not to drown a majestic white albatross as his bird of heaven. He instead wrote of a sharp-winged, predatory seabird—an ideal symbol for Ahab's dark rage against nature and God. Ahab evokes the banished archangel Satan as he shakes his fist against the sun and hunts God's finest white specimen of the largest and deepest-diving of predatory mammals. It was the same type of "black hawk," perhaps even the same "red-billed savage," that had snatched the captain's cap in the approach to the meeting of the White Whale, when Ahab is first hoisted aloft.

Melville's sky-hawk is a frigatebird (*Fregata* spp.). Dr. Bennett penned a full section on "The Frigate-bird" in his *Narrative of a Whaling Voyage*, in which he explained that this bird is also called the "Seahawk," which Ishmael does, too, in the "Epilogue" and "The Hat."[3]

Ornithologists today recognize five separate species of frigatebird, which all look similar yet each have subtle plumage differences in various stages of development, regionally within species, and between males and females. Bennett wrote that frigatebirds are "the most remarkable sea-birds frequenting intertropical regions," with an average six-and-a-half-foot wingspan. The wingspan of the largest of these, the magnificent frigatebird (*Fregata magnificens*), common to coastal South America and the Galápagos, can be as long as eight feet. (See plate 12.) Frigatebirds have the largest wing area to body weight ratio of any bird, at sea or on land, even more than albatrosses. They soar effortlessly on trade winds that take them higher than local thermal breezes. With sharp wings and a scissoring tail, a frigatebird can drop quickly like a hawk and then change direction like a swallow. John James Audubon, who knew them as "Frigate Pelicans," wrote in the 1830s that they have "the power of flight which I conceive superior to that of perhaps any other bird." Audubon described them diving for fish "from on high with the velocity of a meteor."[4] (See fig. 55.)

Frigatebirds earned their name, along with "Man-of-war birds" or "Sky-hawks," because of a practice now called *kleptoparasitism*, a behavior of attacking and scaring other species to give up what they've just caught. Frigatebirds cannot dive under the water. They can barely walk on land and rarely even rest on the surface of the ocean. But they are exceptionally skilled at swooping down and plucking a flying fish out of the air or scaring a tern out of a fish which that bird had just caught. Ishmael alludes to this in "The Pequod meets the Virgin" with the screeching bird trying to escape the "piratical hawks."

This final scene with Tashtego matches Dr. Bennett's own descriptions of the frigatebirds. Bennett wrote that this bird soars higher than any other, until it is "a mere speck in the sky." He told a sailor's yarn about an exceptionally tall landsman who while standing aloft grabbed one of these birds with his hands. "They apparently take a delight," Bennett wrote, "in soaring over the mast-head of a ship, from which they usually tear away the pieces of coloured cloth fixed in the vane."[5]

Melville did not need Bennett, or even Audubon, to be inspired by frigatebirds, however. He had his own experiences. Magnificent frigatebirds were common throughout the southeastern Pacific. Lesser (*F. ariel*, likely named for the character in *The Tempest*) and great frigatebirds (*F. minor*) are found throughout the islands of the South Pacific, where before human settlement, fisheries depletions, and the introduction of rats, cats, and pigs to their island rookeries, seabirds were more prolific than they are today. A couple years after *Moby-Dick*, in "The Encantadas," Melville wrote of the far-ranging "man-of-war hawk" soaring near the Galápagos.[6]

He wasn't always accurate here, though. Despite Ishmael's poetic desire for the blood-red sword imagery, the frigatebird's long thin beak is actually gray, if at most occasionally pinkish. The frigatebird does have a sharp hook at the end of its beak to grasp fish, and during the mating season adult males of all five species have a bright red gular pouch beneath the beak, which puffs out like a balloon as a mating display. This

PLATE CCLXXI

FIG 55. John James Audubon's *Frigate Pelican* in *Birds of America* (1827–1838).

gular pouch is perhaps what Melville confuses or misremembers as its red beak. To be fair, however, even his *Penny Cyclopædia* described the frigatebird's bill as red.[7]

Regardless, frigatebirds had been on Melville's poetic palate since the opening scenes of his very first novel *Typee*:

> As we drew nearer the land, I hailed with delight the appearance of innumerable sea-fowl. Screaming and whirling in spiral tracks, they would accompany the vessel, and at times alight on our yards and stays. That piratical-looking fellow, appropriately named the man-of-war's-hawk, with his blood-red bill and raven plumage, would come sweeping round us in gradually diminishing circles, till you could distinctly mark the strange flashings of his eye; and then, as if satisfied with his observation, would sail up into the air and disappear from the view.[8]

So even in his first book, Melville was thinking about the black criminal frigatebird violently ousting the white harmless albatross for poetic immortality. His sky-hawk is crucified and drowned in *Moby-Dick* with the Native American of the *Pequod*.

Today, a black seabird, grounded, soaked, instead evokes another type of wreck: the oil spill.

Kiribati

Ch. 31

ISHMAEL

Blue Environmentalist and Climate Refugee

> Now small fowls flew screaming over the yet yawning gulf; a sullen white surf
> beat against its steep sides; then all collapsed, and the great shroud of the sea
> rolled on as it rolled five thousand years ago.
> Ishmael, "The Chase—Third Day"

As part of a crew of twenty-four students, a captain, a chief scientist, and ten professional mariners and assistant scientists, I sailed aboard the SSV *Robert C. Seamans* one summer, twenty-five years after my first voyage around the Pacific aboard the *Concordia*. This time I regularly stood aloft looking for whales in the equatorial Pacific, where *Moby-Dick* ended.

Our voyage this time was a round-trip out of American Samoa, an island that was once a provisioning stop for American whalemen and is now a harbor dominated by the StarKist tuna cannery. We traveled from there up to one of the equatorial archipelagoes of the vast island nation of Kiribati, within a day from the navel of the planet where the dateline and equator intersect. Our primary mission, as we taught university students about the ocean, was to traverse and survey the Phoenix Islands Protected Area. This is one of the largest marine reserves on

Earth, spanning 119,000 square nautical miles, nearly the area of the entire state of California.

Sailors have always thought of home while out at sea, but I found myself disappointed during this particular voyage in my inability to fully separate myself from life on land, to fully slow down my mind and forget the "carking cares of Earth." Some of that is simply the reality of our world today. The *Robert C. Seamans* has an engine and modern technologies, which means the program can have a tighter schedule in order to collect more data, faster. Even on a square-rigged sailing ship in the middle of the Pacific, we remained in limited electronic contact with interests ashore.

From my perch aloft, about one hundred feet above the surface, I could theoretically see whales over eleven miles away on a clear day. I calculated this distance from a formula in the ship's copy of *Bowditch*. Ishmael muses in "The Mast-Head" about how easily one could step off the crosstrees and fall to one's death, but I was clipped in with two separate carabiners tethered to a four-point harness. I did notice, though, that during my shifts aloft the ship's rig began to disappear. Standing on those spreaders and looking forward, even with the wire of the topsail lift angled in my view or the tip of the foremast just above, I felt as if I were standing on the ocean itself. The sea, especially during a calm day with few whitecaps, appeared simply, mundanely, like a flecked carpet. On days of low or moderate breezes, the ocean's color from aloft appeared nearly uniform, uninterrupted. I was so high that depth perception was blurred. It felt as if I could step right onto this blue mat and in a few long strides lean over to the edge of the horizon, peel it up, and sweep the clouds back underneath. In other words, standing up at the masthead for hours at a time, I loosened my grasp on scale.

While I looked for whales, I sifted what I'd learned over the course of this natural history of *Moby-Dick*. The novel offers a benchmark for how Americans understood the ocean in the mid-nineteenth century. Melville wrote *Moby-Dick* out of his vast experience at sea, as well as from his reading of popular science books, encyclopedias, narratives of sailors and naturalists, and even the scientific papers of his time. His narrator Ishmael cares about scientific thought and inquiry, values it, even makes an occasional earnest correction or adjustment to the knowledge of his day, but he also questions when this inquiry becomes too distant from the observable, emotional, or spiritual world of humans and God. When in doubt, he preferenced the viewpoint of whalemen who had seen the wonders of the ocean so intimately. Though *Moby-Dick* is a gloriously messy book filled with contradictions, mistakes, and digressions, Melville was surprisingly careful with his marine biology, oceanography, geology, meteorology, and navigation to what seems to be the best of his ability as a self-taught naturalist. He crafted a presciently varied viewpoint of marine life, which included an awareness of human impact on ocean animals, both as species and as individuals—including the qualification that we shouldn't be too hypocritical in how we judge the people out at sea killing in order to satisfy human demands for food and products. In many ways, in our twenty-first century, we may justly read *Moby-Dick* with a three-faceted environmental moral: a story that is proto-Darwinian, proto-environmentalist, and one in which Ishmael serves us now as the symbol of a climate refugee.

MOBY-DICK AS A PROTO-DARWINIAN STORY

At least twice a day during our voyage toward the equator that summer, we conducted a range of biological and oceanographic sampling as part of the ongoing monitoring of the Phoenix Islands Protected Area. As required for the safety of the ship, we posted lookouts on deck twenty-four hours a day. From sunrise to sunset, an additional student stood lookout for fifteen minutes of each hour to record seabirds or other visible fauna. Every day that we were underway, usually midmorning and usually at some peril to my breakfast, I went aloft. Each day I stayed up there for two-hour shifts, which, by the time we returned to the dock in American Samoa, totaled forty-eight hours in the rig searching for whales. When not aloft I searched the horizon for marine life, and, Ahab-like, I ate all my meals on deck. In other words, if anything flew, floated, or dove within several miles of our vessel during daylight hours, someone aboard was going to see it. If it was at night—there was a likely chance someone would hear it.

Alas, though, we had only a few sightings of whales. The voyage had started off fortuitously. Shortly after leaving American Samoa, we saw a few dolphins and then a humpback whale. But for the rest of the voyage we saw only one minke whale and a few other shoals of dolphins, perhaps bottlenose dolphins and long-nosed spinner dolphins (*Stenella longirostris*). When we returned a month later, we saw one last puff, likely a humpback, off American Samoa.

I had wanted to see sperm whales in particular, of course, but recent surveys suggested that we should not have expected to see any large whales at all. The reserve has dozens of steep seamounts and trenches that presumably support enormous fish and squid populations, just like that of Kaikōura Canyon, New Zealand, but while we were drinking kava under a small thatched meeting hut on Kanton Island, the current reserve manager, Tuake Teema, told me that he did not think many whales transited through this region. Five separate surveys within the Phoenix Islands conducted by the New England Aquarium between 2006 and 2013 observed only one whale, which they believed to be a beaked whale.[1]

Yet I had gone into the trip with some reason to hope to see sperm whales: during a similar voyage of the *Robert C. Seamans* two years earlier, the crew saw a unit of some thirty to forty sperm whales, which included females with at least two calves.[2] Historical accounts also recorded sperm whales in this area. Though the *Charles W. Morgan* cruised the equator around here in the early 1850s without much luck, Maury and Wilkes both mapped this area with high sperm whale abundance. Tim Smith and his colleagues of the World Whaling History Project, in their map that compiles whaling activity from the late 1700s to the 1920s, revealed concentrations of sperm whales at the north end of the reserve in our month of July. This was also one of the regions where Captain Lawrence aboard the *Commodore Morris* drew his clumps of whale flukes. When he sketched his cruise track on that Norie chart that's held at Mystic Seaport, he drew his ship's track back and forth across the exact waters in which we sailed. Lawrence's crew caught several sperm whales among the "shoals" of them. He sighted dozens if not hundreds more than he killed. Yet despite this historic precedence, we still saw no whales beside the one minke in this equatorial ocean.[3]

It's an important challenge for modern marine biologists, oceanographers, and environmental historians to determine with confidence whether the nineteenth-century sea, Ahab's sea, was indeed more prolific, more "lively," than it is today—or even as compared to "Noah's time." Professional research reveals more and more that the answer is *yes*: the global ocean in the mid-nineteenth century simply had more and larger life in it. This is based on mariner's journals and published narratives by sailors and naturalists and passengers, as well as anthropological evidence and foundational recent studies in genetics and modeling. In the 1790s Captain James Colnett regularly wrote of abundant fish, birds, and dolphins in the eastern Pacific. He wrote off Mocha Island of "the sea being then covered" with sperm whales. A half-century later Dr. Frederick Bennett wrote of the vast single schools of sperm whales in the Pacific, numbers of sperm whales "beyond all reasonable conception." Recent scholars have applied the concept of shifting baselines to the ocean environment to reevaluate what we might consider normal to see at sea. We should not just brush off as exaggerations or propaganda the reports by early explorers and whalemen who reported voluminous schools of large fish, thousands of enormous sharks, larger and longer marine mammals, and cloud-like colonies of seabirds by comparing these reports to what seems credible to us today, especially to those of us who don't actually spend much time on the water. Douglas McCauley and his colleagues have found that globally since the 1970s, marine vertebrates in the ocean have declined on average by over twenty percent. Fish in particular have declined by nearly forty percent and some of the baleen whales by between eighty and ninety percent. Seabirds as an ecological group are the most endangered of all birds on Earth.[4]

I'm not trying to suggest that the fact we didn't see whales in the Phoenix Islands region of the equatorial Pacific during July of one recent year is any indicator that the sperm whale will perish as a species. Nor does the gradual decline of sperm whales off Kaikōura in recent years necessarily signal problems for sperm whales worldwide. If our ship had traveled just another day north to straddle the equator, our odds would have increased tremendously based on primary productivity and historical data. We also sailed during a neutral year between El Niño and La Niña phases, in which the waters were a bit warmer than normal for July. The plankton diversity and concentrations were light compared to other years when the ship conducted the exact same cruise track, suggesting we sailed in a less productive moment for squid and fish and thus for sperm whales.[5]

In our lifetime and our children's, sperm whales will almost certainly not be significantly diminished by hunting. There is too little demand for their meat and oil. *Moby-Dick* and other works of art have greatly helped raise the status of sperm whales all over the world. The majority of humankind has elevated the sperm whale, along with all whales, to the point where they are icons of conservation. Only the governments of Norway, Japan, and Iceland sanction any killing of large whales for food for humans and farmed animals. The United States, Russia, Canada, and Greenland (Denmark) also allow a small number of cetaceans to be killed by their indigenous Arctic populations. Japan is the only country that has declared rights to a quota of sperm whales as part of a hunt they say is for scientific purposes. Though they have reported killing other species of whales, Japanese whalemen have not reported killing a sperm whale since 2013. Unlike right whales and other baleen whales, the habits of sperm whales tend to keep them out of the path of ship strikes and only occasionally in conflict with fishing gear. Among the other large whales, the southern right whales and minkes, for example, seem to have been steadily recovering since the late 1980s after the international moratorium on industrial-scale hunting. But the North Atlantic right whales and North Pacific right whales remain in grave danger, on the precipice of extinction, with each hovering around only five hundred individuals remaining on Earth.[6]

I do not want to suggest that sperm whales as a species are safe, either.

Sperm whale populations seem to have been recovering slowly after the moratorium — but in truth it's hard to know. For example, sperm whales in the Caribbean, the clans that are perhaps the best studied, do not seem to be recovering as well as was expected. *Physeter macrocephalus* might be eradicated entirely by the same slow motion anthropogenic vectors that could kill us all off. These are the insidious threats: the harpoon lines we hurl at the ocean ecosystem that were beyond Melville's imagination in 1851 and even beyond my own imagination in 1993 when I first climbed aloft in the rigging of the *Concordia*. The prevalence and increase of manmade ocean noise — naval exercises, seismic survey blasts, wind turbines, and the noise of engines across oceans — affect sperm whale behavior and perhaps have even caused the death of individuals and caused mass strandings. Anthropogenic ocean noise is perhaps even more hazardous to smaller, more coastal toothed cetaceans, such as dolphins and beaked whales. More significantly, we now understand that the Ahabian bights of rope that we regularly string around our necks could harm the large whales, too: our addiction to fossil fuels, the prevalence of plastics in the ocean, the seeping of industrial pollution into seaways, and the man-made shifts in ocean chemistry that might begin to crash primary and secondary productivity up and down pelagic and coastal food webs.[7] (See fig. 56.)

More and more, we are learning how essential the large whales are to the ocean environment, as consumers and predators, as movers and distributors of nutrients, and even how much the whales provide in their death to support micro-ecosystems and carrion habitat. Sperm whales with no human predators and no real ocean-dwelling competition other than an occasional killer whale, were limited only by their own species and cephalopod and fish abundance. Today, their primary food of deep-sea squid has little global market for humans, yet Whitehead believes that sperm whales as a population, even at their reduced twenty-first-century populations, probably eat in one year roughly about the same biomass as *all* of the world fisheries conducted by humans.[8]

In *Moby-Dick*, on the third and final day of the chase, after Ahab stands aloft and takes in his final vision of the sea — the "old, old sight"

FIG 56. Graffiti at Skegness, England, on one of a pair of a total of seventeen sperm whales that beached on the coasts of the North Sea in 2016. Scientists remain uncertain of how common mass strandings were before the Anthropocene.

that is the same to Noah as it is to him—the captain sinks another harpoon into the White Whale. Moby Dick rolls and nearly capsizes the small craft. The animal snaps the line and in his pain rushes not back toward Ahab's boat but instead turns to charge at the nearby *Pequod* itself. Ishmael declares: "Retribution, swift vengeance, eternal malice were in his whole aspect." The sperm whale with his "predestinating head" bashes into the broad starboard bow of the whaleship. Seawater rushes in. The White Whale surfaces back near Ahab's boat. Ahab stabs the animal once again, a third time, but ends up killing himself with the line of his own harpoon:

> The harpoon was darted; the stricken whale flew forward; with igniting velocity the line ran through the groove;—ran foul. Ahab stooped to clear it; he did clear it; but the flying turn caught him round the neck, and voicelessly as Turkish mutes bowstring their victim, he was shot out of the boat, ere the crew knew he was gone.[9]

(This was a common accident. For example, during the first lowering for a whale at the start of the 1859 voyage of the *Charles W. Morgan*, a man named Francis Leacock was dragged out of a whaleboat by the line and drowned.[10])

After the White Whale smashes into the *Pequod*, the ship sinks quickly. All three of the harpooners stand atop each masthead. Tashtego and the frigatebird, flailing at the tallest main mast, are the final living things dragged down with the ship. Ahab, the demagogue, has driven a floating nation of men of all races and walks of life to their death so that he could try to fulfill a personal and existential vendetta. He kills himself and causes the death of everyone else. That is, except for Ishmael, who, flung out of Ahab's boat and beyond the sinking *Pequod*, now watches in horror from afar as he clings to Queequeg's coffin. Ishmael is now a Pip-like castaway.[11]

While I stood aloft aboard the *Robert C. Seamans*, I often thought about the closing drama and what a floating Pip or Ishmael might look like when first coming into view on the horizon. One afternoon only a few days north of Samoa, we conducted our first man-overboard drill. We recovered a fender that Captain Chris Nolan threw over the side. Nolan, a former officer of the US Coast Guard, explained to us just how quickly our little coconut heads would disappear in the waves, particularly if any kind of sea was running. For part of his career, Nolan had coordinated search and rescue efforts in a sector of ocean covering more than twelve million square miles, a gargantuan triangle between Samoa, Guam, and Hawaii. The Coast Guard uses computer models with which they punch in local conditions to predict the drifting of a floating object—or human—on any given day and place.[12]

"Survivability is governed mostly by water temperature," Nolan told us. "In Alaska, the North Pacific, the North Atlantic, you lose dexterity in about fifteen minutes."

I thought about Melville on his way to London in 1849, watching that man let go of the rope and drift astern—with a grin.

"In really cold water you're dead in pretty much an hour," Nolan said. "When we were searching for people, if we were confident that they were not in a boat or in a survival suit, the search didn't tend to last too, too long. But in warmer waters, you can sometimes survive up to forty-eight hours, or even longer. Even in ninety-degree waters, like right around here, you're still losing heat and you'll get hypothermia. I've certainly heard of people surviving for days—the *will to live* is big in that. But if you're in the water, you are going to die."

After the drill, I asked Nolan about sharks—whether that was something for which the Coast Guard plans.

"No. There's no *biological* discussions in that training," he said. "I think the sharks are more about curiosity. They are scavengers around bodies. I've never heard of any rescue of someone alive in the water where the rescuers had to fend off sharks."

Our narrator of *Moby-Dick* is not the only survivor of the wreck: Ish *and* fish survive. Melville wrote nothing to suggest the White Whale dies, despite at least five new harpoons in his body from the previous three days—and whatever physical trauma he sustained from thumping his spermaceti melon against the boats and the oak hull of the *Pequod*. In "The Fossil Whale" Ishmael had envisioned that whales will live on "after all humane ages are over."[13]

It's worth reemphasizing that it is not the White Whale that kills Ahab. The last words we hear Starbuck say to his captain are: "See! Moby Dick seeks thee not. It is thou, thou, that madly seekest him!" Then Ahab hangs himself with his own rope in the act of attacking the animal.[14]

Thus one of the potential readings, one of the morals to be derived today from *Moby-Dick* as advanced by scholars such as Bert Bender, Eric Wilson, and Dean Flower is that the story prefigures *On the Origin of Species*—or at least anticipates a modern ecological sensibility. Melville demonstrated two major drivers of evolution by natural selection that Darwin was still percolating in 1851: fitness and chance. Moby Dick, the sperm whale, uses his strength and cunning to survive to swim away and potentially pass on his genes, while Ishmael, the human, by dumb luck beyond his control, drifts alone and survives to one day potentially pass on his genes. (Darwin or Melville didn't mention *genes*, of course:

no one knew the mechanism of inheritance at the time, even as Mendel was beginning to putter away in his monastery garden in the 1850s.)

"The unharming sharks," Ishmael says, "they glided by as if with padlocks on their mouths; the savage sea-hawks sailed with sheathed beaks." The next day, after nearly twenty-four hours in the water, he is rescued by the whalemen of the *Rachel*, who are looking for the captain's son but instead find Ishmael.[15]

Melville wrote in his journal when sailing across the Atlantic before composing *Moby-Dick* about that new friend with Coleridgean sensibilities—a man who accepted the Scripture and the Divine, but was willing to consider scientific advances. So it goes with Ishmael, who in retelling his story, opens himself up to the transmutation of species, to the interconnectedness and irrationality and unimportance of all species, and to the powerful, cannibalistic, seeming immortality of the far larger, longer, deeper, and older ocean. *Moby-Dick* is a novel after the eco-philosopher's heart: a proto-Darwinian fable.

ISHMAEL AS A PROTO-ENVIRONMENTALIST

The steel-hulled two-masted *Robert C. Seamans* is equipped with a diesel engine, single propeller, radar, GPS, depthsounders, diesel generators, refrigerators, freezers, and two reverse osmosis units to convert seawater to potable water. The ship has scientific equipment that is equivalent, if at a smaller scale, to the most advanced oceanographic research vessels operating today. Although we never sent it down this far, the *Seamans* can lower nearly 9,000 feet of wire. We daily attached to this wire a carousel of pressure-activated bottles to capture water samples from different depths. We had computerized sonar equipment, the CHIRP, which pings from the ship's steel hull what sounds remarkably similar to the clicking of a sperm whale. This equipment recorded our depth at all times, tracing and mapping the profile of Ishmael's "undiscoverable bottom." Another suite of equipment, the Acoustic Doppler Current Profiler or ADCP, constantly recorded surface and subsurface currents—and this, too, uses acoustic signals to collect its information. We replicated the sampling sites from previous voyages to the Phoenix Islands. The students worked in groups to answer their own specific questions. Three students, for example, under the guidance of Chief Scientist Deb Goodwin, sampled plankton from surface tows and monitored ocean chemistry for pH and aragonite saturation in order to learn more about how global and local ocean acidification is affecting the growth and resilience of coral reefs and calcifying zooplankton, such as pteropods and foraminifera. Ocean acidification also affects the shell-building ability of larger invertebrates, such as oysters, mussels, and even the paper nautiluses, which scientists believe might be exceptionally vulnerable to climate change, even to the point of extinction because of their thin, brittle shell. When the *Commodore Morris*, *Charles W. Morgan*, and the *Acushnet* sailed these waters in the 1840s and 1850s, the ocean seems to have been less acidic than it was when we dragged our plankton nets across the seas around the Phoenix Islands. The burning of fossil fuels and the industrial removal of the rain forests over the past dozen human generations has apparently accelerated a shift in ocean chemistry so quickly and dramatically that paleo-oceanographers believe this current rate of acidification has not been seen on Earth since fifty-six million years ago, conditions unknown to the evolution of most living marine organisms.[16]

We anchored off three island atolls. We snorkeled the seemingly pristine reefs and lagoons, within an area that scientists and managers believe to be one of the few remaining "intact" coral archipelago ecosystems, even possibly the "last coral wilderness on Earth." The Phoenix Islands were originally settled sparsely by nomadic groups of Polynesians and Micronesians who seem to have rarely stayed on these islands for more than a few years due to lack of fresh water. Most of the islands of Kiribati were first definitively recorded and identified for Western mariners' charts by American and English whalemen in the early 1800s, who named collections of islands here as part of the Kings Mill Group and the Gilbert Islands. One island to the south was named Starbuck Island. Our first anchorage, near where we saw a floating large squid at the surface, was at Enderbury Island, named in 1823 for the English whalemen Samuel Enderby, whom Ishmael mentions and who helped sponsor Colnett's voyage. In the lagoon of Kanton, an island named after a New Bedford whaleship that wrecked there in 1854, we snorkeled over enormous corals, Pip's "colossal orbs." In a small boat over these reefs, I thought of the American whalemen rowing over these corals, also looking wide-eyed over the gunwale. In recent decades the reefs of Kiribati have suffered from two massive bleaching events due to unusually warm El Niño events. Scientists have been amazed to see, however, that the corals seem to be showing signs of recovery far faster than expected.[17]

One day at anchor the captain and I were scouting out snorkeling sites for the following morning. As I was swimming back to the boat, Nolan said there were three large whitetip reef sharks (*Triaenodon obesus*) circling around me as I obliviously swam along at the surface. The coastal sharks around these islands are still recovering from a few months of shark-finning by a single shark long-lining boat in 2001 that nearly decimated the entire local population.[18]

Whenever I had free time at anchor, I explored the beaches. I wanted to find a whale skeleton or a lump of ambergris. On Nikumaroro, the atoll on which scholars believe Amelia Earhart and her navigator crash-landed and died as castaways, I turned a corner and saw a beach of plastic bottles. I saw other plastic trash, too: flip-flops, fishing floats, and polypropylene rope. It's hard to express how remote the islands of Kiribati are. Next time you have a globe—your flat phone won't achieve the proper effect—try to find these islands and just look how far away they are from anywhere. On the windward side of Nikumaroro, plastic bottles at various sun-brittled stages will crumble into smaller and smaller pieces and degrade. At sea, the plastics tend to sink eventually and then persist at the sea bottom, we think, for centuries. On that windward stretch of Nikumaroro, plastic bottles sat on the sand on average every two or three paces, more plentiful than coconuts.[19]

We collected a few garbage bags full of plastic trash, but we did not make a dent. Nearly all of this plastic trash must have floated from dis-

FIG. 57. A sperm whale calf playing, perhaps dangerously, with a plastic bucket, as featured in the documentary film series *Blue Planet II* (2017).

cards thousands of miles away. There was just too much plastic on these desolate beaches for it to be otherwise.

About six months earlier and over a thousand miles to the west in a bay of the Philippines, a 38-foot juvenile male sperm whale had washed ashore. Biologists believed the cause of death to be his stomach crammed with plastic products, fish hooks, rope, wood with nails, and steel wire. In the spring of 2018 another male sperm whale washed ashore on the coast of Spain, with a likely cause of death a clog in his stomach and intestines of some sixty-four pounds of garbage, primarily of plastic bags and rope.[20] (See fig. 57.)

Once on another voyage of the *Robert C. Seamans*, I sailed across the Pacific Garbage Patch in the North Pacific to the east of Hawaii. Our plankton nets dragged up microplastics invisible to the eye, while often we watched larger pieces of garbage drift past, including buckets, Styrofoam flats, and children's toys. One day our ship was entirely, dangerously disabled when a clump of stray fishing gear wrapped around our propeller, requiring two of our crew to scuba dive mid-ocean to cut it free.[21]

The pervasiveness of ocean plastic, yet another derivative of petroleum and thus a contributor to excess carbon dioxide in our atmosphere, is altering the global ocean and its coastlines even in ways that are entirely unexpected. After the 2011 earthquake and subsequent tsunami that obliterated the northeast coasts of Japan, killing more than 15,890 people, melting down a nuclear power plant, and at least partially destroying one million buildings, scientists found that marine species that never would have otherwise been able to survive rode on plastic substrate for trips of months and years across the Pacific. Some species survived on floating docks and debris and settled on the coast of North America. The impacts of these invasive species are still playing out. Invasives used to travel among the weedy hulls of whaling ships, as well as in the storage holds down below, and then later in twentieth-century tanks of ballast water, but this is an entirely new and surprising vector.[22]

Herman Melville never put his fingers on anything during his lifetime, from 1819 to 1891, that was not made from something that occurred in nature—iron, clay, wood, minerals, and rock. It wasn't until the early twentieth century that humans invented an entirely synthetic polymer. Though Melville was not what we call an environmentalist in the sense of the word today, he did see habitat degradation, knew of extinctions, witnessed human populations booming around him, and lived in areas where the forests had been stripped bare. Melville watched and read as railroads unrolled across marshes and prairies. He bought a farmhouse and moved up near the Berkshire Mountains in the middle of his writing of *Moby-Dick* surely, at least in part, to get away from the rush of New York City and the expansion of the Industrial Age.

So another moral of *Moby-Dick* for the twenty-first-century reader is the novel's prescience as a blue fable that decenters man, revealing how messing with the forces of the natural ocean world will end poorly for humans. Ahab's hubris seems mostly a personal, fatal madness, but his actions can be a symbol for any broad anti-environmental human drive. As scholar Elizabeth Schultz explained in 2000, the death of Ahab anticipates "the desire, if not the design, of twentieth-century environmental activists for the annihilation of forces antagonistic toward marine conservation." Ahab on his ivory leg serves as a symbol for modern consumerism and the trampling of nature that can never curb its sharkish, sea-hawkish appetite. More specifically, Ahab easily stands in for Big Oil, the ceaseless quest for fossil fuels. *Moby-Dick* as a proto-environmentalist fable came up especially in the wake of the Deepwater Horizon oil spill in the Gulf of Mexico in 2010. This was the largest spill in the history of the United States, caused by technologies digging deeper and farther to extract oil at sea. Even as oil-soaked seabirds dominated the photography of news media and the oil continued to flow into the ocean, the *New York Times* ran an article on Ahab, *Moby-Dick*, and "its themes of hubris, destructiveness and relentless pursuit."[23]

Do you want to know what happens now on these ocean trips when a thoughtful young person climbs aloft on a school ship and inhales what Melville wrote in *Redburn* as "the very breath that the great whales respire"? The student is now, say, a week from land in the middle of the North Atlantic or the South Pacific or anywhere else on the global ocean. She is finally away from the eco-guilt and everything else she's run away from ashore. She looks at the horizon, all of its 360° of boundlessness. The sea to her seems immortal, timeless, the same to her as Noah's, everything she's been taught to imagine. After some peaceful time, she sees a speck off the bow. It floats half-submerged on the surface. At first she thinks it is the slippery back of a whale or the mottled shell of a sea turtle. She wants to shout out to her shipmates. She berates herself for not bringing up her phone to take a photograph. As it approaches or they are sailing toward it—she can't quite tell which—she then identifies that it is in fact but a sun-faded, falsely pink, coffin-like Styrofoam cooler, half-sunk. It is out of the reach of a boat hook. The captain denies her shouted request to launch a small boat to fish it out.

Does this sight shatter everything she thought about the wild and pristine sea beyond the hand of *Homo sapiens*? I'm telling you that it does: it does, it does, it does.

ISHMAEL AS A CLIMATE REFUGEE

As I cited to open this study, Lewis Mumford predicted in 1929 at the start of the twentieth-century revival of Melville's works: "Each age, one may predict, will find its own symbols in *Moby-Dick*. Over that ocean the clouds will pass and change, and the ocean itself will mirror back those changes from its own depths."[24]

Herman Melville died in obscurity. His masterpiece was a flop. Yet now as we celebrate the bicentennial of his birth in 2019, his novel inspires new readers each year with a thriving, continually growing core of professional scholars and recreational Moby-Dickheads. *Moby-Dick* is now in every Great American Novel discussion. All types of readers in dozens of languages have found in the story rebellions and truths about narrative form, about God, about Melville's parents, children, lovers, about Freudian obsessions, about American democracy, about the Civil War, about labor, about fascism, about communism, about racism, about cultural relativism, about gender, about homosexuality, about ecofeminism, about American economic history, and on and on and on. As early as 1950, the phrase "Moby-Dickering"—meaning too much time spent trying to find the meaning in the novel—appeared in the *New York Times*. This project here before you is no better and hopefully no worse. In 1994 literature professor Paul Lauter wrote: "As is perhaps always the case, what the critics of the 1920s made of Melville tells us more about them than about him." Lauter's point extends far beyond the ivory towers. *Moby-Dick* survives, as strong as ever. Aside from the stories of the Bible, and now perhaps *Harry Potter*, I cannot think of another written work that is so widely known and referenced in American popular culture, even if only a small percentage have actually read the novel in full. This is similar to the poem "The Rime of the Ancient Mariner" or even the phrase "sea change" from *The Tempest*. Newspaper and magazine articles regularly compare obsessions and distant and difficult goals to the White Whale and compare whoever is doggedly in power, wherever, to Ahab.[25]

When I stood aloft in the middle of the South Pacific aboard the *Robert C. Seamans*, I imagined Ishmael floating, thirsty, hypothermic, perhaps resigned, still passive, still inquiring, still philosophical. He is now a lesson for environmental conservation. Though open-minded, scholarly, and often funny, Ishmael is not a typical environmental hero.

Ishmael is not the Lorax. Ishmael advocates no position. Ishmael takes no action to try to correct injustice. In her fable "The Afterlife of Ishmael," Margaret Atwood wrote that Ishmael had sinned by doing nothing. Ishmael, however, chosen by chance—but taking on the challenge in full—does, at least, bear witness to the tragic events. A new ancient mariner, he tells his tale, preaching against arrows, against harpoons, slung at the ocean and at its animals. Perhaps Ishmael convinces us to love all creatures great and small, from sperm whale to brit. Perhaps it really doesn't even matter whether you believe these creatures to be the creations of God or not.[26]

Sometimes during my hours aloft looking for whales, students came up to talk with me, or we had conversations looking over the rail or in our classes on the quarterdeck. Since my first teaching position aboard the *Concordia*, undergraduate perceptions seem, taken crudely on the whole, to have shifted toward focusing activist energies on issues of social justice, not so much on the innate rights of animals or the type of environmentalism of the Earth Day variety with which I was raised. After watching and participating in and occasionally organizing over twenty years of in-class debates about commercial and indigenous whaling, it seems the rising generation sees the "Save the Whales" movements as mostly quaint, even nearing naïve, despite research that continues to reveal even more intelligence in marine mammals, advanced age, and evidence of social behaviors that we can only name as aspects of a culture. Saving whales purely for their own sake, for their own intrinsic individual worth, is not the driving force of the newest environmentalism. For most of the students I've met, it's usually not enough to make economic or policy sacrifices to save a whale for the animal's life or well-being alone. Yes, it's awful to murder the whales in *Moby-Dick*, but what also about the cruelty imposed on Tashtego, as a Native American; Daggoo, as an African; Fedallah, as a Southeast Asian/Middle-Eastern fantastical devil; or the human hero in the story, Queequeg, the Pacific Islander? The face of environmentalism in *Moby-Dick* for the next generation is not the sperm whale but Pip or the harpooners. Students today are more inclined to support indigenous rights to hunt

whales, for example, and to embrace Iceland or Norway or Japan's claim
to hunt whales *if* these nations can defend their claims to cultural necessity. Just as Ishmael teaches, the new generation has learned to tune in
to the gray areas, to not rank humans or nonhuman animals, and, as a
colleague who regularly runs these debates puts it, "students reel a bit
from the cultural imperialism." The rising generation wants a collective
environmentalism without one culture telling another what to do. So
in many ways today the face of the climate change crisis and the ocean
is often not the polar bear or even the harp seal, but, *and*, the Inuit way
of life.[27]

Though we wanted the waters and beaches of Kiribati to be an untouched antediluvian Eden, we learned that the people throughout
the island nation of Kiribati see themselves as the frontline of climate
change. The approximately 113,000 citizens of Kiribati, most of whom
live on the island of Tarawa, have had absolutely nothing to do with
creating this global crisis. The republic is made up now of thirty-three
islands. Two small, uninhabited islands effectively disappeared under
the sea in 1999. Most of the islands are less than 1.2 statute miles wide,
with an average height of six feet above sea level.[28]

The US Global Change Research Program predicted that with "the
lowest emissions scenarios" and with no major loss of ice from the polar
citadels of Greenland and Antarctica, sea level rise will go up at least
another foot by 2099. Their high-end projection is possibly an additional four feet of sea level rise by the end of this century. The melting
polar ice, the rising water, and the associated rising temperatures will
mean not only a loss of physical land in Kiribati, but also continued
damage to the coral reefs that encircle the atolls, upon which so much
of their sustenance and economy depends. Before the waves completely
submerge the sand and soil, it's more likely the islands will first become
uninhabitable for humans because of a lack of freshwater.[29]

In 2017, referencing the famous line from Coleridge's "Ancient
Mariner," the World Bank published an article titled, "Water, Water,
Everywhere, but Not a Drop to Drink: Adapting to Life in Climate-

Change-Hit Kiribati." The author explained that this island nation is one of the most vulnerable to anthropogenic climate change. The shallow water sources are already at risk of sewage contamination. The citizens of Kiribati will almost certainly, in our children's lifetimes, represent one of the first, if not the first, entire nations to be without a claim to land. The people of Kiribati will lose their physical home entirely, with no legal claim to sovereignty on any terra firma. For the first time in recorded human history, a people, an entire culture, will be a stateless republic of climate refugees: of Ishmaels.[30]

The former president of Kiribati, Anote Tong, supported the establishment of the Phoenix Islands Protected Area for conservation measures. Evidence from other reserves around the world suggests that the ocean ecosystem both inside and outside the reserve will improve enormously and quickly. President Tong saw an opportunity to elevate awareness of his country on the world stage and to increase revenue from fishing rights that will likely become more valuable just outside the reserve boundaries. In addition, Tong allocated a large portion of his nation's budget to purchase land in Fiji to be reserved for future emigration. "Migration with dignity" were his rallying words. He led a variety of initiatives, such as planting mangroves around their coasts and creating economic and educational incentives for young I-Kiribati to establish roots in countries abroad. Tong wrote: "Yet the people of Kiribati recognize that they are still stewards of the ocean that now threatens their way of life."[31]

This essential idea would be foreign to Melville or Ahab or Ishmael or Noah: that the ocean in all its fury and force would need or welcome human stewardship.

During our voyage we learned what we assumed was excellent news: that the president of Fiji had declared officially that his nation would welcome those from Kiribati. One of our students aboard was Kareati Waysang, a citizen of Kiribati who was studying biology in New Zealand. "The Fijians look down on us," she explained. She had visited Fiji several times. If in a restaurant Fijians learn she is from Kiribati they serve her differently. The international attention that Kiribati has received as a face of climate change, as victims, has turned out to be a mixed blessing. "I don't want to go live in Fiji," Waysang told me.[32]

Up at the masthead, I began to think of Ishmael in *Moby-Dick* as a climate refugee, of which there are now tens of millions of people on Earth. In the Bible, Ishmael is an orphan, an unwanted castaway, who goes on to found a new religion. Groups of people from Louisiana to Papua New Guinea have already been permanently relocated because of sea level rise.[33]

During our voyage to the equatorial Pacific, as during every long trip at sea, seabirds were the most visible fauna. We observed and counted boobies, tropic birds, and terns. It was the frigatebirds, however, which were by far the most dynamic fauna that we saw consistently throughout the trip. Even over Pago Pago harbor in American Samoa, frigatebirds soared over the water and up over the volcanic peaks. Sometimes near the islands frigatebirds swooped threateningly low and close, while other times when I was aloft I saw them in the distance at truly stunning heights, mere specks flying higher than I'd ever seen any bird, floating up at elevations, it appeared, far into the clouds. While at anchor off Nikumaroro we saw one frigatebird swoop down on a sooty tern (*Sterna fuscata*), stealing its fish midair. When I was aloft at the masthead one morning near Enderbury Island, a frigatebird swooped down and pecked at the radio antennae at the tip of our foremast, only an arm's length away from me. I could've grabbed him if I had the courage—and a helmet.

The nation of Kiribati declared independence from British rule in 1979. They designed a flag that consists of a blue sea, a stylized yellow island on a red sky, and a bird, which is meant to be a frigatebird. According to I-Kiribati historians it is "a symbol of our old people and our dance patterns" that in legend, like the whalers' albatross, carries messages from island to island.[34]

In 1978 an I-Kiribati poet wrote "The Song of the Frigatebird," verse that was eerily prophetic. The song is about a mother frigatebird who flies away to find food for her young. When she returns, her island is underwater. It translates to:

I am searching for my home
I call you by name—Kiribati
Where are you?
Hear my call—hear my song
I have no one to help me
I have been alone for so long
I have no one to help me
I have been alone for so long
Rise up—you, the centre of the world
Rise up from the depths of the sea
So, you may be seen from afar
Rise up! Rise up![35]

Melville never knew or imagined any of this—the first concerns of the global effects of atmospheric carbon were not raised until the 1890s, and sea level rise as a result of global warming wasn't in the public sphere until the 1960s.[36] Yet this song is inseparable for me now when I read *Moby-Dick*. The sky-hawk goes down, drowns, with Tashtego and the *Pequod*. Then another frigatebird hovers, leaving Ishmael in peace to float alone on the surface.

IN CONCLUSION, OR THE IMMORTAL SEA—THE SAME AS AHAB'S WITH A TWENTY-FIRST-CENTURY FLIP SIDE

After Tashtego and the frigatebird sink and drown, Ishmael ends the final chapter: "Now small fowls flew screaming over the yet yawning gulf; a sullen white surf beat against its steep sides; then all collapsed, and the great shroud of the sea rolled on as it rolled five thousand years ago." An additional twenty-first-century moral of *Moby-Dick*, along with the proto-environmentalism, proto-Darwinism, and the image of

Ishmael as a climate refugee, seems to be that no matter the rage and folly of humankind, it will be nature, the ocean, that will look after its own residents. Nature will restore itself to an unspoiled state. The American nineteenth-century sea, harkening back to that of Noah, is immortal and indifferent to man's small, temporary endeavors.[37]

A century after *Moby-Dick*, in her first book about the ocean, Rachel Carson wrote of the ecological lives of ocean animals by way of marine and planetary rhythms rather than human linear storytelling. She ends *Under the Sea-Wind* with the same final message as *Moby-Dick*: "For once more the mountains would be worn away by the endless erosion of water and carried in silt to the sea, and once more all the coast would be water again, and the places of its cities and towns would belong to the sea." Carson, even in 1941, was only just beginning to imagine a world where the ocean would need human stewardship, or at least some self-management.[38]

Today, even though we recognize all of the damage that *Homo sapiens* has inflicted on the sea, we still know and understand that the ocean is bigger and longer-lived than us, if not as a collection of individual animals and plants, then certainly as a body of water and as our planet's most dominant ecosystem. The sea is still rolling into the land, whether slowly by sea level rise or catastrophically with tsunami, hurricanes, and typhoons. We live now with a profound paradoxical relationship with the ocean. On one side of the ship, the sea is immortal and overwhelming and sharkish, while on the other side, we feel we must care for the sea and its inhabitants, which are vulnerable and fragile.

In 2012, Hurricane Sandy, the most powerful hurricane on record to hit the American northeast, flooded New York City, entirely submerging the docks and streets where Ishmael roamed in the opening scenes of *Moby-Dick*. During Sandy, cars floated down Wall Street. Up to six feet of water surrounded the site on Pearl Street where Melville was born. Reasonable projections for 2099 have lower Manhattan as its own separate little island during major storm events.[39]

Thankfully only a few of us are Ahab. But most of us are Ishmaels or Starbucks or Stubbs or Flasks or Pips in our complicity or fear or powerlessness to take individual or collective action aboard our human ark. Most of us are unable to inspire or organize or even participate in the type of social mutiny that might cease our Western societies' desire to hurl harpoon after harpoon after harpoon after harpoon, putting at risk and even killing our future human generations. I'm no better. We've now got solar panels and an electric car, but I still fly all over the place and the computer on which I type this is loaded with mined metals and plastics from all over the world, not to mention my plastic pen, and my coffee—albeit fair-trade organic and in a ceramic mug—is still flown from another hemisphere. I try to eat vegetarian, but I do love a tuna-fish sandwich, and on and on and on with a carking eco-guilt that makes a person just want to go out to sea and get away from it all.

Which brings me back to standing up at the masthead at the top of the foremast of the *Robert C. Seamans*. I was about as far away on this blue Earth as I possibly could be from my home in Mystic, Connecticut, from those hoops aloft of the last American wooden whaleship, which is still tied up to the dock there.

What I've aimed to learn in this natural history is how today we might read *Moby-Dick* to reveal how our American perceptions of the ocean have changed—and how they have not. From experience and from his research, Melville knew a great deal about the marine sciences. Tens of thousands of nineteenth-century American whalemen, including young Melville, had a hunter's intimate knowledge of the ocean. Few people even approach this knowledge today, beyond some fishermen, like Linda Greenlaw, or whalewatch captains, like J. J. Rasler, or the rare scientists like Hal Whitehead or Marta Guerra who are able to spend day after day out on the water—even though those two have never actually placed their palm on a living whale. We may now read in *Moby-Dick* a watery reflection of our American society, living in the slow-motion crisis of climate change. Ishmael floats alone, rescued, and then, like the Ancient Mariner, compels us to sit us down and listen to his story. If we can bless the sperm whale and appreciate its honest wonders—or revel in the ecological role of the right whale, the cormorant, the albatross, the swordfish, the copepod, the giant squid, the cyamid, the barnacle, the shark, the coral, the storm petrel, the sea lion, the paper nautilus, and the frigatebird—if we could spend more patient time in awe of all this, might we be just a little better off?

Melville ended *Moby-Dick* with a nod to Noah, dating his age of the Earth based on the Bible. He knew enough geology, however, to know that the Earth was at least millions of years old. Melville wrote his masterwork on the ocean a full century before radiocarbon dating, before the understanding of plate tectonics, and before Tharpe and Heezen's map of the global ocean bottom. Melville wrote *Moby-Dick* a century before the detonation of the atomic bomb on Hiroshima and the development of commercial underwater color photography, both of which revealed in very different ways just how capable we are of eradicating our own species as well as doing immense, irreversible harm to other extraordinary life on the planet. Melville wrote *Moby-Dick* a century and a quarter before December 7, 1972, when astronauts from Apollo 17 sent down photographs of the Earth from space, an image now known as the "Blue Marble," a photograph that had a profound impact on how we see our world as one dominated by saltwater. This inspired the cartoon on the Earth Day T-shirt that I wore as a child in the 1970s.

Yet even with all these social, technological, and cultural shifts in our perception of the watery natural world, Herman Melville's understanding of the ocean in *Moby-Dick* is, as Ahab puts it standing aloft looking on the waters of the equatorial Pacific: "the same to Noah as to me." Despite our intellectual understanding of our significant negative impacts on the sea and all our new knowledge and access to its ecology, physical characteristics, and weather, we still emotionally, existentially—when we stand on the beach, on a dock, or on the deck of a ship—fear the ocean as immortal, all-powerful, and indifferent to our petty human endeavors. Even when we see that plastic coffin of a cooler float by, we know the ocean will roll on just as it will roll long after the ecological reign of our species, post-Anthropocene. I try to find a comfort in that—as does Ishmael.

ACKNOWLEDGMENTS

This natural history of *Moby-Dick* is barely a sprinkle upon the enormous research of generations of Melville scholarship. The chapter notes and selected bibliography name those deep-diving scholars and writers who have been most helpful for this synthesis.

A few portions of this book have been published in different forms in *Sea History* magazine and in *The Hungry Ocean: The Sea and Nineteenth-Century Anglophone Literary Culture* (2016). For these, I thank Dee O'Regan, Steve Mentz, Marty Rojas, and an anonymous peer reviewer.

Throughout the writing of this book, I've been unreasonably fortunate to benefit from the generosity of experts from so many fields, so many people who gave up their time. Some experts are featured overtly as characters in the narrative here, while others contributed as much or more by helping behind the scenes. Specialists and writer friends read individual chapters and gave me invaluable feedback. Others provided images and still others helped with a range of specialized details. I'm limited here to simply, genuinely, say thank you to Jon Ablett, Ray Atos, Scott Baker, Bert Bender, Lori Beraha, Claudio Campagna, Jim Carlton, the crew of *Catch It!*, Oliver Crimmen, Bryan Donaldson, David Ebert, Chris Fallows, Ewan Fordyce, Glenn Gordinier, Linda Greenlaw, Steven Haddock, Christy Hudack, Stefan Huggenberger,

Terry Hughes, Ken Findlay, Julian Finn, Jennifer Jackson, Bob Kenney, David Laist, Bob Madison, Richard Malley, Abby McBride, Steve Miller, Grace Moore, Catherine Naum, Rob Nawojchik, Michelle Neely, Sandy Oliver, Steven Olsen-Smith, Mark Omura, Hershel Parker, Martina Pfeiler, Charles Paxton, Douglas Perrine, Paul Ponganis, Hugh Powell, Elliot Rappaport, JJ Rasler, Justin Richard, Matt Rigney, Shef Rogers, Clyde Roper, Nancy Shoemaker, Mia Sigler, Hjörtur Gísli Sigurdsson, Elizabeth Schultz, Tim Smith, Hal Whitehead, and Tony Wu.

The staff of several academic institutions and museums have been enormously generous with their resources and space. Thank you to the Maritime Studies Program of Williams College and Mystic Seaport, and all the staff, students, and fellow faculty there who taught me so much over two decades, especially my longtime mentors in literature of the sea: Mary K. Bercaw Edwards, Dan Brayton, and Susan Beegel, the last of whom introduced me to eco-criticism and so many of the titles I now love, as well as inspiring this book with her "A Guide to the Marine Life in Ernest Hemingway's *The Old Man and the Sea*" (*Resources for American Literary Study*, 2005). Several Williams-Mystic research assistants helped with the project at various stages, including Beryl Manning-Geist, Ben Seretan, and Steve Telsey, and Leila Crawford, Rachel Earnhardt, and Emma McCauley both researched and served as valuable editors. Thank you to the Sea Education Association, Mystic Seaport Museum, Mystic Aquarium, the New Bedford Whaling Museum, the University of California-Santa Cruz (especially Kresge College and Rachel Carson College), and the Moss Landing Marine Laboratories. I completed the book as a guest of the University of Otago's English and Linguistics Department and the Science Communication Programme, both of whom could not have been more generous with work space, resources, and academic collegiality.

I'm indebted to the librarians and research staff and curators in the libraries and collections at the following institutions: Maribeth Belinski, Louisa Watrous, Richard Malley, Krystal Rose, Fred Calabretta, and Paul O'Pecko at Mystic Seaport's Blunt White Library; Meg Costello at the Falmouth Historical Society; Mary Sears and Elizabeth Meyer at the Ernst Mayr Library of the Museum of Comparative Zoology (Harvard University); Mark Procknik and Michael Dyer at The New Bedford Whaling Museum library; the generous Interloans staff at the University of Otago library; and, once again as ever, the amazing Alison O'Grady and her assistants at Williams College.

Two classes of undergraduates were kind to allow me to be a part of their academic communities during the practical research for this book: the Marine Mammals class from the University of Rhode Island at Mystic Aquarium, led by Justin Richard and MaryEllen Mateleska, and the S-274 Phoenix Islands Protected Area class from the Sea Education Association aboard the *Robert C. Seamans*, led by Captain Chris Nolan, Chief Scientist Deb Goodwin, and Professor Jeff Wescott. In the final stages of this book, the S-283 class from the Sea Education Association worked in particular with the logbook of the *Commodore Morris*, and I was fortunate to learn more from this group, both in Woods Hole and at sea off New Zealand.

Professor Steve Dawson led me to the experience of a lifetime as a guest of Marta Guerra, who took me out to study sperm whales on Kaikōura Canyon with her assistant Rebecca Bakker. Thank you to Marta and the marine mammal team at Otago for their collegiality and generosity with information and resources.

Thank you to two outside readers and their thoughtful and helpful feedback of the manuscript. Thank you to my agent, Russel Galen, and to Christie Henry for her enthusiastic acquisition of this project. Thank you to Karen Merikangas Darling and her thoughtful editing, kind feedback, and patient shepherding. Susannah Engstrom provided careful, good-humored, and tireless logistical help, and copy editor Mary Corrado, designer Isaac Tobin, and production controller Skye Agnew brought the book smoothly to publication with great care, knowledge, and talent.

Thank you to my brother Seth for his cartoon of Melville and the creation of the "Fish Documents" infographic, which was designed by Skye Moret. Thank you to my daughter, Alice, for her encouragement, suggestions, and artwork, and, most importantly, thank you to my partner, patron, editor, expert resource, and love of my life, Lisa Gilbert, to whom this book is dedicated.

NOTES

NOTES TO INTRODUCTION

1   Lewis Mumford, *Herman Melville* (New York: Literary Guild of America, 1929), 194. This quotation was introduced to me as a lead by Jennifer Baker.

2   Herman Melville, *Moby-Dick or The Whale*, ed. Harrison Hayford, Hershel Parker, and G. Thomas Tanselle (Evanston: Northwestern University Press and The Newberry Library, 1988), 565.

3   On Melville and marine life, see, e.g., Bert Bender, *Sea-Brothers: The Tradition of American Sea-Fiction from "Moby-Dick" to the Present* (Philadelphia: University of Pennsylvania Press, 1988), vii, 19.

4   *Moby-Dick*, 64, 273, 424. Throughout this natural history I'll refer to Ishmael as the narrator of *Moby-Dick*. In other words, it will be "Melville wrote" (past tense) and "Ishmael says" (present tense). Certainly there are gray areas, especially after the story departs Nantucket and heads out to sea, and the chapters and scenes and even footnotes seem to be narrated by someone we might call Hermael or Ishmelville. But the storyteller, who's asked us to call him Ishmael, is, like all fictional characters, a creation of the author at a single point in his life, even if the tone feels distant or omniscient or even if this voice refers to himself sitting at his writing desk. On this, see Robert Zoellner, *The Salt-Sea Mastodon: A Reading of Moby-Dick* (Berkeley: University of California Press, 1973), xi; and Maurice S. Lee, "The Language in *Moby-Dick*: 'Read It If You Can,'" in *A Companion to Herman Melville*, ed. Wyn Kelley (West Sussex, UK: Wiley-Blackwell, 2015), 395–96.

5   NOAA Carbon Dioxide Information Analysis Center as displayed in "Climate Change: Seven Things You Need to Know," *National Geographic* 231, no. 4 (April 2017): 31–32; John Walsh and Donald Wuebbles, et al., "Ch. 2: Our Changing Climate," *Climate Change Impacts in the United States: The Third National Climate Assessment*, ed. J. M. Melillo, Terese (T. C.) Richmond, and G. W. Yohe (US Global Change Research Program, 2014), 20–21, 44–45. See also fig. 1, "Global Average Absolute Sea Level Change, 1880–2015," Climate Change Indicators: Sea Level, United States Environmental Protection Agency, August 2016, www.epa.gov

/climate-indicators/climate-change-indicators-sea-level; and S. Jevrejeva, J. C. Moore, A. Grinsted, A. Matthews, and G. Spada, "Trends and Acceleration in Global and Regional Sea Levels since 1807," *Global and Planetary Change* 113 (2014): 11–22.

6   Transportation Safety Board of Canada, "Marine Investigation Report, M10F0003: Knockdown and Capsizing, Sail Training Yacht *Concordia*" (Minister of Public Works and Government Services Canada, 2011).

7   Greg Dobie, ed., "Safety and Shipping Review 2016," Allianz Global Corporate & Specialty SE (March 2016), 4; See also George Michelson Foy, *Run the Storm* (New York: Scribner, 2018).

8   *Moby-Dick*, 273.

NOTES TO CHAPTER ONE

1   NYC Department of City Planning, "Total and Foreign-Born Population, New York City, 1790–2000," accessed 31 January 2019, www1.nyc.gov/site/planning/data-maps/nyc-population/historical-population.page; Jean-Paul Rodrigue, "World's Largest Cities, 1850," accessed 31 January 2019, https://transportgeography.org/?page_id=4976. See also Richard F. Selcer, *Civil War America, 1850 to 1875* (New York: Facts on File, 2006), 271.

2   Margaret S. Creighton, *Rites and Passages: The Experience of American Whaling, 1830–1870* (Cambridge: Cambridge University Press, 1995), 67, 129; Williams A. Allen, "25 November 1842," Logbook of the *Samuel Robertson*, 1841–46, New Bedford Whaling Museum Log ODHS 1040; also Richard Henry Dana Jr., *Two Years before the Mast: A Personal Narrative of Life at Sea* (New York: Harper and Bros., 1840), 44–45.

3   *Moby-Dick*, 524.

4   Pip as a figure to consider environmental justice was introduced to me by Dana Luciano, "Love and Death in the Anthropocene: Geologic Time, Genre, *Moby-Dick*," Lecture at Connecticut College, 21 April 2016.

5   *Moby-Dick*, 370–71; Thomas Beale, *The Natural History of the Sperm Whale* (London: John Van Voorst, 1839), 44, 161; Randall R. Reeves, Brent S. Stewart, Phillip J. Clapham, James A. Powell, and Pieter A. Folkens, *Guide to Marine Mammals of the World* (New York: Knopf, 2002), 21; Stephanie L. Watwood, et al., "Deep-Diving Foraging Behavior of Sperm Whales (*Physeter macrocephalus*)," *Journal of Animal Ecology* 75, no. 3 (May 2006): 814–25.

6   Helen M. Rozwadowski, *Fathoming the Ocean: The Discovery and Exploration of the Deep Sea* (Cambridge, MA: Belknap Press, 2005), 74–75; Matthew Fontaine Maury, *Explanations and Sailing Directions to Accompany the Wind and Current Charts*, 3rd ed. (Washington, DC: C. Alexander Printer, 1851), 62–63.

7   The lecture Melville heard was one of Emerson's in a series of five titled "Mind and Manners in the Nineteenth Century." See William Braswell, "Melville as a Critic of Emerson," *American Literature* 9, no. 3 (November 1937): 317; Herman Melville, "To Evert A. Duyckinck, 3 March 1849, Boston," *Correspondence*, ed. Lynn Horth (Evanston: Northwestern University Press and The Newberry Library, 1993), 121.

8   Herman Melville, "Hawthorne and His Mosses," *The Piazza Tales and Other Prose Pieces, 1839–1860*, ed. Harrison Hayford, Alma A. MacDougall, G. Thomas Tanselle,

et al. (Evanston: Northwestern University Press and The Newberry Library, 1987), 242.

9   This interview with Mary K. Bercaw Edwards was originally conducted on 11 May 2016, then revised in collaboration.

10   The *Acushnet* was 359 tons and the *Charles W. Morgan* was 351 tons, both ship-rigged in 1841. Alexander Starbuck, *History of the American Whale Fishery from Its Earliest Inception to the Year 1876* (Waltham, MA: self-published, 1878), 372, 376.

11   Meredith Farmer, "Herman Melville and Joseph Henry at the Albany Academy; or, Melville's Education in Mathematics and Science," *Leviathan* 18, no. 2 (June 2016): 4–28. See also Laurie Robertson-Lorant, "A Traveling Life," in *A Companion to Herman Melville*, 3–18; and Tyrus Hillway, "Melville's Education in Science," *Texas Studies in Literature and Language* 16, no. 3 (Fall 1974): 411–25; Jay Leyda, *The Melville Log: A Documentary Life of Herman Melville, 1819–1891*, vol. 1 (New York: Harcourt, Brace, 1951), 43, 45, 52; Wilson Heflin, *Herman Melville's Whaling Years*, ed. Mary K. Bercaw Edwards and Thomas Farel Heffernan (Nashville: Vanderbilt University Press, 2004), 4–5.

12   Leyda, 110; Andrew Delbanco, *Melville: His World and Work* (New York: Vintage, 2006), 35; Merton M. Sealts Jr., *Melville's Reading: Revised and Enlarged Edition* (Columbia: University of South Carolina Press, 1988), 19–22; Hershel Parker, *Herman Melville: A Biography, Vol. 1, 1819–1851* (Baltimore: Johns Hopkins University Press, 1996), 110, 181, etc.

13   J. Ross Browne, *Etchings of a Whaling Cruise*, ed. John Seelye (Cambridge, MA: Belknap Press, 1968), 193.

14   *Moby-Dick*, 156.

15   Heflin, 59, 67, 69, 106, 110–15. They might have also anchored in Independencia Bay, Peru. See a helpful summary of all Melville's ocean passages in R. D. Madison, ed., *The Essex and the Whale: Melville's Leviathan Library and the Birth of Moby-Dick* (Santa Barbara, CA: Praeger, 2016), 264.

16   John James Audubon, "To Daniel Webster, New York, 8 September 1841," in *The Audubon Reader*, ed. Richard Rhodes (New York: Everyman's Library, 2006), 570.

17   Mary K. Bercaw Edwards, *Cannibal Old Me: Spoken Sources in Melville's Early Works* (Kent: Kent State University Press, 2009), 1–23; Heflin, 161–70.

18   He dubiously claimed later that he was a harpooner on board, also known as the boatsteerer, who actually threw the first harpoon at a whale. Herman Melville, "To Richard Bentley, 27 June 1850, New York," *Correspondence*, 163.

19   Herman Melville, *Journals*, ed. Howard C. Horsford and Lynn Horth (Evanston: Northwestern University Press and The Newberry Library, 1989), 4–5.

20   Herman Melville, "To R. H. Dana, Jr., 1 May 1850, New York," *Correspondence*, 162. Madison suggests convincingly that Melville was not *actually* halfway through the novel. See "Introduction: Swimming through Libraries," *The Essex and the Whale*, xx–xxii.

21   Herman Melville, "To Nathaniel Hawthorne [1 June?] 1851, Pittsfield," *Correspondence*, 191. On this letter, see Samuel Otter, *Melville's Anatomies* (Berkeley: University of California Press, 1999), 7.

22   *Moby-Dick*, 112; Harold J. Morowitz, "Herman Melville, Marine Biologist," *Biological Bulletin* 220 (April 2011): 83.

23   See also Laurie Robertson-Lorant, "A Traveling Life," 5.

24   Robert Madison, 10 June 2016, pers. comm.

25   I've estimated conservatively over 8,000 masthead standers per year by multiply-
ing 15 sailors by 550 ships. Charles Wilkes, for example, states, "Between fifteen and
sixteen thousand of our countrymen are required to man these [675] vessels." Not all
men on a ship took regular tricks aloft to look for whales. Charles Wilkes, *Narrative
of the United States Exploring Expedition*, vol. 5 (London: Wiley and Putnam, 1845),
485–86. See also Starbuck, 98; on reading cultures, Hester Blum, *The View from the
Masthead* (Chapel Hill: University of North Carolina Press, 2008), 5.

26   James C. Osborn, Logbook of the *Charles W. Morgan*, 1841–45, Mystic Seaport
Log 143. See also Hester Blum, "A List of Books that I Did Not Read on the Voy-
age," *Leviathan* 17, no. 1 (March 2015): 129–32; Herman Melville, *White-Jacket, or
The World in a Man-of-War*, ed. Harrison Hayford, Hershel Parker, and G. Thomas
Tanselle (Evanston: Northwestern University Press and The Newberry Library,
1988), 167; Mary K. Bercaw [Edwards], *Melville's Sources* (Evanston: Northwestern
University Press, 1987), 85; Tyrus Hillway, "Melville's Education in Science," 417;
on natural history and sailors' journals, D. Graham Burnett, *Trying Leviathan: The
Nineteenth-Century New York Court Case That Put the Whale on Trial and Challenged
the Order of Nature* (Princeton: Princeton University Press, 2007), 110; Michael
Dyer, 6 Friday 2017, pers. comm.

27   Herman Melville, *Typee: A Peep at Polynesian Life*, ed. Harrison Hayford, Hershel
Parker, and G. Thomas Tanselle (Evanston: Northwestern University Press and The
Newberry Library, 1968), 3.

28   Herman Melville, "Review of *Etchings of a Whaling Cruise* and *Sailors' Life and
Sailors' Yarns*," *Piazza Tales*, 206. See Beale, 3.

29   John F. Leavitt, *The Charles W. Morgan* (Mystic: Mystic Seaport, 1973), 35, 37.

30   *Moby-Dick*, 159.

NOTES TO CHAPTER TWO

1   *Moby-Dick*, 136.

2   *Moby-Dick*, 135, 443; Sumner W. D. Scott, "The Whale in *Moby Dick*," PhD diss.
(Chicago: University of Chicago, 1950), 6–27.

3   Madison, *The Essex and the Whale*, 169–76; Parker, vol. 1, 723–24; Vincent, 128–35;
Kendra Gaines, "A Consideration of an Additional Source for *Moby-Dick*," *Melville
Society Extracts* 29 (1977): 6–12; Mary K. Bercaw Edwards, "The Infusion of Useful
Knowledge: Melville and *The Penny Cyclopædia*," *Melville Society Extracts* 70 (1987):
9–13; Hal Whitehead, *Sperm Whales: Social Evolution in the Ocean* (Chicago: Univer-
sity of Chicago Press, 2003), 16.

4   For brief biographies of Beale and Bennett, see Ian A. D. Bouchier, "Some Experiences of Ships' Surgeons during the Early Days of the Sperm Whale Fishery,"
British Medical Journal 285 (18–25 December 1982): 1811–13; and Honore Forster,
"British Whaling Surgeons in the South Seas, 1823–1843," *Mariner's Mirror* 74, no. 4
(1988): 401–15.

5   Frederick Bennett, *Narrative of a Whaling Voyage Round the Globe* (London: Richard
Bentley, 1840), vol. 1, 118–19.

6   *Moby-Dick*, 265; Beale, 33; Beale, Melville's Marginalia Online, ed. Steven Olsen-
Smith, Peter Norberg, and Dennis C. Marnon, melvillesmarginalia.org.

7   "Herman Melville's Moby Dick," *Southern Quarterly Review* 5, New Series
     (Charleston: Walker and Richards, January 1852), 262.

NOTES TO CHAPTER THREE

1   Burnett, *Trying Leviathan*, 97. Burnett's *Trying Leviathan* is the definitive source on
     this case. See also for a brief summary "Maurice v. Judd," *Historical Society of the New
     York Courts*, accessed 31 January 2019, www.nycourts.gov. The case was published as
     a pamphlet in the summer of 2019 by the successful prosecuting attorney. William
     Sampson, *Is a Whale A Fish? An accurate report of the case of James Maurice against
     Samuel Judd* ... (New York: C. S. Van Winkle, 1819).
2   Bennett, vol. 2, 148–49.
3   *Moby-Dick*, 136, 307, 370; The Society for the Diffusion of Useful Knowledge, *The
     Penny Cyclopædia*, vol. 27, "Wales—Zygophyllaceæ," ed. George Long (London:
     Charles Knight and Co., 1843), 272; Gaines, 6. "Penem intrantem feminam mam-
     mis lactantem" refers to the penis entering the female and the female feeding with
     breast milk; "ex lege naturæ jure meritoque" refers to natural law and merit, which
     doesn't have much to do with biology. On the biblical groupings, Burnett, *Trying
     Leviathan*, 20–23.
4   Burnett, *Trying Leviathan*, 14.
5   *Moby-Dick*, 305; Tyrus Hillway, "Melville as Critic of Science," *Modern Language
     Notes* 65, no. 6 (June 1950): 411–14.
6   Browne, 59.
7   Beale, e.g. 15, 18, 106; Bennett, vol. 2, 145; John Mason Good, *The Book of Nature*
     (Hartford: Belknap and Hamersley, 1837), 192–93; "Whales," *Penny Cyclopædia*, vol.
     27, 273.
8   *Moby-Dick*, 135, 262; on the "squash," Stuart M. Frank, *Herman Melville's Picture
     Gallery* (Fairhaven, MA: Edward J. Lefkowicz, 1986), 34–35; Good, 192.
9   *Moby-Dick*, xxiii, 137; "Whales," *Penny Cyclopædia*, vol. 27, 273; Sealts, 170; Baron
     Georges Cuvier, *The Class Pisces*, with supplementary editions by Edward Griffith
     and Charles Hamilton Smith, vol. 10 of *The Animal Kingdom* (London, Whittaker
     and Co., 1834), 27; Cuvier, Melville's Marginalia Online. For further discussion on
     the history of the whales v. fish debate, see Burnett, *Trying Leviathan*, 211–22.
10   Robert Nawojchik, 14 September 2016, Lecture, Mystic Aquarium.
11   *Moby-Dick*, 306; Christoph Irmscher, *Louis Agassiz: Creator of American Science*
     (Boston: Houghton Mifflin, 2013), 64–84; David Dobbs, *Reef Madness: Charles
     Darwin, Alexander Agassiz, and the Meaning of Coral* (New York: Pantheon Books,
     2005), 31–36; on Ishmael's "vast floating icebergs" scraping New England rocks, and
     Melville's reading of Lyell, Agassiz, et al., see Elizabeth S. Foster, "Melville and
     Geology," *American Literature* 17, no. 1 (March 1945): 61.
12   Louis Agassiz and A. A. Gould, *Principles of Zoology*, rev. ed. (Boston: Gould and
     Lincoln, 1851), 210.
13   This interview with Robert Nawojchik was originally conducted on 26 May 2017,
     then revised in collaboration.
14   Beale, 10–12; Beale, Melville's Marginalia Online. For more on how the "Cetology"
     chapter mocks the scientific community, see, e.g. J. A. Ward, "The Function of

the Cetological Chapters in *Moby-Dick*," *American Literature* 28, no. 2 (May 1956), 176–77.

15   Agassiz and Gould, *Principles of Zoology*, 84, 95.

16   Ewan Fordyce, 15 March 2018, pers. comm.; Ewan Fordyce, "Cetacean Evolution," *Encyclopedia of Marine Mammals [EMM]*, 3rd ed., ed. Bernd Würsig, J. G. M. Thewissen, and Kit M. Kovacs (London: Academic Press, 2018), 180.

17   Charles Darwin, *On the Origin of Species by Means of Natural Selection*, ed. William Bynum (New York: Penguin, 2009), 169; Fordyce, "Cetacean Evolution," *EMM*, 3rd ed., 182; Reeves, et al., *Guide*, 12–13; Janet Browne, *Charles Darwin: The Power of Place*, vol. 2 (London: Pimlico, 2003), 99. See also Annalisa Berta, "Pinniped Evolution," *EMM*, 3rd ed., 712–22.

18   Annalisa Berta and Thomas A. Deméré, "Baleen Whales, Evolution," *EMM*, 3rd ed., 70–72; David W. Laist, *North Atlantic Right Whales: From Hunted Leviathan to Conservation Icon* (Baltimore: Johns Hopkins University Press, 2017), 61.

19   *Moby-Dick*, 140.

20   Melville would be pleased to know that until quite recently there has been some haggling over the full scientific title. *Physeter catadon*, referring to the teeth, was in fashion for a while. *Physeter* is Greek for the blow-hole and was first given by Linnaeus in 1758. See Hal Whitehead, "Sperm Whale," *EMM*, 3rd ed., 919; and A. A. Berzin, *The Sperm Whale (Kashalot)*, ed. A.V. Yablokov, trans. E. Hoz and Z. Blake (Jerusalem: Israel Program for Scientific Translation, 1972), 7–13, 16.

21   Agassiz and Gould, 25–26.

22   See David W. Sisk, "A Note on *Moby-Dick*'s "Cetology" Chapter," *ANQ: A Quarterly Journal of Short Articles, Notes, and Reviews* 7, no. 2 (April 1994): 80–82.

23   *Moby-Dick*, 138.

24   *Moby-Dick*, 138; Bennett, vol. 2, 154; Beale, 1, 15–16; Reeves, et al., *Guide*, 234, 241.

25   Charles M. Scammon, *The Marine Mammals of the North-western Coast of North America* (New York: Dover Publications, 1968), 52; and John Jones, *Meditations from Steerage: Two Whaling Journal Fragments*, ed. Stuart M. Frank (Sharon, MA: Kendall Whaling Museum, 1991), 21; Laist, 23–30. For right whale split as considered in 1851 see Matthew Fontaine Maury, "The Whale Fisheries . . . ," *New York Herald*, 29 April 1851, 3. For current taxonomy and genetics, Scott D. Kraus and Rosalind M. Rolland, eds., *The Urban Whale* (Cambridge, MA: Harvard University Press, 2007), 9; Reeves, et al., *Guide*, 190; Robert D. Kenney, "Right Whales," *EMM*, 3rd ed., 817–18.

26   Phillip J. Clapham and Jason S. Link, "Whales, Whaling, and Ecosystems in the North Atlantic Ocean," in *Whales, Whaling, and Ocean Ecosystems*, ed. James A. Estes, et al. (Berkeley: University of California Press, 2006), 314–16; Laist, 101–3, 109–10, 165–69. Accounts today about the name of the right whale sometimes suggest that right whales did not sink when killed, but they did, and it was a problem. See Henry T. Cheever, *The Whale and His Captors; or, The Whaleman's Adventures*, ed. Robert D. Madison (Hanover, NH: University Press of New England, 2018), 33.

27   *Moby-Dick*, 139. Wallace argued that Melville purposefully did not name Gray so he could skewer his findings: Robert K. Wallace, "Melville, Turner, and J. E. Gray's Cetology," *Nineteenth-Century Contexts* 13, no. 2 (Fall 1989): 155–64.

28   *Moby-Dick*, 360.

29   For similar common names in the nineteenth century, see Scoresby in the 1820s,

whalemen Dean C. Wright, James Osborn, the logkeeper of the *Commodore Morris* in the 1840s, and William Scammon in the 1870s. Cheever wrote of razor-backs that sounded more like a blue whale by size (*The Whale and His Captors*, 43–44), as did Bennett (vol. 2, 154). The fin whale's dorsal fin is about a foot or foot and a half tall. Rob Nawojchik, pers. comm.; OBIS SEAMAP, "Fin Whale—*Balaenoptera physalus*," Marine Geospatial Ecology Lab, Duke University, accessed 31 January 2019, http://seamap.env.duke.edu/species/180527/html; Ishmael seems to conveniently fudge the characteristics of the fin whale spout, however, to confuse the fin whale with a sperm whale at the close of "The Pequod Meets the Virgin" (360). Reeves, et al., *Guide*, 184, 208; A. G. Bennett, "On the Occurrence of Diatoms on the Skin of Whales," *Proceedings of the Royal Society B* 91, no. 641 (15 November 1920): 352–57. For more on historical common names, see Lance E. Davis, Robert E. Gallman, and Karin Gleiter, *In Pursuit of Leviathan: Technology, Institutions, Productivity, and Profits in American Whaling, 1816–1906* (Chicago: University of Chicago Press, 1997), 53–56.

30   Reeves, et al., *Guide*, 226–27.

31   For example, Jones on the whaleship *Eliza Adams* wrote in 1852 off Cape Horn of "one little fin back," swimming around the ship (*Meditations from Steerage*, 18). It seems unlikely this was a juvenile finback if all alone. Perhaps this was a minke, just as Scammon's "Sharp-headed Finner Whale (Balænoptera Davidsoni, *Scammon.*) [*sic*]" was almost certainly a minke (49–51).

32   Osborn, e.g. "15 October 1841"; Dean C. Wright, *Meditations from Steerage*, 2.

33   Good, 192; Robert Hamilton, *The Naturalist's Library: Mammalia. Whales, &c.*, vol. 26, ed. William Jardine (Edinburgh: W. H Lizards, 1843), 228–30; Michael Dyer, "Whalemen's natural history observations and the Grand Panorama of a Whaling Voyage Round the World," *New Bedford Whaling Museum Blog*, 29 March 2016, whalingmuseumblog.org; Dan Bouk and D. Graham Burnett, "Knowledge of Leviathan: Charles W. Morgan Anatomizes His Whale," *Journal of the Early Republic* 27 (Fall 2008): 453; Charles W. Morgan, "Address before the New Bedford Lyceum," Charles Waln Morgan Papers, 1796–1861, Mss 41, Sub-group 1, Series Y, Folder 1, New Bedford Whaling Museum (1830/37), 12. In 1912 naturalist Robert Cushman Murphy found that the men referred to grampus as beaked whales; his captain called them "algerines," in Robert Cushman Murphy, *Logbook for Grace: Whaling Brig Daisy, 1912–1913* (Chicago: Time-Life Books, 1982), 146; *Typee*, 10; Cheever, *The Whale and His Captors*, 43, 55–56, 173. See also William Scoresby Jr., *An Account of the Arctic Regions with a History and Description of the Northern Whale-Fishery* vol. 1 (Edinburgh: Archibald Constable and Co., 1820), 474.

34   *Moby-Dick*, 261, 282. For examples of the dolphin fish in sea literature see Dana's *Two Years* and William Falconer's "The Shipwreck" (1762). See Bennett's discussion of the dying "dolphin," too; he puts the word in quotes, since it's the sailor's word (his *Coryphaena hippuris*), vol. 1, 8.

35   Bernd Würsig, "Bow-Riding," *EMM*, 3rd ed., 135–36. See also Francis Allyn Olmsted, *Incidents of a Whaling Voyage* (Rutland, VT: Charles E. Tuttle Co., 1970), 90–92.

36   On "algerines," see Dyer, "Whalemen's natural history observations;" Sisk, 81; and, again, Murphy, *Logbook for Grace*, 146.

37   Bolster, 72; Paul Dudley, "An Essay upon the Natural History of Whales, with a particular Account of the Ambergris found in the *Sperma Ceti* Whale," *Philosophical*

Transactions 33 (1724–25): 258; Laist, 18; see also Burnett, *Trying Leviathan*, 39–40; and M. E. Bowles, "Some Account of the Whale-Fishery of the N. West Coast and Kamschatka," *Polynesian* (4 October 1845): 83.

38   Darwin, *On the Origin of Species*, 427.

39   Eric Wilson, "Melville, Darwin, and the Great Chain of Being," *Studies in American Fiction* 28, no. 2 (Autumn 2000): 132. See also Wyn Kelley, "Rozoko in the Pacific: Melville's Natural History of Creation," in *"Whole Oceans Away": Melville and the Pacific*, ed. Jill Barnum, Wyn Kelley, and Christopher Sten (Kent: Kent State University Press, 2007), 139–52; and Karen Lentz Madison and R. D. Madison, "Darwin's Year and Melville's 'New Ancient of Days,'" in *America's Darwin: Darwinian Theory and U. S. Literary Culture*, ed. Tina Gianquitto and Lydia Fisher (Athens: University of Georgia Press, 2014), 86–103.

NOTES TO CHAPTER FOUR

1   Ralph Waldo Emerson, *The Journals and Miscellaneous Notebooks of Ralph Waldo Emerson, Volume IV, 1832–1834*, ed. Alfred R. Ferguson (Cambridge, MA: Belknap Press, 1964), 265.

2   *Moby-Dick*, 204–5; Howard P. Vincent, *The Trying Out of Moby-Dick* (Carbondale: Southern Illinois University Press, 1965), 169, 174; Frank Luther Mott, *A History of American Magazines 1741–1850* (Cambridge, MA: Harvard University Press, 1966), 607–11; Jeremiah N. Reynolds, "Mocha Dick; or The White Whale of the Pacific: A Leaf from a Manuscript Journal," *Knickerbocker* 12 (May 1839) in Madison, *The Essex and the Whale*, 64; "Five Wicked Whales: A Quintet of Leviathans Well Known to All Whalers," *Chicago Daily Tribune*, 3 April 1892, 35.

3   See Walter Harding, "A Note on the Title 'Moby-Dick,'" *American Literature* 22, no. 4 (January 1951): 501; Ben J. Rogers, "Melville, Purchas, and Some Names for 'Whale' in *Moby-Dick*," *American Speech* 72, no. 3 (Autumn 1997): 333, 335; Ben Rogers, "From Mocha Dick to Moby Dick: Fishing for Clues to Moby's Name and Color," *Names: A Journal of Onomastics* 46, no. 4 (1998): 263–76.

4   Parker, vol. 1, 696; Madison, *The Essex and the Whale*, 9–10, 61; Owen Chase, et al., *Narratives of the Wreck of the Whale-ship Essex* (New York: Dover, 1989); Leyda, 119; Harrison Hayford and Lynn Horth, "Melville's Memoranda in Chase's *Narrative of the Essex*," in *Moby-Dick*, 971–95; *Moby-Dick*, 203–5. Melville extracted the note on the white baleen whale for *Moby-Dick*, xxiii, from John Harris's *Compleat Collection of Voyages* (1705); see *Moby-Dick*, 820; Bennett, vol. 2, 220. Whaling historian Michael Dyer found in only one logbook a reference to a lancing of "Old Zeek," and not a single mention of a named whale in an article in the *New Bedford Whalemen's Shipping List* (Michael Dyer, 15 February 2018, pers. comm.); Michael P. Dyer, *"O'er the Wide and Tractless Sea": Original Art of the Yankee Whale Hunt* (New Bedford: Old Dartmouth Historical Society/New Bedford Whaling Museum, 2017), 304. See also the logbook by whaleman-artist John F. Martin, who wrote in 1843 that his shipmates named three right whales that they chased all day, but were unable to kill: "The largest of the whales we chased to day the men gave the name of Old Sorrel. The next, which was spotted, they called Stewball, & the smallest one, Betz." This seems more like something to pass the time that single day, however. John F.

Martin, *Around the World in Search of Whales: A Journal of the Lucy Ann Voyage, 1841–44*, ed. Kenneth R. Martin (New Bedford: Old Dartmouth Historical Society/New Bedford Whaling Museum, 2016), 126.

5   Ralph Waldo Emerson, "The Uses of Natural History (1833–35)," *The Selected Lectures of Ralph Waldo Emerson*, ed. Ronald A. Bosco and Joel Myerson (Athens: University of Georgia Press, 2005), 8. On historic Boston temperatures: Richard B. Primack, *Walden Warming: Climate Change Comes to Thoreau's Woods* (Chicago: University of Chicago Press, 2015), 50.

6   For more on natural theology here, see Wilson, 134; and Bruce A. Harvey, "Science and the Earth," in *A Companion to Herman Melville*, 73.

7   *Moby-Dick*, 208; Jennifer J. Baker, "Dead Bones and Honest Wonders: The Aesthetics of Natural Science in *Moby-Dick*," in *Melville and Aesthetics*, ed. Samuel Otter and Geoffrey Sanborn (New York: Palgrave Macmillan, 2011), 85–86; Sealts, 122, 171; Parker, vol. 1, 267, 499, 724.

8   *Moby-Dick*, 183.

9   Dagmar Fertl and Patricia E. Rosel, "Albinism," *EMM*, 3rd ed., 20. See also Bennett, vol. 1, 157–58.

10   *Moby-Dick*, 306–7.

11   Beale, 31.

12   Whitehead, "Sperm Whales," *EMM*, 3rd ed., 920; Berzin, 25, 27; Bennett, vol. 2, 155; Beale, 31. See also Scoresby, *Arctic Regions*, vol. 1, 459; Richard Ellis, *The Great Sperm Whale: A Natural History of the Ocean's Most Magnificent and Mysterious Creature* (Lawrence: University Press of Kansas, 2011), 97; Reeves, et al., *Guide*, 240; Whitehead, *Sperm Whales: Social Evolution*, 6–7.

13   Dagmar Fertl, et al., "An Update on Anomalously White Cetaceans, Including the First Account for the Pantropical Spotted Dolphin (*Stenella Attenuata Graffmani*)," *Latin American Journal of Aquatic Mammals* 3, no. 2 (July/Dec 2004): 163; Chester Howland, "The Real Moby Dick?" as collected in *Yankees Under Sail: A Collection of the Best Sea Stories from Yankee Magazine*, ed. Richard Heckman (Dublin, NH: Yankee, 1968), 184–85; Curatorial display, New Bedford Whaling Museum, 2017; Berzin, 27–28; S. Ohsumi, "A Descendant of Moby Dick, or a White Sperm Whale," *Scientific Reports of the Whales Research Institute* 13 (September 1958): 207–9; Tim Severin, *In Search of Moby Dick: The Quest for the White Whale* (New York: Da Capo, 2000), 181–86; Charles "Flip" Nicklin, with K. M. Kostyal, *Among Giants: A Life with Whales* (Chicago: University of Chicago Press, 2011), 12–13; Flip Nicklin, 24 July 2018, pers. comm. via Chris Carey/Minden; Hiroya Minakuchi, 26 July 2018, pers. comm. via Chris Carey/Minden; Helmut Corneli, Alamy Stock Photo, 2016.

14   Cheever, *The Whale and His Captors*, 39; Olmsted, 62; Bennett, vol. 2, 165; New Bedford Whaling Museum, e.g., S-1918 (labeled sperm whale), KWM-686, and S-1890 from a voyage on the barque *Newton*, 1846–49.

15   *Moby-Dick*, 337; Berzin, 38–41.

16   *Moby-Dick*, 307. On blubber see Whitehead, *Sperm Whales: Social Evolution*, 32, 62.

17   Emerson, "The Uses of Natural History," 16. *Moby-Dick*, 312; Ralph Waldo Emerson, *Nature*, new ed. (Boston: James Munroe & Co., 1849), 30. See also Parker, vol. 1, 776; Blum, *The View from the Masthead*, 121–23; Bercaw [Edwards], *Melville's Sources*, 79–80.

18   *Moby-Dick*, 307.

NOTES TO CHAPTER FIVE

1   "Southern Pacific Ocean," J. W. Norie (1825), Mystic Seaport No. 1942.1082.

2   The chart was donated in 1943 with a set of several other charts by a wealthy descendent of a whaling family named Edward Howland Greene, the same man who gave Mystic Seaport the *Charles W. Morgan*; Lewis H. Lawrence, captain, Logbook of the *Commodore Morris*, 1849-1853, Falmouth Historical Society, 2006.044.002; Logkeeper, Logbook of the *Commodore Morris*, 1845-1849 (capt. Silas Jones), Falmouth Historical Society, 2013.076.09.

3   Chase, 24; *Moby-Dick*, 504, 513-14. Other factors to consider to argue for the destination to the east include that the seals encountered in "The Life-Buoy" suggest Galápagos-like rocks (see Olmsted, 177), and Maury's notice, as discussed below, for the region of the *Essex*'s sinking lists the best time for whaling as February. On the names of whaling grounds, see John L. Bannister, Elizabeth A. Josephson, Randall R. Reeves, and Tim D. Smith, "There She Blew! Yankee Sperm Whaling Grounds, 1760-1920," in *Oceans Past: Management Insights from the History of Marine Animal Populations*, ed. David J. Starkey, Poul Holm, and Michaela Barnard (London: Earthscan, 2008), 116-18.

4   Logbook of the *Commodore Morris*, 1849-1853; Logbook of the *Commodore Morris*, 1845-1849. See also Wilkes, vol. 5, 487; Cheever, *The Whale*, 165, 186; Beale, 189-91; Browne, 548-58.

5   Vincent, 180-85; D. Graham Burnett, "Matthew Fontaine Maury's 'Sea of Fire': Hydrography, Biogeography, and Providence in the Tropics," in *Tropical Visions in the Age of Empire*, ed. Felix Driver and Luciana Martins (Chicago: University of Chicago Press, 2014), 113-14; Samuel Otter, "Reading *Moby-Dick*," in *The New Cambridge Companion to Herman Melville*, ed. Robert S. Levine (Cambridge: Cambridge University Press, 2014), 73.

6   *Moby-Dick*, 199.

7   Heflin, 39-40, 48-49, 59-62; Wilkes, vol. 5, 470; Nathaniel Philbrick, *In the Heart of the Sea: The Tragedy of the Whaleship Essex* (New York: Viking, 2000), 63-64, 77-78; Matthew Fontaine Maury, "On the Navigation of Cape Horn," *American Journal of Science and Arts (Silliman's Journal)* 26, no. 1 (1834): 54-63.

8   Braswell, 329; Ralph Waldo Emerson, *Essays: First Series*, new ed. (Boston: James Munroe and Co., 1847), 216.

9   *Moby-Dick*, 104, 237. According to whaling historian Alexander Starbuck, of the ninety-four whaleships that left New Bedford and Fairhaven in 1841, including the *Acushnet*, twenty-one whaled most significantly in the Indian Ocean, although it's not stated if they went by the way of Good Hope (Starbuck, 372-77). For 25% see Raymond A. Rydell, *Cape Horn to the Pacific: The Rise and Decline of an Ocean Highway* (Berkeley: University of California Press, 1952), 66.

10 Scott, "The Whale in *Moby Dick*," 157-58; Wilkes, vol. 5, 483. See also *Mardi*, 4.

11 Douglas Stein, "Paths through the Sea: Matthew Fontaine Maury and His Wind and Current Charts," *The Log of Mystic Seaport* 32, no. 3 (1980): 99-101.

12 *Moby-Dick*, 199. The quotation in Melville's footnote appears in Maury, *Explanations and Sailing Directions*, 3rd ed., 207; Vincent, 184-85; Maury, "The Whale Fisheries," 3.

13   *Moby-Dick*, 198; "Abstract Log of the *Acushnet*, 1841–1844," The Maury Abstract
      Logs, 1796–1861, US National Archives and Records Service, record group 27;
      Heflin, xvii. Melville does not mention whale stamps in *Moby-Dick*, which would
      seem a delicious metaphor for Ishmael. These stamps were in use in the 1840s and
      '50s and served as an easy reference for whalemen to see what had been killed and
      where over the course of a previous voyage.

14   Rozwadowski, *Fathoming the Ocean*, 71; Andrew W. German and Daniel V. McFad-
      den, *The Charles W. Morgan: A Picture History of an American Icon* (Mystic: Mystic
      Seaport, 2014), 11; Davis, Gallman, and Gleiter, 6; Starbuck, 98.

15   Marc I. Pinsel, "The Wind and Current Chart Series Produced by Matthew Fon-
      taine Maury," *Journal of the Institute of Navigation* 28, no. 2 (Summer 1981): 125, 130,
      137; Vincent, 184; Maury, "The Whale Fisheries," 3.

16   John Leighly, "Introduction," in Matthew Fontaine Maury, *The Physical Geography
      of the Sea and Its Meteorology* (Cambridge, MA: Belknap Press, 1963), xxiv–xxvii;
      Burnett, "Maury's 'Sea of Fire,'" 116, 127–28; Matthew Fontaine Maury, *The Physi-
      cal Geography of the Sea* (New York: Harper and Brothers, 1855), 152–53, 167–70. See
      also on Maury the special issue of the *International Journal of Maritime History* 28,
      no. 2 (2016).

17   Rozwadowski, *Fathoming the Ocean*, 74; Maury, "The Whale Fisheries . . . ," 3.

18   See, for example, Nathaniel Philbrick, *Sea of Glory: America's Voyage of Discovery,
      The U.S. Exploring Expedition, 1838–1842* (New York: Penguin, 2004), 108–11.

19   For Ahab as inspired by Wilkes, see David Jaffé, *The Stormy Petrel and the Whale:
      Some Origins of Moby-Dick* (Washington, DC: University Press of America, 1982).

20   Wilkes, vol. 5, 457.

21   Wilkes, vol. 5, 482–84; explaining otherwise was Beale, 34–35, 61–62. On Wilkes
      and Melville, Jason Smith, "Charles Wilkes," *Searchable Sea Literature*, 2013, http://
      sites.williams.edu/searchablesealit/w/wilks-charles; Parker, vol. 1, 456; Bercaw
      [Edwards], *Melville's Sources*, 130. See also Anne Baker, *Heartless Immensity: Lit-
      erature, Culture, and Geography in Antebellum America* (Ann Arbor: University of
      Michigan Press, 2006), 30–43.

22   Saana Isojunno, Manuel C. Fernandes, and Jonathan Gordon, "Effects of Whale
      Watching on Underwater Acoustic Behaviour of Sperm Whales in the Kaikōura
      Canyon Area," in *Effects of Tourism on the Behaviour of Sperm Whales Inhabiting the
      Kaikōura Canyon*, ed. Tim M. Markowitz, Christoph Richter, and Jonathan Gordon
      (PACE-NZRP, 2011), 83.

23   Tim D. Smith, 12 December 2016, interview conducted by Skype and then revised
      in collaboration.

24   Tim D. Smith, Randall R. Reeves, Elizabeth A. Josephson, and Judith N. Lund,
      "Spatial and Seasonal Distribution of American Whaling and Whales in the Age of
      Sail," *PLOS One* 7, no. 4 (April 2012): 1–25.

25   Hal Whitehead, "Sperm Whales in Ocean Ecosystems," in *Whales, Whaling, and
      Ocean Ecosystems*, 325; Whitehead, *Sperm Whales: Social Evolution*, 98–100.

26   Whitehead, "Sperm Whale," *EMM*, 3rd ed., 922; Beale, 51.

27   D. Graham Burnett, *The Sounding of the Whale: Science and Cetaceans in the Twentieth
      Century* (Chicago: University of Chicago Press, 2013), 153–71.

28   M. Guerra, et al. "Diverse Foraging Strategies by a Marine Top Predator: Sperm
      Whales Exploit Pelagic and Demersal Habitats in the Kaikōura Submarine

Canyon," *Deep-Sea Research Part 1* 128 (2017): 99, 101; Ladd Irvine, Daniel M. Pala-
cios, Jorge Urbán, and Bruce Mate, "Sperm Whale Dive Behavior Characteristics
Derived from Intermediate-Duration Archival Tag Data," *Ecology and Evolution*
7 (2017): 7822–23.

29   Maury, "The Whale Fisheries," 3.

30   Staff, "Nomad Whale Spotted Back in the Strait of Gibraltar," *Gibraltar Chronicle*,
20 July 2016, http://chronicle.gi/2016/07/nomad-whale-spotted-back-in-the-strait
-of-gibraltar/; Ruth Esteban, 13 March 2018, pers. comm.; Renaud de Stephanis,
24 March 2018, pers. comm. See also E. Carpinelli, et al., "Assessing Sperm Whale
(*Physeter macrocephalus*) Movements within the Mediterranean Sea through Photo-
Identification," *Aquatic Conservation: Marine and Freshwater Ecosystems* (special
issue) 24, no. S1 (July 2014): 23–30.

NOTES TO CHAPTER SIX

1   The Pythagorean maxim includes *not* eating beans, which seems to have been a
known producer of flatulence as far back as the ancient Greeks. Melville's use of
"head winds" is a nice pun, too, as this refers to winds at the bow, but also, of course,
a toilet at sea. Stubb makes another fart joke later (352–53).

2   Lawrence, "17 September 1849," Logbook of the *Commodore Morris*.

3   Dana, 262–64. He back-pedaled on whalemen as poor sailors in a later edition.

4   Vincent, 220.

5   Maury, *The Physical Geography of the Sea* (1855), 69–74; Anders O. Perrson, "Hadley's
Principle: Understanding and Misunderstanding the Trade Winds," *History of
Meteorology* 3 (2006): 17, 30. As early as 1735, Englishman George Hadley had
put forth a paper for the Royal Society titled "On the Cause of the General Trade
Winds," the first coherent explanation as to why these winds blow in the direction
they do, but he couldn't quite get at it either.

6   *Moby-Dick*, 557, 564.

7   *Moby-Dick*, 564. Melville seems to have had little knowledge or interest in
nineteenth-century debates over the physics and causes of trade winds, yet Ahab
does mention here the poles in relation to wind.

NOTES TO CHAPTER SEVEN

1   For more on Melville and Romanticism, see Rachela Permenter, "Romantic Philosophy, Transcendentalism, and Nature," in *A Companion to Herman Melville*,
266–81. See also Bender, *Sea-Brothers*.

2   Samuel Taylor Coleridge, "The Rime of the Ancient Mariner (1834)," in *The Annotated Ancient Mariner*, ed. Martin Gardner (Amherst, NY: Prometheus Books,
2003), 47.

3   Melville did not mention sea snakes in *Moby-Dick*, but he might have seen them
himself during his Pacific travels, and he surely, like Coleridge, read about sailors
catching and eating them. See, e.g., James Colnett, *A Voyage to the South Atlantic and
Round Cape Horn Into the Pacific Ocean, for the Purpose of Extending the Spermaceti*

Whale Fisheries... (London: W. Bennett, 1798), 124; Bennett, vol. 2, 68; and John Livingston Lowes, *The Road to Xanadu: A Study in the Ways of the Imagination* (Boston: Houghton Mifflin, 1927), 49–50, 478–79.

4   *Moby-Dick*, 549; Lawrence, "12 Nov 1849," and "15 Jan 1850," Logbook of the *Commodore Morris*.

5   *Moby-Dick*, 180–81; *Typee*, 223.

6   On the biology of gulls and interactions with humans, see John Eastman, *Birds of Field and Shore: Grassland and Shoreline Birds of Eastern North America* (Mechanicsburg, PA: Stackpole Books, 2000), 80–81, 88–89, 95–96; Kenn Kaufman, "Herring Gull, *Larus argentatus*" and "Ring-billed Gull, *Larus delawarensis*," Audubon Guide to North American Birds, accessed 31 January 2019, http://www.audubon.org/field -guide/bird/herring-gull and http://www.audubon.org/field-guide/bird/ring-billed -gull.

7   *Moby-Dick*, 234.

8   *Moby-Dick*, 234. John Milton, *The Poetical Works of John Milton*, vol. 1 (Boston: Hilliard, Gray & Co., 1834), 118; Bercaw [Edwards], *Melville's Sources*, 103. In "The Encantadas" (1854), published only a few years after *Moby-Dick*, Melville quoted a line from Edmund Spenser's *Faerie Queene* (1590) that included "the cormoyrants with birds of ravenous race." See *Prose Pieces*, 133; "Cormorant, n.," *Oxford English Dictionary*, 2nd ed. (1989), www.oed.com/view/Entry/41582.

9   *Moby-Dick*, 190.

10 Derek Onley and Paul Scofield, *Albatrosses, Petrels & Shearwaters of the World* (Princeton: Princeton University Press, 2007), 32–34, 122, 124; Graham Barwell, *Albatross* (London: Reaktion, 2014), 14; "Northern Royal Albatross," New Zealand Birds Online (accessed 16 April 2018), www.nzbirdsonline.org.nz/species/northern -royal-albatross; Robert Cushman Murphy, *Oceanic Birds of South America*, vol. 1 (New York: MacMillan/American Museum of Natural History, 1836), 541–43.

11 Lowes, 222–28; Coleridge, "Mariner," 47–48; George Shelvocke, *A Voyage Round the World by the Way of the Great South Sea* (London: J. Senex, 1726), 72–74.

12 Lowes, 227; Samuel Taylor Coleridge, *Biographia Literaria* (New York: Leavitt, Lord & Co., 1834), 174. Emphasis mine; it's a pun, people!

13 *Moby-Dick*, 190; Martin, 45.

14 Herman Melville, *Omoo: A Narrative of Adventures in the South Seas*, ed. Harrison Hayford, Hershel Parker, and G. Thomas Tanselle (Evanston: Northwestern University Press and The Newberry Library, 1968), 74; Gabrielle A. Nevitt, Marcel Losekoot, and Henri Weimerskirch, "Evidence for Olfactory Search in Wandering Albatross, *Diomedea exulans*," *Proceedings of The National Academy of Sciences of the USA* 105, no. 12 (25 March 2008): 4578–79; Murphy, *Oceanic Birds of South America*, vol. 1, 544; Barwell, 95–96; William M. Davis, *Nimrod of the Sea; or, The American Whaleman* (New York: Harper and Brothers, 1874), 42.

15 Murphy, *Oceanic Birds of South America*, vol. 1, 546. See also "An Albatross Brought the News," *Sailors' Magazine and Seaman's Friend* 63, no. 7 (July 1891): 216–17.

16 Dana, 43. His comment on "Rime" in the revised edition was in Richard Henry Dana Jr., *Two Years before the Mast: A Personal Narrative of Life at Sea*, English copyright ed. (London: Sampson Low, Son, & Marston, 1869), 35; Browne, 207; Olmsted, 101–2, 111. For Melville's use (or not) of Olmsted, see Vincent, 210; Madison, *The Essex and the Whale*, 89; Bercaw [Edwards], *Melville's Sources*,107. See also

Gurdon Hall, "1 December 1842," Log of the *Charles Phelps* 1842–1844, Mystic Seaport Log 141. Albatrosses do have a gizzard—a grinding area of their stomach—but it's just a relatively small one.

17   "Gony, n.," *Oxford English Dictionary*, 2nd ed. (1989), www.oed.com/view/Entry /79920; Mary Brewster, *"She Was a Sister Sailor": The Whaling Journals of Mary Brewster, 1845–1851*, ed. Joan Druett (Mystic Seaport: Mystic, 1992), 31.

18   Martin Walters, *Bird Watch: A Survey of Planet Earth's Changing Ecosystems* (Chicago: University of Chicago Press, 2011), 123–24; Kevin T. Fitzgerald, "Longline Fishing (How What You Don't Know Can Hurt You)," *Topics in Companion Animal Research* 28 (2013): 151; see Orea R. J. Anderson, et al., "Global Seabird Bycatch in Longline Fisheries," *Endangered Species Research* 14, no. 2 (2011): 91–106. On the current status of the 22 species they evaluate, search "Albatross," in the IUCN Red List of Threatened Species, Version 2018-1, accessed 15 September 2018.

19   Scott Shaffer, "Albatross Flight Performance and Energetics," in *Albatross: Their World, Their Ways*, ed. Tui De Roy, Mark Jones, and Julian Fitter (Auckland: David Bateman, 2008), 152–53; Carl Safina, "On the Wings of the Albatross," *National Geographic* 212, no. 6 (December 2007): 91–92, 97.

20   Henry T. Cheever, *The Island World of the Pacific* (New York: Harper & Brothers, 1851), 52–63. Madison argues that Melville's note in *Moby-Dick* is lifted directly from Cheever's account, in Cheever, *The Whale and His Captors*, 213.

21   Robert Louise Chianese, "The Tales We All Must Tell," *American Scientist* 104, no. 4 (July-August 2016): 212, www.americanscientist.org/article/the-tales-we-all-must -tell.

22   Melville, *Journals*, 4.

NOTES TO CHAPTER EIGHT

1   *Moby-Dick*, 237. Herman Melville, *Mardi, and A Voyage Thither*, ed. Harrison Hayford, Hershel Parker, and G. Thomas Tanselle (Evanston: Northwestern University Press and The Newberry Library, 1970), 55. See his later "The Maldive Shark" (1888), in Herman Melville, *Published Poems*, ed. Robert C. Ryan, Harrison Hayford, Alma MacDougall Reising, and G. Thomas Tanselle (Evanston: Northwestern University Press and The Newberry Library, 2009), 236, 739–40; Vincent, 210–11; Olmsted, 95, 146; Bennett, vol. 2, 274–77.

NOTES TO CHAPTER NINE

1   *Moby-Dick*, xxv; Elizabeth Oakes Smith, "The Drowned Mariner," *Western Literary Messenger: A Family Magazine of Literature, Science, Art, Morality, and General Intelligence*, 6 (Buffalo: Clement and Faxon, 1846), 344.

2   Charles Darwin, *Journal of Researches into the Natural History and Geology . . .* , vol. 1 (New York: Harper & Brothers, 1846), 207, 208. Melville purchased this edition in 1847 (Sealts, 171).

3   Edith A. Widder, "Bioluminescence and the Pelagic Visual Environment," *Marine and Freshwater Behaviour and Physiology* 35, no. 1 (March 2002): 2–4; *"Noctiluca scin-*

tillans," Phyto'Pedia, accessed 31 January 2019, https://www.eoas.ubc.ca/research /phytoplankton/dinoflagellates/noctiluca/n_scintillans.html; Steve Miller, 8 August 2018, pers. comm.

4   E. Newton Harvey, *A History of Luminescence: From the Earliest Times Until 1900* (Philadelphia: American Philosophical Society, 1957), 534–37; Widder, 1–26; Julian C. Partridge, "Sensory Ecology: Giant Eyes for Giant Predators?" *Current Biology* 22, no. 8 (2012): R269; Séverine Martini and Steven H. D. Haddock, "Quantification of Bioluminescence from the Surface to the Deep Sea Demonstrates Its Predominance as an Ecological Trait," *Scientific Reports* 7, no. 45750 (2017): 1–11. For more on early exploration of bioluminescence, see, e.g., Arthur Hassall, "Note on Phosphorescence," *Annals and Magazine of Natural History* 9, no. 55 (London: R. and J.E. Taylor, 1842), 78.

5   Beale, 205–6; Olmsted, 72–73, 159–60; Bennett, vol. 2, p. 320–21. Bennett's below-decks whale was probably covered in bioluminescent bacteria.

6   *Moby-Dick*, 193.

7   Matthew Fontaine Maury, *Explanations and Sailing Directions to accompany the Wind and Current Charts*, 7th ed. (Philadelphia: E. C. and J. Biddle, 1855), 174–75. The substance of these milky seas wasn't hypothesized until 1985, when a research vessel sampled a glowing swath of the Arabian Sea and found that in addition to copepods and dinoflagellates and dozens of other zooplankton species there were numerous luminous bacteria *Vibrio harveyi* that had colonized on decaying organic matter floating at the surface. See David Lapota, et al., "Observations and Measurements of Planktonic Bioluminescence in and around a Milky Sea," *Journal of Experimental Marine Biology and Ecology* 119 (1988): 55. For the first image of milky seas, see Steven D. Miller, Steven H. D. Haddock, Christopher D. Elvidge, and Thomas F. Lee, "Detection of a Bioluminescent Milky Sea from Space," *Proceedings of the National Academy of Sciences of the United States of America* 102, no. 40 (4 October 2005): 14181–84; and P. J. Herring and M. Watson, "Milky Seas: A Bioluminescent Puzzle," *Marine Observer* 63 (1993): 22–30.

8   *Moby-Dick*, 234; Coleridge, "Mariner," 57, 75.

9   Melville, *White-Jacket*, 106; Herman Melville, *Redburn: His First Voyage*, ed. Harrison Hayford, Hershel Parker, and G. Thomas Tanselle (Evanston: Northwestern University Press and The Newberry Library, 1969), 244; *Mardi*, 121–23; *Moby-Dick*, 193–94, 309. See also Melville's end to "Commemorative of a Naval Victory" (1866); and J. E. Bowman, "Remarks on the Luminosity of the Sea," *Magazine of Natural History* 5, no. 23 (January 1832): 1–3.

NOTES TO CHAPTER TEN

1   *Moby-Dick*, 243.

2   Interview with Linda Greenlaw conducted on 8 May 2017.

3   *Mardi*, 104–5. See also E. W. Gudger, "The Alleged Pugnacity of the Swordfish and the Spearfishes as Shown by Their Attacks on Vessels," *Memoirs of the Royal Asiatic Society of Bengal* 12, no. 2 (Calcutta: Royal Asiatic Society of Bengal, 1940), 215–315; Cuvier, *The Class Pisces*, Pl. 27, 187–88, 351–52, 642; Sealts, 170; Cuvier, Melville's Marginalia Online; Parker, 146; Bennett, vol. 1, 272; Heflin, 137, 204; Harry L.

Fierstine and Oliver Crimmen, "Two Erroneous, Commonly Cited Examples of 'Swordfish' Piercing Wooden Ships," *Copeia* 2 (1996): 472–75; John Cruickshank, "Letter to Peter Amber, 29 October 1832," Natural History Museum, London, courtesy of Oliver Crimmen; Parker, vol. 1, 679, 685; Henry G. Clarke, *The British Museum: Its Antiquities and Natural History, a Hand-book Guide for Visitors* (London: H. G. Clarke and Co., 1848), 21.

4   Matt Rigney, *In Pursuit of Giants: One Man's Global Search for the Last of the Great Fish* (Hanover: University Press of New England, 2017), 247, 312; Hsing-Juin Lee, Yow-Jeng Jong, Li-Min Chang, and Wen-Lin Wu, "Propulsion Strategy Analysis of High-Speed Swordfish," *Transactions of the Japan Society for Aeronautical and Space Sciences* 52, no. 175 (2009): 11; B. Collette, et al., "*Xiphias gladius*," The IUCN Red List of Threatened Species, 2011/2016, www.iucnredlist.org/details/23148/0.

5   Cuvier, *The Class Pisces*, 351. Melville did not comment on this in his copy, which makes you wonder if he ever read it; Osborn, "19 January 1844"; Logkeeper, "26 March 1846," Logbook of the *Commodore Morris*, 1845–1849.

6   Beale, 48–50; Robert Weir, "1 August 1856," Journal aboard the *Clara Bell* 1855–1858, Mystic Seaport Log 164.

7   Whitehead, *Sperm Whales: Social Evolution*, 59; Richard Ellis, *Swordfish: A Biography of the Ocean Gladiator* (Chicago: University of Chicago Press, 2013), 39–40, 61–90. On a billfish stuck into a pipeline, Christopher Helman, "Swordfish Attacks BP Oilfield in Angola," *Forbes* (14 March 2014), www.forbes.com/.

8   *Moby-Dick*, 282; Callum Roberts, *The Unnatural History of the Sea* (Washington, DC: Island Press, 2007), 276–78, 323; Ransom A. Myers and Boris Worm, "Rapid Worldwide Depletion of Predatory Fish Communities," *Nature* 423, no. 6937 (15 May 2003): 280–84; Julia Lajus, "Understanding the Dynamics of Fisheries and Fish Populations: Historical Approaches form the 19th Century," in *Oceans Past*, 175–87.

9   Beale, 211–12.

10 Beale, 212; David R. Callaway, *Melville in the Age of Darwin and Paley: Science in Typee, Mardi, Moby-Dick, and Clarel* (Binghamton: State University of New York, 1999), 151.

11 Linda Greenlaw, *Seaworthy: A Swordboat Captain Returns to the Sea* (New York: Penguin, 2011), 106, 111.

12 *Moby-Dick*, 206; Walter Scott, *The Antiquary*, vol. 1 (New York: Van Winkle and Wiley, 1816), 124. See also Cheever, *The Whale and His Captors*, 76.

NOTES TO CHAPTER ELEVEN

1   Starbuck, 377.

2   *Moby-Dick*, 272.

3   *Moby-Dick*, 210 (Melville's italics), 334; Noah Webster, *The American Dictionary of the English Language*, ed. Chauncey A. Goodrich (New York: Harper and Brothers, 1848), 150; Samuel Johnson, H. J. Todd, Alexander Chalmers, and John Walker, *Johnson's English Dictionary* (Philadelphia: Griffith & Simon, 1844), 156.

4   "Krill, n.," *Oxford English Dictionary*, 2nd ed. (1989), www.oed.com/view/Entry /104465; "Plankton, n." *Oxford English Dictionary*, 2nd ed. (1989), www.oed.com /view/Entry/145123.

5   Mathew Fontaine Maury, *Explanations and Sailing Directions to Accompany the Wind and Current Charts*, 4th ed. (Washington, DC: C. Alexander, 1852), 251; Isaac J. Sanford, "8 December 1840," Log of the *Jasper* 1840-1841, New Bedford Whaling Museum ODHS Log 359; Elihu Gifford, "8 February, 1836," Log of the *America*, 1835-1838, New Bedford Whaling Museum, Log 933, *America*, 1836. Two other journals of Gifford's, which are bound with this one, have at the end of each of their last entries: "Copied for Lieut Maury." New Bedford Whaling Museum, Log 933, *America*. For other uses of brit, see Martin, 41, 172; Wright, *Meditations from Steerage*, 7; Scammon, 54; Dudley, "An Essay," 262; and Joel S. Polack, *New Zealand: Being a Narrative of Travels and Adventures* [. . .], vol. 2 (London: Richard Bentley, 1838), 402.

6   John P. Harrison, ed., "An 1849 Statement on the Habits of Right Whales by Captain Daniel McKenzie of New Bedford," *American Neptune* 14, no. 2 (1954): 140; Charles Haskins Townsend, "Chart C. Distribution of Northern and Southern Right Whales Based on Logbook Records Dating from 1785 to 1913," in "The Distribution of Certain Whales as Shown by Logbook Records of American Whaleships," *Zoologica* 19, no. 1 (New York: Society of the Zoological Park/Kraus Reprint Co., 3 April 1935), 53. The abstract log of the *Acushnet* makes no mention of seeing any baleen whales on their way around and up to the Galápagos, however. See also Starbuck, 376-77, and Heflin, 67-68.

7   Robert Kenney, 7 January 2017, pers. comm.; the sei whales are the only baleen whales that employ both strategies; Takahisha Nemoto, "Feeding Pattern of Baleen Whales in the Ocean," *Marine Food Chains*, ed. J. H. Steele (Berkeley: University of California Press, 1970), 241-52. Laist (33-34) divides baleen whales into three feeding strategies: continuous ram feeding (skimming), lunge feeding (gulping), and suction feeding (a sort of benthic vacuum, unique to gray whales). For discussion of the physics and current theories regarding the evolution of right whale skim feeding, see Laist 34-51.

8   *Moby-Dick*, 334.

9   Reeves, et al., *Guide*, 194, 198; Scoresby, *Arctic Regions*, vol. 1, 457, vol. 2, 416.

10   Robert Weir, "24 November 1855." For more on Weir and his journal, see Dyer, *Tractless Sea*, 247-56.

11   *Moby-Dick*, 334. See Scoresby, *Arctic Regions*, vol. 1, 457; Vincent, 256; Rebecca Kessler, "Written in Baleen," *Aeon* (28 September 2016), https://aeon.co/essays /the-natural-history-of-whales-is-written-in-their-baleen; Kathleen E. Hunt, et al., "Baleen Hormones: A Novel Tool for Retrospective Assessment of Stress and Reproduction in Bowhead Whales (*Balaena mysticetus*)," *Conservation Physiology* 2, no. 1 (1 January 2014), doi.org/10.1093/conphys/cou030.

12   This interview with Robert Kenney was originally conducted on 7 March 2017, then revised in collaboration. See also D. E. Gaskin, *The Ecology of Whales and Dolphins* (London: Heinemann Educational Books, 1982), 67; and Laist, 45, 49.

13   J. Mauchline, *The Biology of Calanoid Copepods*, in *Advances in Marine Biology*, vol. 33 (San Diego: Academic Press, 1998), 1; Stephen Nicol and Yoshinari Endo, "Introduction to Euphasiids or Krill," *Krill Fisheries of the World* (Rome: FAO, 1997), www .fao.org.

14   Christy Hudak, 26 April 2017, pers. comm. For common colors of plankton, E. J. Slijper, *Whales*, 2nd ed., trans. A. J. Pomerans (Ithaca, NY: Cornell University Press, 1979), 255-57.

15   Darwin, *Journal of Researches*, vol. 1, 21. See also Cheever, *The Whale and His Captors*, 29-30; and "Whales," *Penny Cyclopædia*, vol. 27, 296.

16   Beale, 61, 189; Beale, Melville's Marginalia Online. See also Cheever, *The Whale and His Captors*, 44.

17   Scoresby, *Arctic Regions*, vol. 2, plate 16.

18   Scoresby, *Arctic Regions*, vol. 1, 469.

19   William Scoresby Jr., *Journal of a Voyage to the Northern Whale-Fishery* (Edinburgh: Archibald Constable and Co., 1823), 353.

20   W. A. Watkins and W. E. Schevill, "Right Whale Feeding and Baleen Rattle," *Journal of Mammalogy* 57, no. 1 (1976): 62-63.

21   Scoresby, *Arctic Regions*, vol. 1, 179-80. See also Bennett, vol. 2, 175-76.

22   Bennett, vol. 2, 183.

23   *Moby-Dick*, 64; on Americans living on farms, Jimmy M. Skaggs, *The Great Guano Rush: Entrepreneurs and American Overseas Expansion* (New York: St. Martin's Griffin, 1995), 11; Obed Macy, *The History of Nantucket* (Boston: Hilliard, Gray, and Co., 1835), 33 (Macy's italics); Jack Scherting, "Tracking the *Pequod* along *The Oregon Trail*: The Influence of Parkman's Narrative on Imagery and Characters in *Moby-Dick*," *Western American Literature* 22, no. 1 (Spring 1987): 3-15.

24   In 2011 Matt Kish illustrated "Brit" brilliantly in his *Moby-Dick in Pictures: One Drawing for Every Page* (Portland: Tin House Books, 2011), 266. For a period comparison of the whale to the elephant that Melville read, see Cheever, *The Whale and His Captors*, 54. For the deception of the surface and Transcendentalism, see Ward, 182.

NOTES TO CHAPTER TWELVE

1   Herman Melville, "To Nathaniel Hawthorne [17?] November 1851, Pittsfield," *Correspondence*, 213.

2   *Moby-Dick*, 276.

3   Clyde F. E. Roper and Elizabeth K. Shea, "Unanswered Questions about the Giant Squid *Architeuthis* (Architeuthidae) Illustrate Our Incomplete Knowledge of Coleoid Cephalopods," *American Malacological Bulletin* 31, no. 1 (2013): 109-22; Vincent, 223-27; *Moby-Dick*, 274.

4   Richard Ellis, *The Search for the Giant Squid* (New York: Penguin, 1999), 257-65; Jonathan Ablett, "The Giant Squid, *Architeuthis dux* Steenstrup, 1857 (Mollusca: Cephalopoda): The Making of an Iconic Specimen," *NatSCA News* 23 (2012): 16.

5   All quotations in this chapter are from the interview with Jon Ablett during my visit to the Natural History Museum between 31 May to 1 June, 2016, then revised in collaboration.

6   Julian C. Partridge, "Sensory Ecology: Giant Eyes for Giant Predators?" *Current Biology* 22, no. 8 (2012): R268-70; T. Kubodera and K. Mori, "First-Ever Observations of a Live Giant Squid in the Wild," *Proceedings of the Royal Society B* 272 (2015): 2585; N. H. Landman, et al. "Habitat and age of the giant squid (*Architeuthis sanctipauli*) inferred from isotopic analyses," *Marine Biology* 144 (2004): 685.

7   See K. S. Bolstad and S. O'Shea, "Gut Contents of a Giant Squid *Architeuthis dux* (Cephalopoda: Oegopsida) from New Zealand Waters," *New Zealand Jour-*

nal of Zoology 31, no. 1 (2010): 15, 18; Ronald B. Toll and Steven C. Hess, "A Small, Mature Male *Architeuthis* (Cephalopoda: Oegopsida) with Remarks on Maturation in the Family," *Proceedings of the Biological Society of Washington* 94, no. 3 (1981): 756; and Roper and Shea, 114–15.

8   C. G. M. Paxton, "Unleashing the Kraken: On the Maximum Length in Giant Squid (*Architeuthis* sp.)," *Journal of Zoology* 300, no. 2 (2016): 82; Charles Paxton, 21 September 2016, pers. comm.; Michael J. Sweeney, "Records of *Architeuthis* Specimens from Published Reports," National Museum of Natural History, Smithsonian (2001): 1–131; Charles Paxton, 12 September 2016, pers. comm. See also Clyde F. E. Roper, et al. "A Compilation of Recent Records of the Giant Squid, *Architeuthis dux* (Steenstrup, 1857) (Cephalopoda) from the Western North Atlantic Ocean, Newfoundland to the Gulf of Mexico," *American Malacological Bulletin* 33, no. 1 (2015): 78–88; Ellis, *The Search for the Giant Squid*, 80–92; Richard Ellis, "*Architeuthis*—The Giant Squid: A True Explanation for Some Sea Serpents," *Log of Mystic Seaport* (Autumn 1994): 35–36; Kubodera and Mori, "First-Ever Observations," 2583–84; Mark Schrope, "Giant Squid Filmed in Its Natural Environment," *Nature News* (14 January 2013), doi:10.1038/nature.2013.12202.

9   Roper, et al. "A Compilation of Recent Records of the Giant Squid," 80; and Roper and Shea, 114.

10   Nelson Cole Haley, *Whale Hunt: The Narrative of a Voyage by Nelson Cole Haley, Harpooner in the Ship Charles W. Morgan, 1849–1853*, 3rd ed. (Mystic: Mystic Seaport, 1990), 219–23.

11   Lukas Rieppel, "Albert Koch's *Hydrarchos* Craze: Credibility, Identity, and Authenticity in Nineteenth-Century Natural History," in *Science Museums in Transition: Cultures of Display in Nineteenth-Century Britain and America*, ed. Carin Berkowitz and Bernard Lightman (Pittsburgh: University of Pittsburgh Press, 2017), 144–47; Eugene Batchelder, *A Romance of The Sea-Serpent, or The Ichthyosaurus*, 4th ed. (Cambridge: John Bartlett, 1850), 127, 135–37; Parker, vol. 1, 746; Clyde Roper, 10 January 2017, pers. comm.; Ellis, *The Search for the Giant Squid*, 10–30; Charles G. M. Paxton, "Giant Squids Are Red Herrings: Why *Architeuthis* Is an Unlikely Source of Sea Monster Sightings," *Cryptozoology Review* 4, no. 2 (Autumn 2004): 10–16.

12   Clyde Roper, 10 January 2017, pers. comm.

13   *Moby-Dick*, 277. See Pontoppidan as quoted, abridged, in John Knox, ed., *A New Collection of Voyages, Discoveries and Travels*, vol. 4 (London: J. Knox, 1767), 100–102; and Erich Pontoppidan, *The Natural History of Norway* (London: A. Linde, 1755), 212–13; Herman Melville, *Moby-Dick*, 3rd ed., edited by Hershel Parker (New York: W. W. Norton & Co., 2018), 216n4. See also Paxton, "Giant Squids Are Red Herrings," 13.

14   See, for example, Colnett, 171. So far I've only found one illustration of a squid in a logbook or journal by a nineteenth-century whaleman, but this was of a small flying squid. Martin, 67 (in the back of the original journal), 175.

15   *Moby-Dick*, 282.

16   Cheever, *The Whale and His Captors*, 51; Beale, 8, 66. See also Olmsted, 156, and Logbook of the *Annawan* 1848–1850, New Bedford Whaling Museum Log 16; Karen Evans and Mark A. Hindell, "The Diet of Sperm Whales (*Physeter macrocephalus*) in Southern Australian Waters," *ICES Journal of Marine Science* 61, no. 8

(2004): 1313-29; and M. R. Clarke and P. L. Pascoe, "Cephalopod Species in the Diet of a Sperm Whale (*Physeter catodon*) Stranded at Penzance, Cornwall," *Journal of the Marine Biological Association of the United Kingdom* 77, no. 4 (1997): 1255-58.

17   Bennett, vol. 2, 175; Olmsted, 156; Beale, 63; Joseph Banks, *Journal of the Right Hon. Sir Joseph Banks*, ed. Sir Joseph D. Hooker (London: MacMillan and Co., 1896), 65. The Banks specimen is now classified as the Dana octopus squid (*Taningia danae*). You can see this beak today in an old bottle of preservative on display at the Hunterian Museum in London.

18   Beale, 59-60; Clyde Roper, 10 January 2017, pers. comm.

19   A. Fais, et al., "Sperm Whale Predator-Prey Interactions Involve Chasing and Buzzing, but No Acoustic Stunning," *Scientific Reports* 6, no. 28562 (24 June 2016): 1-13. Clyde Roper is also dubious of this theory.

20   Clarke and Pascoe, 1256.

NOTES TO CHAPTER THIRTEEN

1   *Moby-Dick*, 110, 169, 542.

2   *Moby-Dick*, 127, 190, 308-9, 498. Melville read of this derivation in part in Cuvier, *The Class Pisces*, 632-33.

3   *Moby-Dick*, 293.

4   *Moby-Dick*, 293. See also "Brit," 273, for the first comparison of sharks to dogs.

5   *Moby-Dick*, 293.

6   *Moby-Dick*, 302.

7   *Moby-Dick*, 302. See especially Edgar Allan Poe, *Arthur Gordon Pym: or, Shipwreck, Mutiny, and Famine*... (London: John Cunningham, 1841), 40-41. See also Parker, vol. 1, 370; Bercaw [Edwards], *Melville's Sources*, 110.

8   *Moby-Dick*, 573.

9   *Moby-Dick*, 321.

10   Olmsted, 5. His dissertation was titled "Dissertation on the Use of Narcotics in the Treatment of Insanity" (New Haven: Yale University, 1844), Archives T113 Y11 1844.

11   Michael P. Dyer, "Francis Allyn Olmsted," *Searchable Sea Literature*, 2000, sites.williams.edu/searchablesealit/o/olmsted-francis-allyn/.

12   Olmsted, 170, 355; W. Storrs Lee, "Preface to the New Edition," in Olmsted, vii-x.

13   Olmsted, 146.

14   Olmsted, 184.

15   Robert Weir, "22 August 1855." On 1 Aug 1856 Weir drew a "shovel-nosed shark," meaning a hammerhead, and a "bone shark," meaning a whale or basking shark. He described both species, including that the hammerhead is "said to be perfectly harmless though it has a loathsome appearance."

16   Beale, 276; Colnett, 89.

17   Browne, 130-31.

18   Cheever, *The Whale and His Captors*, 76; Brewster, 343; Davis, 248-49; William B. Whitecar Jr., *Four Years Aboard the Whaleship* (Philadelphia: J. B. Lippincott, 1864), 146-47.

19   The Society for the Diffusion of Useful Knowledge, "Squalidæ," *The Penny Cyclopædia*, vol. 22 (London: Charles Knight and Co., 1842), 391-93; Samuel Maunder, *The Treasury of Natural History, or A Popular Dictionary of Animated Nature*, 3rd ed.

(London: Longman, Brown, Green, and Longmans, 1852), 605; Good, 11, 185; Todd Preston, "*Moby-Dick* and John Singleton Copley's *Watson and the Shark*," *Melville Society Extracts* 129 (July 2005): 3–4. On the longest white shark, David A. Ebert, Sarah Fowler, and Marc Dando, *A Pocket Guide to Sharks of the World* (Princeton: Princeton University Press, 2015), 110.

20 Cuvier, *The Class Pisces*, 634.

21 *Mardi*, 39.

22 *Mardi*, 40. David Ebert, "Help Crowdfund Shark Research: Jaws, Lost Sharks, and the Legacy of Peter Benchley," *Southern Fried Science* (22 June 2016), www.southern friedscience.com; Ebert, et al., 5; David Ebert, 16 November 2017, pers. comm.; Bennett, vol. 1, 165.

23 Ebert, pers. comm.; Marcus Rediker, "History from below the Water Line: Sharks and the Atlantic Slave Trade, *Atlantic Studies* 5, no. 2 (2008): 291–92.

24 All quotations in this chapter are from the interview on 16 December 2017 with David Ebert at Moss Landing, CA, with subsequent correspondence for clarification and fact-checking.

25 In August 1912, naturalist Robert Cushman Murphy observed in the Atlantic sharks around a dead sperm whale: "Some were washed back into deep water badly mutilated but still able to swim, and these, even though their entrails were hanging out of the side of a body that had been cut away, would turn again toward the whale, bury their teeth, twist, yank, and swallow. I believe, though I am not quite certain, that one or more of these insensate fish were taking food in at the mouth and immediately losing it through a stomach that had been severed by a blubber spade." Murphy, *Logbook for Grace*, 71.

26 Bolster, 101; J. D. Stevens, R. Bonfil, N. K. Dulvy, and P. A. Walker, "The Effects of Fishing on Sharks, Rays, and Chimaeras (Chondrichthyans), and the Implications for Marine Ecosystems," *ICES Journal of Marine Science* 57 (2000): 476–77. On historic reports see, e.g., Roberts, 73–75, 242–57.

27 Greenlaw, *Seaworthy*, 153.

28 Roberts, 283; Myers and Worm, 282; Ewa Magiera and Lynne Labanne, "A Quarter of Sharks and Rays Threatened with Extinction," IUCN, 21 January 2014, www .iucn.org/content/quarter-sharks-and-rays-threatened-extinction; J. Stevens, "*Prionace glauca*," The IUCN Red List of Threatened Species, 2009, www.iucnredlist.org /details/39381/0; I. Fergusson, L. J. V. Compagno, and M. Marks, "*Carcharodon carcharias*," The IUCN Red List of Threatened Species, 2009, www.iucnredlist.org /details/3855/0; J. Denham, et al., "*Sphyrna mokarran*," The IUCN Red List of Threatened Species, 2007, www.iucnredlist.org/details/39386/0.

29 Mary Schley, "Off-Duty Deputies Save Shark Bite Victim," *Carmel Pine Cone*, 1 December 2017, 5A, 23A.

30 Chris Fallows, Austin J. Gallagher, and Neil Hammerschlag, "White Sharks (*Carcharodon carcharias*) Scavenging on Whales and Its Potential Role in Further Shaping the Ecology of an Apex Predator," *PLOS One* 8, no. 4 (2013): 5; Sheldon F. J. Dudley, Michael D. Anderson-Reade, Greg S. Thompson, and Paul B. McMullen, "Concurrent Scavenging off a Whale Carcass by Great White Sharks, *Carcharodon carcharias*, and Tiger Sharks, *Galeocerdo cuvier*," *Fishery Bulletin* 98, no. 3 (2000): 646–47.

31 Chris Fallows, 4 December 2017, pers. comm.

32 Zoellner, 220.

33 John Stauffer, "Melville, Slavery, and the American Dilemma," in *A Companion to Herman Melville*, 218.

34 *Moby-Dick*, 60, 89, 143, 321.

35 Zoellner, 220–25.

36 *Moby-Dick*, 295.

37 Rediker, 286–88, 291–95. Melville likely read of this painting, but never saw it. See Robert K. Wallace, *Melville and Turner: Spheres of Love and Fright* (Athens: University of Georgia Press, 1992), 63, 212, 588, 608.

38 Cuvier, *The Class Pisces*, 632. See also Ishmael's line in "The Whiteness of the Whale" about "giving the white man ideal mastership over every dusky tribe" (189). See analysis on Melville and race, particularly in relation to Pip and Queequeg, in Christopher Freeburg, *Melville and the Idea of Blackness: Race and Imperialism in Nineteenth-Century America* (Cambridge: Cambridge University Press, 2012).

39 *Moby-Dick*, 492.

NOTES TO CHAPTER FOURTEEN

1 *Moby-Dick*, 298.

2 *Moby-Dick*, 445.

3 Sandy Oliver, 13 December 2017, pers. comm.; Creighton, 125–26. For a humorous account of foul bread see Henry Eason, "28 May 1860," Logbook of the USS *Marion* 1858–60, Mystic Seaport Log 902; Mary K. Bercaw Edwards, Lecture, 5 April 2016, Mystic Seaport; *Typee*, 3; Simon Spalding, *Food at Sea: Shipboard Cuisine from Ancient to Modern Times* (London: Rowman and Littlefield, 2015), 88–89; Sandra L. Oliver, *Saltwater Foodways* (Mystic: Mystic Seaport Museum, 1995), 105–6; *Omoo*, 202.

4 *Moby-Dick*, 262. On Colnett's whale illustration see Frank, *Herman Melville's Picture Gallery*, 26–27. For Melville's reading of Colnett, see Madison, *The Essex and the Whale*, 221; Parker, vol. 1, 723; Bercaw [Edwards], *Melville's Sources*, 70.

5 "Giant Tortoises," Galapagos Conservancy, www.galapagos.org/about_galapagos /about-galapagos/biodiversity/tortoises/. See, for example, P. P. van Dijk, A. G. J. Rhodin, L. J. Cayot, and A. Caccone, "*Chelonoidis niger*," The IUCN Red List of Threated Species, 2017, www.iucnredlist.org/details/9023/0; *Moby-Dick*, 242; Heflin, 100–102; Philbrick, *In the Heart of the Sea*, 72–75; Colnett, 73, 158.

6 Colnett, 107–8, 123.

7 Lawrence, 4 April 1850, 14 August 1850, 6 December 1850, etc., Logbook of the *Commodore Morris* 1849–1853.

8 Creighton, 127; Osborn, "15 October, and 4, 10, and 12 November, 1851, et al."

9 Cheever, *The Whale and His Captors*, 44; Weir, "16 April 1856"; Jones, *Meditations from Steerage*, 16; *Moby-Dick*, 144, 298; Olmsted, 92.

10 *Moby-Dick*, 299; Browne, 62–63. See also Olmsted, 65–66.

11 Scoresby, *Arctic Regions*, vol. 1, 463, 475–76; Bennett, vol. 2, 167, 232; *Moby-Dick*, 388; Martin, 79.

12 Colnett, 80; Murphy, *Logbook for Grace*, 72.

13 Severin, 158–162; Nancy Shoemaker, "Whale Meat in American History," *Environmental History* 10, no. 2 (April 2005): 273. For images of the current sperm whale processing at Lamalera, see "Dividing the Whale among the Villagers," Photo-

voices International, www.photovoicesinternational.org. For more on the historical accuracy of "The Whale as a Dish," see *Moby-Dick*, ed. Luther S. Mansfield and
Howard P. Vincent (New York: Hendricks House, 1952), 757.

14   Shoemaker, 278–79. For an example of early twentieth-century whalemen eating
whale meat see Murphy, *Logbook for Grace*, 118, etc., and see Olmsted, 92–93, for
Ishmael-like ideas about prejudices against whale meat; *Moby-Dick*, 30. Also, 284.

15   Henry David Thoreau, *Cape Cod* (New York: Penguin, 1987), 166; Shoemaker,
276–77.

16   *Moby-Dick*, 143–44.

17   Diane L. Beers, *For the Prevention of Cruelty: The History and Legacy of Animal Rights
Activism in the United States* (Athens: Ohio University Press, 2006), 22; "Status of
Whales," and "Population Estimates," International Whaling Commission (2018),
https:/iwc.int/status; J. G. Cook, "*Balaenoptera acutorostrata*," The IUCN Red List
of Threatened Species, 2018, www.iucnredlist.org/species/2474/50348265.

18   *Moby-Dick*, 157; "Whale Friendly Restaurants," IceWhale, icewhale.is; "Visiting
Iceland? Help us Keep Whales off the Dinner Menu," Whale and Dolphin Conservation, us.whales.org/campaigns/visiting-iceland-help-us-keep-whales-off-dinner
-menu.

19   Jonny Zwick, director, *Breach* (Side Door Productions, 2015). For more on this
debate, see Karen Oslund, "Of Whales and Men: Images of Iceland and the North
Atlantic in Contemporary Whaling Politics," in *Images of the North: Histories—
Identities—Ideas*, ed. Sverrir Jakobsson (Amsterdam: Rodopi, 2009), 91–101; Larissa
Kyzer, "Whale Meat Popular among Tourists," *Iceland Review* (15 June 2017),
icelandreview.com/news.

NOTES TO CHAPTER FIFTEEN

1   *Moby-Dick*, 134, 266. See also Frank, *Melville's Picture Gallery*, 73, 77.

2   *Moby-Dick*, 333–34.

3   Scoresby, Cheever, and Beale, for example, all do not use the term callosity or callosities. See also "Callosity, n.," *Oxford English Dictionary*, 3rd ed. (2016), www.oed
.com/view/Entry/26461; Laist, 273; Kenney, "Right Whales," *EMM*, 3rd ed.,
817–18; Victoria J. Rowntree, "Callosities," *EMM*, 3rd ed., 157–58.

4   Bennett, vol. 2, 169; Cheever, *The Whale and His Captors*, 60; Jon Seger and Victoria J. Rowntree, "Whale Lice," *EMM*, 3rd ed, 1051–54; Juan Antonio Raga, Mer-
cedes Fernández, Juan A. Balbuena, and Francisco Javier Aznar, "Parasites," *EMM*,
3rd ed, 685.

5   Dagmar Fertl and William A. Newman, "Barnacles," *EMM*, 3rd ed., 75–77;
James E. Scarff, "Occurrence of the Barnacles *Coronula Diadema, C. Reginae*, and
Cetopirus Complanatus (Cirripedia) on Right Whales," *Scientific Reports of the Whales
Research Institute* 37 (1989): 130. Neither Bennett or Darwin use the *Coronula* scien-
tific name in the books that Melville read before 1851.

6   Fertl and Newman, "Barnacles," *EMM*, 3rd ed., 75; Kenney, "Right Whales,"
EMM, 2nd ed., 964; Rowntree, "Callosities," *EMM*, 3rd ed., 158; Bennett, vol. 2,
164–65.

7   Weir, "9 December 1855." See also Martin, 79, and Cheever, *The Whale and His Captors*, 186.

NOTES TO CHAPTER SIXTEEN

1   Justin T. Richard, et al., "Testosterone and Progesterone Concentrations in Blow
    Samples Are Biologically Relevant in Belugas (*Delphinapterus leucas*)," *General and
    Comparative Endocrinology* 246 (2017): 183–84.
2   Bennett, vol. 2, 151; Beale, 16–17.
3   Justin Richard, 26 October 2016, pers. comm., with details refined in subsequent
    discussions and emails.
4   *Moby-Dick*, xxii; Bennett, vol. 2, 174; Vincent, 292–93. See also Barbara Todd,
    Whales and Dolphins of Kaikōura, New Zealand (Nelson, NZ: Nature Down Under,
    2007), 20.
5   Burnett, *Trying Leviathan*, 125–26.
6   *Moby-Dick*, 222. See also Cheever, *The Whale and His Captors*, 92.
7   *Moby-Dick*, 139, 162; Eschricht as cited in Burnett, *Trying Leviathan*, 126.
8   J. J. Rasler, 9 August 2016, 7 October 2017, pers. comm.
9   *Moby-Dick*, 232–33.
10 *Moby-Dick*, 372; Annalisa Berta, James L. Sumich, and Kit M. Kovacs, *Marine
    Mammals: Evolutionary Biology*, 2nd ed. (Boston: Elsevier, 2006), 155; Reeves, et al.,
    Guide, 28; J. G. M. Thewissen, John George, Cheryl Rosa, and Takushi Kishida,
    "Olfaction and Brain Size in the Bowhead Whale (*Balaena mysticetus*)," *Marine
    Mammal Science* 27, no. 2 (April 2011): 282–94; Takushi Kishida, et al., "Aquatic
    Adaptation and the Evolution of Smell and Taste in Whales," *Zoological Letters* 1,
    no. 9 (2015): 1–10.
11 *Moby-Dick*, 370.
12 Beale, 43–44; "Whales," *Penny Cyclopædia*, vol. 27, 294 (rewritten from Beale). See
    also Wright, *Meditations from Steerage*, 6; William A. Watkins, et al. "Sperm Whale
    Dives Tracked by Radio Tag Telemetry," *Marine Mammal Science* 18, no. 1 (January
    2002), 55; Ladd Irvine, Daniel M. Palacios, Jorge Urbán, and Bruce Mate, "Sperm
    Whale Dive Behavior Characteristics Derived from Intermediate-Duration Archi-
    val Tag Data," *Ecology and Evolution* 7 (2017): 7834; Hal Whitehead and Luke Ren-
    dell, *The Cultural Lives of Whales and Dolphins* (Chicago: Chicago University Press,
    2015), 152.
13 Colin D. MacLeod, "Beaked Whales," *EMM*, 3rd ed., 82; Peter L. Tyack, et al.,
    "Extreme Diving of Beaked Whales," *Journal of Experimental Biology* 209 (2006):
    4238, 4246–48; George Edgar Folk Jr., Marvin L. Riedesel, and Diana L. Thrift,
    Principles of Integrative Environmental Physiology (Bethesda, MD: Austin & Win-
    field, 1998), 400; Scoresby, *Arctic Regions*, vol. 2, 249–50. See also Charles W. Mor-
    gan, "Address before the New Bedford Lyceum," 19, who repeated Scoresby's calcu-
    lations directly; Beale, 92; Paul J. Ponganis and Gerald Kooyman, "How Do Deep
    -Diving Sea Creatures Withstand Huge Pressure Changes?," *Scientific American*
    (2 May 2002), www.scientificamerican.com/article/how-do-deep-diving-sea-cr/.
14 *Moby-Dick*, 357, 371; Paul Ponganis, 8 February 2018, pers. comm.; Paul J. Ponganis,
    "Circulatory System," *EMM*, 3rd ed., 192–93; Berta, Sumich, and Kovacs, *Marine
    Mammals*, 244; "Whales," *Penny Cyclopædia*, vol. 27, 284–85; Beale, 103–5, 106.
15 William E. Damon, *Ocean Wonders: A Companion for the Seaside* (New York:
    D. Appleton and Co., 1879), 175–76. See also Philip Hoare, *The Whale* (New York:
    Ecco, 2010), 12.

16   *Moby-Dick*, 310, 340.

17   *Moby-Dick*, 449; Bennett, vol. 2, 167–69. See also Haley, 250.

18   J. B. S. Jackson, "Dissection of a Spermaceti Whale and Three Other Cetaceans," *Boston Journal of Natural History* 5, no. 2 (October 1845): 10–171.

19   On "sea canaries," see Henry Lee, "The White Whale" (London: R. K. Burt & Co., 1878), 3. Oddly, Melville does not mention belugas in *Moby-Dick* at all—perhaps the whiteness was too close to home—unless he was referring to them as the "Iceberg Whale" in "Cetology." Melville would never have seen a beluga in person by 1851, but he definitely saw an illustration of this species in Scoresby's narrative as well as in the "Whales" entry of his *Penny Cyclopædia*.

20   Stefan Huggenberger, Michel André, and Helmut H. A. Oelschläger, "The Nose of the Sperm Whale: Overviews of Functional Design, Structural Homologies, and Evolution," *Journal of the Marine Biological Association of the United Kingdom* 96, no. 4 (2016): 783–84, 793.

21   Bennett, vol. 2, 224; Dale W. Rice, "Spermaceti," in *Encyclopedia of Marine Mammals*, ed. William F. Perrin, Bernd Würsig, and J. G. M. Thewissen, 2nd ed. (New York: Academic Press, 2009), 1098–99.

22   Bouk and Burnett, 436; David Littlefield with Edward Baker, "Oil from Whales," in Heflin, 231–40.

23   *Moby-Dick*, 338.

24   Beale, 54.

25   Gerhard Neuweiler, *The Biology of Bats*, trans. Ellen Covey (Oxford: Oxford University Press, 2000), 140–41; Wright, *Meditations from Steerage*, 5.

26   Woods Hole Oceanographic Institution, "Bill Schevill," *Woods Hole Currents* 1, no. 2 (Spring 1992): 6–7; Berta, Sumich, and Kovacs, *Marine Mammals*, 279; B. Mohl, et al. "The Monopulsed Nature of Sperm Whale Clicks," *Journal of The Acoustical Society of America* 114, no. 2 (August 2003): 1143; Huggenberger, et al., 790, 795.

27   Richard Sears and William F. Perrin, "Blue Whale," *EMM*, 3rd ed., 113.

28   Although the term "carpenter fish" is often written, I've yet to find any nineteenth-century source by or about whalemen that actually uses this phrase or even mentions sounds through the hull.

29   *Moby-Dick*, 331; Bennett, vol. 2, 159, 180; and Beale, 114–15, but none of these sources mention any plug or membrane in baleen whales.

30   *Moby-Dick*, 283; on whalemen and sperm whale vision, see Davis, 169–70; and Wright, *Meditations from Steerage*, 6; Marta Guerra, 2 January 2008, pers. comm.; Severin, 198.

31   See Huggenberger, et al., 794–96.

32   *Moby-Dick*, 379.

NOTES TO CHAPTER SEVENTEEN

1   Adrienne Wilber, 5 August 2017, pers. comm.; Riley Woodford, "Sperm Whales Awe and Vex Alaska Fisherman," *Alaska Fish and Wildlife News* (August 2003), http://www.adfg.alaska.gov/index.cfm?adfg=wildlifenews.view_article&articles _id=61; BBC (film), "Sperm Whales and Fishing Boats," *Alaska: Earth's Frozen Kingdom*, 27 January 2015, http://www.bbc.co.uk/programmes/p02htrqb.

2 Whitehead and Rendell, 3–7, 11–12, 153–57, 252–54.

3 Kathleen Dudzinski, Lecture, 8 December 2016, Mystic Aquarium; Jason N. Bruck, "Decades-Long Social Memory in Bottlenose Dolphins," *Proceedings of the Royal Society B* 280, no. 1768 (7 October 2013): 1–6.

4 *Moby-Dick*, 199, 348–50, 549; Beale, 29–30, 79. For other whalemen who saw sperm whales as more intelligent, see, for example, Wright, *Meditations from Steerage*, 6.

5 Susanne Shultz, "Whales and dolphins have rich cultures—and could hold clues to what makes humans so advanced," *The Conversation*, 18 October 2017, https://theconversation.com/whales-and-dolphins-have-rich-cultures-and-could-hold-clues-to-what-makes-humans-so-advanced-85858; Kieran C. R. Fox, Michael Muthukrishna, and Susanne Shultz, "The Social and Cultural Roots of Whale and Dolphin Brains," *Nature Ecology and Evolution* 1 (2017): 1699–1705; Whitehead, "Sperm Whales," *EMM*, 3rd ed., 919; Burnett, *The Sounding of the Whale*, 622–23. See also Samuel L. Metcalfe, *Caloric: Its Mechanical, Chemical and Vital Agencies in the Phenomena of Nature*, vol. 1 (Philadelphia: Lippincott & Co., 1859), 540–46.

6 Parker, vol. 1, 689; Richard Dean Smith, *Melville's Science: "Devilish Tantalization of the Gods!"* (New York: Garland Publishing, 1993), 127–28; Sealts, 193; Bercaw [Edwards], *Melville's Sources*, 97; Darwin, *Journal of Researches*, vol. 2, 203–4; Janet Browne, *Charles Darwin Voyaging*, vol. 1 (London: Pimlico, 2003), 161; The Darwin Human Nature Project, "Darwin and Phrenology," *Darwin Correspondence Project Blog*, 24 November 2010, https://darwinhumannature.wordpress.com/2010/11/24/darwin-and-phrenology. On phrenology and physiognomy as connected to *Typee*, see Otter, *Anatomies*, 30–38.

7 John Caspar Lavater, *Essays on Physiognomy: Designed to Promote the Knowledge and the Love of Mankind*, trans. by Thomas Holcroft, 5th ed. (London: William Tegg & Co., 1848), 217–18, 304.

8 J. G. Spurzheim, M. D., *Phrenology, or the Doctrine of the Mental Phenomena*, 5th ed. (New York: Harper & Brothers, 1846), 223.

9 On sources, see Cheever, *The Whale and His Captors*, 74; Vincent, 266–67.

10 *Moby-Dick*, 347.

11 *Moby-Dick*, 311–12.

NOTES TO CHAPTER EIGHTEEN

1 *Moby-Dick*, 403. See also Cheever, *The Whale and His Captors*, 34.

2 *Moby Dick*, 407.

3 Beale, 133; Dudley, "An Essay," 262, 265–69; Dale W. Rice, "Ambergris," *EMM*, 2nd ed., 29.

4 *Moby-Dick*, 407–9.

5 Robert Clarke, "The Origin of Ambergris," *Latin American Journal of Aquatic Mammals* 5, no. 1 (June 2006): 11. See also Robert Clarke, "A Great Haul of Ambergris," *Nature* 174, no. 4421 (1954): 155–56; Christopher Kemp, *Floating Gold* (Chicago: Chicago University Press, 2012), 96.

6 Kemp, *Floating Gold*, frontmatter; R. Clarke, "The Origin of Ambergris," 11–13; Dr. [Zabdiel] Boylston, "Ambergris Found in Whales," 33, no. 385 [1724] in *Philosophical Transactions of the Royal Society of London* [1665–1800], vol. 7 (London: C. and R. Baldwin, 1809), 57. The latter was quoted in full in Beale, 131–32.

7   Beale, 135. See also "Proceedings of the Society for the Communication of the Useful Arts in Scotland: Antediluvian Ambergris," *Edinburgh New Philosophical Journal* 15 (April–October 1833): 398; Vincent, 317–26; Kemp, *Floating Gold*, 144; Christopher Kemp, "Ambergris," *EMM*, 3rd ed., 24. See also Bennett, vol. 2, 226.

8   Beale, 132; Beale, Melville's Marginalia Online; Mr. Payne, "Ambergris," *American Journal of Pharmacy* 9, no. 4 (1844): 296 (Payne cites both Beale and Bennett); Kemp, *Floating Gold*, 80. See also Hoare, *The Whale*, 393–99.

9   William H. Griffith, 29 March 1913, Journal aboard the *Charles W. Morgan*, 1911–1913, Log 157, Mystic Seaport; Starbuck, 148.

10   Ambergris NZ, Ltd, accessed 17 January 2019, www.ambergris.co.nz/buy-ambergris.

11   Hasan Shaban Al Lawati, "Omani Fishermen Net a Fortune in a Catch," *Times of Oman* (2 November 2016), http://timesofoman.com/article/95707/OmanPride/.

12   *Moby-Dick*, 409.

NOTES TO CHAPTER NINETEEN

1   William Shak[e]speare, *The Tempest*, in *The Dramatic Works of William Shakespeare*, vol. 1 (Boston: Hilliard, Gray, 1837), Act 1, Sc. 2, 23; Shakespeare, Melville's Marginalia Online; Sealts, 213.

2   See Dan Brayton, *Shakespeare's Ocean: An Ecocritical Exploration* (Charlottesville: University of Virginia Press, 2012), 53–55; and Steve Mentz, *At the Bottom of Shakespeare's Ocean* (London: Continuum, 2009), 1–13.

3   See Herman Melville, "To Evert A. Duyckinck, 24 February 1849, Boston," *Correspondence*, 119–20.

4   On the cabin boy on the *Acushnet*, Parker, vol. 1, 697.

5   *Moby-Dick*, 414.

6   *Moby-Dick*, 295, 522, 535.

7   Dobbs, 145–46; Bennett, vol. 1, 95; Maury, *The Physical Geography of the Sea* (1855), 170.

8   Darwin, *Journal of Researches*, vol. 2, 260–81; Charles Darwin, *The Structure and Distribution of Coral Reefs, being the first part of the geology of the voyage of the Beagle* (London: Smith, Elder, and Co., 1842); Dobbs, 5, 254–56; Janet Browne and Michael Neve, "Introduction," in Charles Darwin, *Voyage of the Beagle* (New York: Penguin, 1989), 20; *Omoo*, 62–63; *Typee*, 155; Harvey, "Science and the Earth," 73–74.

9   Anonymous, "The Drowned Harpooner," *Casket* 2 (February 1827): 65; Martina Pfeiler, *Ahab in Love: The Creative Reception of Moby-Dick in Popular Culture*, Habilitation thesis, TU Dortmund, Germany (May 2017), 85–87.

10   B. D. Emerson, ed., *The First Class Reader* (Philadelphia: Hogan and Thompson, 1843), 56; Emerson, "The Uses of Natural History," 12.

11   *Relics from the Wreck of a Former World; or Splinters Gathered on the Shores of a Turbulent Planet* (New York: Henry Long and Brother, 1847), 8, 12.

12   *Omoo*, 162; J. E. N. Veron and Mary Stafford-Smith, *Corals of the World*, vol. 3 (Townsville: Australian Institute of Marine Science, 2000), 280–91; "National Marine Sanctuary of American Samoa and Rose Atoll Marine National Monument," National Marine Sanctuary Foundation, accessed 31 January 2019, www.marinesanctuary.org/explore/american-samoa/.

13 Damien Cave and Justin Gillis, "Large Sections of Australia's Great Reef Are Now Dead, Scientists Find," *New York Times* (15 March 2017), https://www.nytimes .com/2017/03/15/science/great-barrier-reef-coral-climate-change-dieoff.html; Terry P. Hughes, et al., "Global Warming and Recurrent Mass Bleaching of Corals," *Nature* 543 (16 March 2017): 373; Dennis Normile, "Survey Confirms Worst-Ever Coral Bleaching at Great Barrier Reef," *Science*, 19 April 2016, http:// www.sciencemag.org/news.

14 *Omoo*, 192. Melville seems to have actually learned this from William Ellis's *Polynesian Researches* (1831). See David Farrier, *Unsettled Narratives: The Pacific Writings of Stevenson, Ellis, Melville and London* (New York: Routledge, 2007), 13–15.

NOTES TO CHAPTER TWENTY

1 *Moby-Dick*, 419.
2 *Moby-Dick*, 20.
3 Interview with Michael Dyer was originally conducted on 6 January 2017, then revised in collaboration after later visits and correspondence.
4 Hjörtur Gisli Sigurdsson, 29 March 2017, pers. comm.
5 *Moby-Dick*, 416.
6 *Moby-Dick*, 388; Bennett, vol. 2, 178.
7 Whitehead, *Sperm Whales: Social Evolution*, 271–77; *Moby-Dick*, 391–93; Sarah L. Mesnick and Katherine Ralls, "Mating Systems," and "Sexual Dimorphism," *EMM*, 3rd ed., 587, 590, 848, 1116; Koen Van Waerebeek and Bernd Würsig, "Dusky Dolphin," *EMM*, 3rd ed., 279; Reeves, et al., *Guide*, 24; Kenney, "Right Whales," *EMM*, 3rd ed., 819–820; Dudley, "An Essay," 260.
8 Tom Dalzell, *The Routledge Dictionary of Modern American Slang and Unconventional English* (New York: Routledge, 2009), 283; "Dick, n. 1," *Oxford English Dictionary*, 2nd ed. (1895/1989), www.oed.com/view/Entry/52255.
9 D. H. Lawrence, *Studies in Classic American Literature*, ed. Ezra Greenspan, Lindeth Vasey, and John Worthen (Cambridge: Cambridge University Press, 2003), 354, 294; *Moby-Dick*, 463. See, for example, Herbert N. Schneider and Homer B. Pettey, "Melville's Ichthyphallic God," *Studies in American Fiction* 26, no. 2 (Autumn 1998): 193–212; Harry Slochower, "Freudian Motifs in *Moby-Dick:* The White Whale: The Sex Mystery," in *Moby-Dick as Doubloon: Essays and Extracts, 1851–1970* (New York: W.W. Norton & Co., 1970), 234–37; Leland S. Person, "Gender and Sexuality," in *A Companion to Herman Melville*, 231–246; and for connections of Ahab to the Fisher King myth, see Christopher Sten, *Sounding the Whale: Moby-Dick as Epic Novel* (Kent: Kent State University Press, 1996), 67–68; Robert Shulman, "The Serious Functions of Melville's Phallic Jokes," *American Literature* 33, no. 2 (May 1961): 179–194. See, e.g., *Moby-Dick*, 550.
10 See, for example, Amy S. Greenberg, "Fayaway and Her Sisters," in *"Whole Oceans Away": Melville and The Pacific*, ed. Jill Barnum, Wyn Kelley, and Christopher Sten (Kent: Kent State University Press, 2007), 17–30; Caleb Crain, "Melville's Secrets," *Leviathan* 14, no. 3 (October 2012): 6–24; Matthew Knip, "Homosocial Desire and Erotic *Communitas* in Melville's Imaginary: The Evidence of Van Buskirk," *ESQ* 62, no. 2 (2016): 355–414.

11 Nathaniel Philbrick, *Why Read Moby-Dick?* (New York: Viking, 2011), 88. Shakespeare spelled it "archbishopricke," but it seems ending with a "c" was more common. "Prick, n. 12b," *Oxford English Dictionary*, 3rd ed. (2007), www.oed.com/view/Entry/151146; "Archbishopric, n.," *Oxford English Dictionary*, 3rd ed. (2007), www.oed.com/view/Entry/10308.

NOTES TO CHAPTER TWENTY-ONE

1 This interview with Ewan Fordyce was conducted on 15 March 2018, then revised in collaboration.
2 *Moby-Dick*, 452–53; Reeves, et al., *Guide*, 241.
3 Edward O. Wilson, "Introduction," *The Rarest of the Rare: Stories Behind the Treasures at the Harvard Museum of Natural History*, by Nancy Pick and Mark Sloan (New York: Harper Collins, 2004), 1; Mary Sears, Ernst Mayr Library of the Museum of Comparative Zoology, 23 May 2017, pers. comm.; Invoice to Museum of Comparative Zoology, 17 July 1891, from Ward's Natural Science Establishment, collection of the Museum of Comparative Zoology Archives; Emerson, "The Uses of Natural History," 4. See also Emerson, *The Journals and Miscellaneous Notebooks of Ralph Waldo Emerson, Volume IV, 1832–1834*, 406.
4 Sealts, 222; Henry David Thoreau, *A Week on the Concord and Merrimack Rivers* [1849] (Cambridge: Riverside Press, 1894), 323. Notably different in Melville's quotation is that Thoreau uses "skeleton" while Melville uses "specimen." Melville traveled up to New Hampshire during his honeymoon with Elizabeth Shaw, but there is no record that he stopped here. In 1847 the Manchester museum opened its third floor space, below a theater, for its collections, and then seems to have closed only a few years later. Nobody knows what became of the whale. Jeffrey Barraclough, 8 March 2018, pers. comm.; L. Ashton Thorp, *Manchester of Yesterday: A Human Interest Story of Its Past* (Manchester, NH: Granite State Press, 1939), 332–33. See also Leyda, 255–58; William Henry Flower, "On the Osteology of the Cachalot or Sperm-Whale (*Physeter macrocephalus*)," *Transactions of the Zoological Society of London* 6 (London: Taylor and Francis, 1868): 309–10.
5 *Moby-Dick*, 454; Beale, 80, 82; Beale, Melville's Marginalia Online.
6 Jane Walsh, "From the Ends of the Earth: The United States Exploring Expedition Collections," Smithsonian Libraries, 29 March 2004, www.sil.si.edu/DigitalCollections/usexex/learn/Walsh-01.htm.
7 John Rickards Betts, "P. T. Barnum and the Popularization of Natural History," *Journal of the History of Ideas* 20, no. 3 (June–Sept 1959): 366; J. B. S. Jackson, 138–39.
8 William S. Wall, "History and Description of the Skeleton of a New Sperm Whale" (Sydney: W.R. Piddington, 1851), 2–5.
9 Wall, 5, 33; Beale, 84; Darwin, *On the Origin of Species*, 394.
10 Wall, 21, 23, 24.
11 *The Holy Bible ... Together with the Apocrypha* (Philadelphia: E. H. Butler & Co., 1860), 648. See Janis Stout, "Melville's Use of the Book of Job," *Nineteenth-Century Fiction* 25, no. 1 (June 1970): 69–83. For other biblical connections to "A Bower in the Arsacides," see A. Baker, *Heartless Immensity*, 35–38; Wilkes, vol. 5, 14–15.
12 J. Baker, "Dead Bones," 91–95; Ralph Waldo Emerson, "The Naturalist," in *The*

Early Lectures of Ralph Waldo Emerson, ed. Stephen E. Whicher and Robert E. Spiller, vol. 1, 1833–1836 (Cambridge, MA: Harvard University Press, 1959), 79.

13 Mark D. Uhen, "Basilosaurids and Kekenodontids," *EMM*, 3rd ed., 78–79; Richard Owen, "Observations on the Basilosaurus of Dr. Harlan (*Zeuglodon cetoides*, Owen)," *Transactions of the Geological Society of London*, 2nd series, vol. 6 (London: R. and J. E. Taylor, 1842), 69–79; "Hydrarchos Advertisement," *The Encyclopedia of Alabama*, 2018, http://www.encyclopediaofalabama.org/article/m-8529; Rieppel, 139–161.

14 *Moby-Dick*, 457; Foster, 50–54; R. D. Smith, *Melville's Science*, 136, 301; Harvey, "Science and the Earth," 73–74. For Melville's earlier engagement with geology, see for example, *Typee*, 155, and *Mardi*, 414–18. On Owen and the basilosaurus, see Gaines, 11, and Richard S. Moore, "Owen's and Melville's Fossil Whale," *American Transcendental Quarterly* 26 (1975): 24.

15 *Moby-Dick*, 457.

16 Melville seems to have combined Agassiz's ice age theory with Lyell's one that icebergs deposited and transported large boulders and carved landscapes. Foster, 61–62.

17 Foster, 50.

NOTES TO CHAPTER TWENTY-TWO

1 *Moby-Dick*, 460. Melville took Lacépède's figures not from the original book in French, but from Scoresby (Vincent, 365).

2 "Five Wicked Whales," 35; Beale, 15; Beale, Melville's Marginalia Online; Davis, 188; Philbrick, *In the Heart of the Sea*, 254–55; Stuart Frank, *Ingenious Contrivances, Curiously Carved: Scrimshaw in the New Bedford Whaling Museum* (Boston: David R. Godine, 2012), 54; Clifford W. Ashley, *The Yankee Whaler*, 2nd ed. (London: George Routledge & Sons, 1938), 73. On overestimates of bowhead whale lengths, see Scoresby, *Arctic Journals*, vol. 1, 449–54. Also Tyrus Hillway, "Melville and Nineteenth-Century Science," PhD diss. (New Haven: Yale University, 1944), 100.

3 Bennett, vol. 2, 154; Reeves, et al., *Guide*, 241. See also Ellis, *The Great Sperm Whale*, 97; Whitehead, *Sperm Whales: Social Evolution*, 8, 328; Berzin, 16, 30.

4 Elizabeth Schultz, "Melville's Environmental Vision in *Moby-Dick*," *Interdisciplinary Studies in Literature and Environment* 7, no. 1 (2000): 102.

5 Christopher F. Clements, Julia L. Blanchard, Kirsty L. Nash, Mark A. Hindell, and Arpat Ozgul, "Body Size Shifts and Early Warning Signals Preceded the Historic Collapse of Whale Stocks," *Nature Ecology & Evolution* 1, no. 0188 (2017): 1–6. See also Berzin who in 1971 recognized the mean length was decreasing (30). For debate on the reliability of this study see Phillip J. Clapham and Yulia V. Ivashchenko, "Whaling Catch Data Are Not Reliable for Analyses of Body Size Shifts," *Nature Ecology & Evolution* 2 (May 2018): 756; and the authors' reply: Christopher F. Clements, et al., "Reply to 'Whaling Catch Data Are Not Reliable . . . ,'" *Nature Ecology & Evolution* 2 (May 2018): 757–58. I had help thinking about this from David Laist, 14 August 2018, pers. comm.

6 *Moby-Dick*, 460.

7 On numbers of whales killed per voyage, see, e.g., Wilkes, vol. 5, 493, and the whaleship *Pocahontas* (1844–46) in Dyer, *Tractless Sea*, 70.

8   *Moby-Dick*, 461.

9   *Moby-Dick*, 461-62.

10   *Moby-Dick*, 462; Smith, Reeves, Josephson, and Lund, "Spatial and Seasonal Distribution of American Whaling and Whales in the Age of Sail," 1; Jennifer A. Jackson, et al., "An Integrated Approach to Historical Population Assessments of the Great Whales: Case of the New Zealand Southern Right Whale," *Royal Society Open Science* 3 (2016): 11; Elizabeth A. Josephson, Tim D. Smith, and Randall R. Reeves, "Depletion within a Decade: The American 19th-Century North Pacific Right Whale Fishery," in *Oceans Past*, 133-47; Cheever, *The Whale and His Captors*, 48; Bowles, "Some Account of the Whale-Fishery of the N. West Coast and Kamschatka," 83. I'm not sure where Ishmael got his number of 13,000 North Pacific right/bowhead whales killed each year by American whaleships, but it's not absurd for what was known in the 1840s. Scarff (2001) as cited and discussed in Randall R. Reeves and Tim D. Smith, "A Taxonomy of World Whaling," in *Whales, Whaling, and Ocean Ecosystems*, 91, calculated that numbers of North Pacific right whales killed by all nationalities between 1839 and 1909 to be from 26,500 to 37,000 individuals. This was on top of an estimated 30,000 bowheads killed by American whaleships alone from 1804 to 1909, as Reeves and Smith cite from Best (1987). Bowles (1845), however, as above, estimates 12,000 North Pacific right whales taken a year in just the four-month season. See, also, for barrels per year, Josephson, Smith, and Reeves, as above, 143.

11   Tim Smith, 12 December 2016, pers. comm.; Whitehead, "Sperm Whales in Ocean Ecosystems," 330.

12   *Moby-Dick*, 388; Whitehead, "Sperm Whales," *EMM*, 3rd ed., 923; Whitehead, *Sperm Whales: Social Evolution*, 11-12.

13   *Moby-Dick*, 358, 462; Laist, 64-65; Kenney, "Right Whales," *EMM*, 3rd ed., 820; Cheryl Rosa, et al., "Age Estimates Based on Aspartic Acid Racemization for Bowhead Whales (*Balaena mysticetus*) Harvested in 1998-2000 and the Relationship between Racemization Rate and Body Temperature," *Marine Mammal Science* 29, no. 3 (July 2013), 424. On "fraternal congenerity," see Zoellner, 166-90. See also Bode, 189, and "Melville's Tendency to Lateralize," in Geoffrey Sanborn, "Melville and the Nonhuman World," in *The New Cambridge Companion to Herman Melville*, ed. Robert S. Levine (Cambridge: Cambridge University Press, 2014), 13.

14   Agassiz and Gould, 237-39.

15   "On the Fur Trade, and Fur-bearing Animals," *American Journal of Science and Arts*, ed. Benjamin Silliman, 25:2 (New Haven: 1834), 329.

16   For example, Washington Irving, *Astoria, or Anecdotes of an Enterprise beyond the Rocky Mountains*, vol. 2 (Philadelphia: Carey, Lea, & Blanchard, 1836), 274; Gideon Algernon Mantell, *The Wonders of Geology*, vol. 1 (London: Relfe and Fletcher, 1839), 117; Samuel Gilman Brown, *The Works of Rufus Choate, with a Memoir of His Life*, vol. 2 (Boston: Little, Brown & Co., 1862), 159-60; Primack, 60-61; *Moby-Dick*, 153.

17   Schultz, 107-10; Francis Parkman Jr., *The California and Oregon Trail: Being Sketches of Prairie and Rocky Mountain Life* (New York: George P. Putnam, 1849), 176, 229. By the preface to his 1892 edition, Parkman wrote that all of the buffalo were now gone, "of all his millions nothing is left but bones" (Boston: Little, Brown, and Co., 1892), vii.

18   Beale, 151.

19 Morgan, "Address before the New Bedford Lyceum," 15-16; Wilkes, vol. 5, 493.

20 Schultz, 106.

21 *Moby-Dick*, 4, 380; Schultz, 107.

22 On the monk seal, see Deborah A. Duffield, "Extinctions, Specific," *EMM*, 3rd ed., 344-45; J. A. Allen, "The West Indian Seal (*Monanchus tropicalis* Gray)," in *Bulletin of the American Museum of Natural History*, vol. 2, 1887-90 (New York: AMNH, 1890), 27-28. On salmon, Roberts, 53-56; US Fish and Wildlife Service, "Final Environmental Impact Statement, Restoration of Atlantic Salmon to New England Rivers" (Newton Corner, MA: USFWS, 1989), 11. See Catherine Schmitt, *The President's Salmon: Restoring the King of Fish and Its Home Waters* (Camden: Down East Books, 2015). On North Atlantic right whales, Laist, 165-77, 262. On the gray whale, Clapham and Link, "Whales, Whaling, and Ecosystems in the North Atlantic Ocean," in *Whales, Whaling, and Ocean Ecosystems*, 317-18. On South Shetland fur seals, Roberts, 107; Robert Hamilton, *The Naturalist's Library*, ed. Sir William Jardine, *Mammalia*, vol. 8 (Edinburgh: W. H. Lizars, 1839), 95-96.

23 Douglas J. McCauley, et al. "Marine Defaunation: Animal Loss in the Global Ocean (Review Summary)," *Science* 347, no. 6219 (16 January 2015): 247.

24 On "The Try-Works," see Dean Flower, "Vengeance on a Dumb Brute, Ahab? An Environmentalist Reading of *Moby-Dick*," *Hudson Review* 66, no. 1 (2013): 144-45.

25 Whitehead, *Sperm Whales: Social Evolution*, 20; Whitehead, "Sperm Whale," *EMM*, 3rd ed., 923. On uses of sperm whale oil, see Joe Roman, *Whale* (London: Reaktion, 2006), 131, 144-45, and Rice, "Spermaceti," *EMM*, 2nd ed., 1099.

26 Whitehead, *Sperm Whales: Social Evolution*, 130-31; B. L. Taylor, et al., "*Physeter microcephalus*," IUCN Red List of Threatened Species, 2008, www.iucnredlist.org /details/41755/0.

27 Whitehead and Rendell, 154-55; Whitehead, "Sperm Whale," *EMM*, 3rd ed., 922.

28 Laist, 262.

29 Michael Dyer, "Why Black Whales Are Called 'Right Whales,'" *New Bedford Whaling Museum Blog* (13 September 2016), www.whalingmuseumblog.org; "An 1849 Statement on the Habits of Right Whales," 140; "Active Maps: The First Voyage of the *Charles W. Morgan*, 1841-1845," Mystic Seaport for Educators, 2016, http:// educators.mysticseaport.org/maps/voyage/morgan_first/; Logbook of the *Commodore Morris*, 1845-49.

30 Recent genetic work has put into question how much hunting by early Basques actually affected right whales—they might have been catching more bowheads. See Laist, 136-37; Toolika Rastogi, et al., "Genetic Analysis of 16th-Century Whale Bones Prompts a Revision of the Impact of Basque Whaling on Right and Bowhead Whales in the Western North Atlantic," *Canadian Journal of Zoology* 82 (2004): 1647-54; Richard M. Pace III, Peter J. Corkeron, and Scott D. Kraus, "State-Space Mark-Recapture Estimates Reveal a Recent Decline in Abundance of North Atlantic Right Whales," *Ecology and Evolution* (July 2017): 1; National Marine Fisheries Service, "North Atlantic Right Whale (*Eubalaena glacialis*), 5-Year Review: Summary and Evaluation," (October 2017), 1-33; J. G. Cooke and A. N. Zerbini, "*Eubalaena australis*," The IUCN Red List of Threatened Species, 2018, https://www.iucn redlist.org/species/8153/50354147; J. A. Jackson, N. J. Patenaude, E. L. Carroll, and C. Scott Baker, "How Few Whales Were There after Whaling? Inference from Contemporary mtDNA Diversity," *Molecular Ecology* 17 (2008): 244; Kenney,

"Right Whales," *EMM*, 3rd ed., 818; Richard L. Merrick, Gregory K. Silber, and Douglas P. DeMaster, "Endangered Species and Populations," *EMM*, 3rd ed., 313; M. M. Muto, et al., "North Pacific Right Whale (*Eubalaena japonica*): Eastern North Pacific Stock," NOAA-TM-AFSC-355 (30 December 2016), 1–9.

31 Reeves, et al., *Guide*, 190–93.

32 Joe Roman and Stephen R. Palumbi, "Whales before Whaling in the North Atlantic," *Science* 301 (25 July 2003): 508. See also Stephen R. Palumbi, "Whales, Logbooks, and DNA," in *Shifting Baselines: The Past and the Future of Ocean Fisheries*, ed. Jeremy B.C. Jackson, Karen E. Alexander, and Enric Sala (Washington, DC: Island Press, 2011), 163–73.

33 Peter Kareiva, Christopher Yuan-Farrell, and Casey O'Connor, "Whales Are Big and It Matters," in *Whales, Whaling, and Ocean Ecosystems*, 383.

34 Joe Roman and James J. McCarthy, "The Whale Pump: Marine Mammals Enhance Primary Productivity in a Coastal Basin," *PLOS One* 5, no. 10 (October 2010): 1. See Elizabeth Shultz's poem "Holy Shit!" in *Ishmael on the* Morgan (2015), 17.

35 Thoreau, *Cape Cod*, 219–20.

NOTES TO CHAPTER TWENTY-THREE

1 *Moby-Dick*, 487.

2 Onley and Scofield, 14–15; Good, 137, 192; Robin Hull, *Scottish Birds: Culture and Tradition* (Edinburgh: Mercat Press, 2001), 91.

3 W. B. Alexander, *Birds of the Ocean*, rev. ed. (New York: G. P. Putnam's Sons, 1963), 52; John James Audubon, *Ornithological Biography, or An Account of the Habits of the Birds of the United States of America*, vol. 3 (Edinburgh: Adam and Charles Black, 1835), 486–90.

4 Isaac Jessup, "19 August 1849," Logbook of the *Sheffield* 1849–50, Mystic Seaport Log 351.

5 Melville, "The Encantadas," *The Piazza Tales*, 135–36.

6 Hull, 92–93.

7 Baron Cuvier and Edward Griffith, *The Animal Kingdom*, vol. 8 (London: Whittaker, Treacher, and Co., 1829), 641; Bennett, vol. 1, 12–13.

NOTES TO CHAPTER TWENTY-FOUR

1 *Moby-Dick*, 503. For the smashing of the quarter-boat as bad omen, see Olmsted, 24.

2 *Moby-Dick*, 506–7; John G. Rogers, *Origins of Sea Terms* (Mystic: Mystic Seaport Museum, 1985), 49, 150. See also "Corposant" and "St. Elmo, n.," *Oxford English Dictionary*, 2nd ed. (1989), www.oed.com/view/Entry/41856 and www.oed.com/view/Entry/189733.

3 *Moby-Dick*, 507. See R. D. Smith, *Melville's Science*, 142–44.

4 *Moby-Dick*, 508.

5 "Tropical Cyclones in 1993," Royal Observatory Hong Kong (1995), 14; Associated Press, "Typhoon Kills 4 in Seas off Hong Kong," *Los Angeles Times* (28 June 1993),

http://articles.latimes.com/1993-06-28/news/mn-8022_1_hong-kong; Melville also read of a typhoon off Japan, also not far from the Bonin Islands, in Beale, 269–73.

6   The barometer mention in *Moby-Dick*, 235.

7   "Tropical Cyclones in 1993," 20.

8   *Moby-Dick*, 513, 516.

9   Webster, 491.

10   Scott Huler, *Defining the Wind: The Beaufort Scale, and How a 19th-Century Admiral Turned Science into Poetry* (New York: Three Rivers Press, 2004), 123. On whalemen and the Beaufort Scale, Michael Dyer, 5 April 2018, pers. comm.

11   *Moby-Dick*, 158, 432, 433; Chase, 27; Nathaniel Bowditch and J. Ingersoll Bowditch, *The New American Practical Navigator* (New York: E. & G. W. Blunt, 1851), 117–19, 318; Huler, 121.

12   Huler, 121.

13   On Maury in relation to Franklin, see R. D. Smith, *Melville's Science*, 298–99. On grounding chains on whaleships, Michael Dyer, 5 April 2018, pers. comm.

14   Eason, "16 March 1858."

15   Melville will later continue the exploration of lightning, reason, and authority in his short story "The Lightning-Rod Man" (1854).

16   Melville, *Journals*, 6. On Melville's experience with storms, see Mary K. Bercaw Edwards, "Ships, Whaling, and the Sea," in *A Companion to Herman Melville*, 89–92.

17   Bennett, vol. 1, 4, 190; Wilkes, vol. 2, 159; Kenneth W. Cameron, "A Note on the Corpusants in *Moby-Dick*," *Emerson Society Quarterly* 19 (1960): 22–24; Darwin, *Journal of Researches*, vol. 1, 49.

18   Dana, 434.

19   On storms as objective correlative, Dan Brayton, 1 July 2016, pers. comm.; Sealts, 214, 225; Shak[e]speare, "Tempest," in *Dramatic Works*, 16; Coleridge, "The Ancient Mariner," 56–57.

20   On knowledge of circular storms at the time, see Bowditch and Bowditch (1851), 119.

21   Henry Piddington, *The Sailor's Horn-Book for the Law of Storms: being a Practical Exposition of the Theory of the Law of Storms . . .* (London: Smith, Elder, and Co., 1848).

22   *Moby-Dick*, 158.

23   Weir, "24 April 1856."

24   See Dan Brayton, *Shakespeare's Ocean: An Ecocritical Exploration* (Charlottesville: University of Virginia Press, 2012); Gwilym Jones, *Shakespeare's Storms* (Manchester: Manchester University Press, 2016); and Kris Lackey, "'More Spiritual Tenors': The Bible and Gothic Imagination in *Moby-Dick*," *South Atlantic Review* 52, no. 2 (May 1987): 37–50.

25   Geophysical Fluid Dynamics Laboratory, "Global Warming and Hurricanes," rev. 20 September 2018, www.gfdl.noaa.gov/global-warming-and-hurricanes; Maggie Astor, "The 2017 Hurricane Season Really Is More Intense than Normal," *New York Times* (19 September 2017), www.nytimes.com. See also Richard J. Murnane and Kam-bui Liu, eds., *Hurricanes and Typhoons: Past, Present, and Future* (New York: Columbia University Press, 2004).

NOTES TO CHAPTER TWENTY-FIVE

1　Heflin, 42, 259–60; Starbuck as "patent chronometer" in *Moby-Dick*, 115; Dava Sobel, *Longitude* (New York, Penguin: 1996); Tamara Plakins Thornton, *Nathanial Bowditch and the Power of Numbers* (Chapel Hill: University of North Carolina Press, 2016), 73.
2　Chase, 27. There's been some confusion over the years about the names of these tools. An octant is so named because the tool has an arc that is one-eighth of a circle; it can measures up to 45°. A quadrant is the same size, but with mirrors it can measure 90°. A sextant has an arc that is one-sixth of a circle, but because of the mirrors, can measure 120°. See, for example, Bowditch (1851), 128, 133, Plate 9, incorporated here into fig. 48.
3　*Moby-Dick*, 501.
4　*Moby-Dick*, 501.
5　*Moby-Dick*, 501.
6　On the Chain of Being, Wilson, 136.
7　See R. D. Smith, *Melville's Science*, 301.
8　If Melville did not see a compass flip for himself, he also would have read about this in Scoresby's *Journal of a Voyage to the Northern Whale Fishery*. See R. D. Smith, *Melville's Science*, 144–45.

NOTES TO CHAPTER TWENTY-SIX

1　Colnett, 169.
2　Colnett, 176.
3　*Moby-Dick*, 523–24; Vincent, 382–83.
4　*Moby-Dick*, 524, 532.
5　Claudio Campagna, 11 April 2017, pers. comm.; see also a compelling parallel in Olmsted, 177.
6　*Moby-Dick*, 150, 523; Roberts, 99–113; George W. Peck, *Melbourne, and the Chincha Islands; with sketches of Lima, and a voyage round the world* (New York: Charles Scribner, 1854), 191–92; Mark Bousquet, "Afterword: 'The Cruel Harpoon' and the 'Honorable Lamp': The Awakening of an Environmental Consciousness in Henry Theodore Cheever's *The Whale and His Captors*," in Cheever, *The Whale and His Captors*, 241; Brewster, 389. On whaling and sealing voyages combined, see, for example, log of the ship *Emeline* 1843–44, New Bedford Whaling Museum Log 147; Joshua Drew, et al., "Collateral Damage to Marine and Terrestrial Ecosystems from Yankee Whaling in the 19th Century," *Ecology and Evolution* 6 (2016): 8181–92.
7　Beers, 191; Roman, 157.
8　Brian Clark Howard, "Haunting Whale Sounds Emerge from Ocean's Deepest Point," *National Geographic* (5 March 2016), news.nationalgeographic.com. See also Roger Payne, "Melville's Disentangling of Whales," in *Moby-Dick*, ed. Parker, 3rd ed., 702–4.

NOTES TO CHAPTER TWENTY-SEVEN

1   "To Sophia Peabody Hawthorne, 8 Jan 1852," *Correspondence*, 218–19.
2   *Moby-Dick*, 542.
3   *Moby-Dick*, 190, 191, 274, 393, 497; Whitehead and Rendell, 157.
4   Victor Reinking and David Willingham, "Conversation with Ursula K. Le Guin," in *Conversations with Ursula K. Le Guin*, ed. Carl Freedman (Jackson: University Press of Mississippi, 2008), 118. For a retelling of *Moby-Dick* and the historical era from a woman's point of view, see the novel *Ahab's Wife, or The Stargazer* (1999) by Sena Jeter Naslund.
5   *Moby-Dick*, 397. See Person, "Gender and Sexuality," in *A Companion to Herman Melville*, 231–46.
6   *Moby-Dick*, 544; Rita Bode, "'Suckled by the Sea': The Maternal in *Moby-Dick*," in *Melville and Women*, ed. Elizabeth Schultz and Haskell Springer (Kent: Kent State University Press, 2006), 181–98.
7   *Moby-Dick*, 545.

NOTES TO CHAPTER TWENTY-EIGHT

1   *Mardi*, 179, 283.
2   *Moby-Dick*, 548. Melville mentions tropic birds in *Omoo*, "The Encantadas," and likely *Typee*, and it's nearly the only seabird with long tail plumage like this. See R. D. Madison, "The Aviary of Ocean: Melville's Tropic-Birds and Rock Rodondo—Two Notes and an Emendation," in *This Watery World: Humans and the Sea*, ed. Vartan P. Messier and Nandita Batra (Newcastle upon Tyne: Cambridge Scholars Publishing, 2008), 158–62.
3   *Moby-Dick*, 548.
4   Beale, 60–61; Beale, Melville's Marginalia Online; See Bercaw [Edwards], *Melville's Sources*, 55; Peter Mark Roget, *The Bridgewater Treatises on the Power Wisdom and Goodness of God as Manifested in the Creation: Treatise V, Animal and Vegetable Physiology Considered with Reference to Natural Theology*, vol. 1 (London: William Pickering, 1834), 265–66. On *The Bridgewater Treatises*, see Callaway, 141–46.
5   William Wood, *Zoography, or, The Beauties of Nature Displayed*, 3 vols. (London: Cadell and Davies, 1807), vol. 1: vii–xiv, vol. 2: 579; Roget, 30. Other authors who have used the myths of the paper nautilus for poetic effect include Alexander Pope, Oliver Wendell Holmes, Jules Verne, and Marianne Moore.
6   Bernd Brunner, *The Ocean at Home: An Illustrated History of the Aquarium*, trans. Ashley Marc Slapp (London: Reaktion, 2011), 30–31; A. Louise Allcock, et al., "The Role of Female Cephalopod Researchers: Past and Present," *Journal of Natural History* 49, nos. 21–24 (2015): 1242–43; The Society for the Diffusion of Useful Knowledge, "Paper Nautilus," *The Penny Cyclopædia*, vol. 17 (London: Charles Knight and Co., 1840), 210–15.
7   Brunner, 105–8.
8   *Moby-Dick*, 548.

NOTES TO CHAPTER TWENTY-NINE

1   This interview with Marta Guerra and Rebecca Bakker was originally conducted
    2-5 January 2018, then revised in collaboration.
2   Todd, *Whales and Dolphins of Kaikōura*, 21, 26; Guerra, 7 May 2018, pers. comm.
    Male sperm whale populations have been in decline in the region, and the definition
    between "residents" and "transients" can be fuzzy.
3   For more on hydrophone distance see Whitehead, *Sperm Whales: Social Evolution*,
    144. For a discussion of how clangs or "slow clicks" might relate to foraging, see
    Nathalie Jaquet, Stephen Dawson, and Lesley Douglas, "Vocal Behavior of Male
    Sperm Whales: Why Do They Click?," *Journal of the Acoustical Society of America* 109,
    no. 5, pt. 1 (May 2001): 2254-59.
4   A comparatively similar productive area to Kaikōura Canyon is Bremer Bay, Aus-
    tralia. Kaikōura is productive non-chemosynthetically, meaning it is not a hydro-
    thermally or actively chemically driven ecosystem. See Fabio C. De Leo, Craig R.
    Smith, Ashley A. Rowden, David A. Bowden, and Malcolm R. Clark, "Submarine
    Canyons: Hotspots of Benthic Biomass and Productivity in the Deep sea," *Proceed-
    ings of the Royal Society B* 277 (2010): 2783, 2785; National Institute of Water and
    Atmospheric Research Ltd, "Kaikōura Canyon: Depths, Shelf Texture and Whale
    Dives," NIWA Miscellaneous Chart Series, 1998, teara.govt.nz/en/zoomify/31738
    /kaikoura-canyon-poster.
5   Guerra, et al., "Diverse Foraging Strategies by a Marine Top Predator," 98-108.
6   *Moby-Dick*, 556.
7   Whitehead, "Sperm Whale," *EMM*, 922.
8   Todd, *Whales and Dolphins of Kaikōura*, 7; D. E. Gaskin, "Analysis of Sightings and
    Catches of Sperm Whales (*Physeter catodon* L.), in the Cook Strait Area of New
    Zealand in 1963-4," *New Zealand Journal of Marine and Freshwater Research* 2, no. 2
    (1968): 260.
9   Todd, 23, 26.
10 *Moby-Dick*, 183.
11 Deaths by sperm whale in *Moby-Dick*: 180, 183 (implied head and jaws), 316-17
    (Macey, tail), 257 (Radney, jaws), 438-39 (Boomer, tail).
12 Morgan, "Address before the New Bedford Lyceum," 16; Beale, 3, 5.
13 Bennett, vol. 2, 214.
14 Bennett, vol. 2, 217; Weir, "16 November 1856," and "9 December 1857"; Michael
    Dyer, "Introduction to the Art of the American Whale Hunt," *New Bedford Whal-
    ing Museum Blog* (6 March 2013), https://whalingmuseumblog.org; Dyer, *Tractless
    Sea*, 250-56. See also Creighton, 67.
15 Heflin, 85, 91.
16 Reeves, et al., *Guide*, 242, 256-57, 422-23; Whitehead, "Sperm Whale," *EMM*,
    3rd ed., 1095; Whitehead, *Sperm Whales: Social Evolution*, 194-95; Hidehiro Kato,
    "Observation of Tooth Scars on the Head of Male Sperm Whale, as an indication
    of Intra-sexual Fightings," *Scientific Reports of the Whales Research Institute Tokyo* 35
    (1984): 39-46.
17 Whitehead, *Sperm Whales: Social Evolution*, 193; Ellis, *The Great Sperm Whale*, 98;
    Beale, 36-37.

18   Kazue Nakamura, "Studies on the Sperm Whale with Deformed Lower Jaw with Special Reference to Its Feeding," *Bulletin of Kanagawa Prefecture Museum* 1, no. 1 (March 1968): 13, 17, 19. See also Berzin, 93, 94, 274. Gender is skewed toward males, too, since the study was based on the regions where the whalemen were hunting.

19   Whitehead, *Sperm Whales: Social Evolution*, 45; Dolin, 402.

20   Bennett, vol. 2, 220; Haley, 250.

21   Philbrick, *In the Heart of the Sea*, 81; Parker, vol. 1, 725; Leyda, 411; Cheever, "1853 Additions," in *The Whale and His Captors*, 168–69.

22   Chase, 26–27; Philbrick, *In the Heart of the Sea*, 81. For more on the *Essex*, see Philbrick, *In the Heart of the Sea*; David Dowling, *Surviving the Essex: The Afterlife of America's Most Storied Shipwreck* (Hanover: ForeEdge, 2016); and Madison, *The Essex and the Whale*.

23   "The Whale Fishery," No. 82 (Jan 1834), *North American Review* 38 (Boston: Charles Bowen, 1834), 112; Olmsted, 144–45; Madison, *The Essex and the Whale*, 89; Hal Whitehead and Marta Guerra, 6 February 2018, pers. comm.; Philbrick, *In the Heart of the Sea*, 87, 255–56. For discussion of this and other elements of sperm whale sentience and ramming ships, see also Dowling, 141–65.

24   *Correspondence*, 209; Parker, vol. 1, 878; Sidney Kaplan, "Can a Whale Sink a Ship? The Utica *Daily Gazette* vs the New Bedford *Whalemen's Shipping List*," *New York History* 33, no. 2 (April 1952): 159–63. See also Cheever, "1853 Additions," in *The Whale and His Captors*, 165–68; Starbuck, 123–25, 159; Andrew B. Myers, "Two More Attacks," *Melville Society Extracts* 29 (1977): 12; and the account of the whaleship *Osceola* in which a sperm whale rammed the bow, knocking off the cutwater, and biting at the copper sheathing with his teeth, see A. Howard Clark, "The Whale-Fishery," in *The Fisheries and Fishery Industries of the United States*, ed. George Brown Goode, 5:2 (Washington, DC: Government Printing Office, 1887), 261–62; in addition, see the 1902 sinking of the whaleship *Kathleen*, Captain Thomas H. Jenkins, *Bark Kathleen Sunk by a Whale* (New Bedford: H. S. Hutchinson & Co., 1902). For recent accounts, see Gregory L. Fulling, et al., "Sperm Whale (*Physeter macrocephalus*) Collision with a Research Vessel: Accidental Collision or Deliberate Ramming?," *Aquatic Mammals* 43, no. 4 (2017): 421–29. (Video of the Fulling, et al., event is viewable in "Supplemental Material" here: www.aquaticmammalsjournal .org.)

25   Reeves, et al., *Guide*, 360; David Lusseau, "Why Are Male Social Relationships Complex in the Doubtful Sound Bottlenose Dolphin Population?," *PLOS One* 2, no. 4 (2007): 1–8; Ingrid N. Visser, et al., "First Record of Predation on False Killer Whales (*Pseudorca crassidens*) by Killer Whales (*Orcinus orca*)," *Aquatic Mammals* 36, no. 2 (2010): 195; Whitehead, *Sperm Whales: Social Evolution*, 278–81; Bennett, vol. 2, 218; Whitehead, "Sperm Whale," *EMM*, 3rd ed., 922.

26   See another similar observation in Olga Panagiotopoulou, Panagiotis Spyridis, Hyab Mehari Abraha, David R. Carrier, and Todd Pataky, "Architecture of the Sperm Whale Forehead Facilitates Ramming Combat," *PeerJ* (2016): 3.

27   David R. Carrier, Stephen M. Deban, and Jason Otterstrom, "The Face that Sank the *Essex*: Potential Function of the Spermaceti Organ in Aggression," *Journal of Experimental Biology* 205 (2002): 1755, 1760–62.

28   Carrier, et al., 1762.

29   Burnett, *Trying Leviathan*, 131.
30   "Sunrise or sunset he will invariably die towards the sun," wrote William A. Allen, "27 June 1842," Journal of the *Samuel Robertson* 1841-46. New Bedford Whaling Museum ODHS Log 1040; Beale, 161.
31   *Moby-Dick*, 354-55.
32   *Moby-Dick*, 356-57.
33   Bousquet, 221-222; Enoch Carter Cloud, *Enoch's Voyage: Life on a Whale Ship, 1851-1854*, ed. Elizabeth McLean (Wakefield, RI: Moyer Bell, 1994), 53.
34   Beers, 21, 23; Humphry Primatt, *A Dissertation on the Duty of Mercy and Sin of Cruelty to Brute Animals* (London: R. Hett, 1776), 13, 237, 308-9; Jeremy Bentham, *An Introduction to the Principles of Morals and Legislation* (London: T. Payne and Son, 1789), 309. The translation of Lacépède is in Jacques Cousteau, *Whales* (New York: H. N. Abrams, 1988), 13; Barwell, 57.
35   *Moby-Dick*, 74.
36   *Moby-Dick*, 382, 461.
37   Lori Cuthbert and Douglas Main, "Orca Mother Drops Calf, after Unprecedented 17 Days of Mourning," *National Geographic* (13 August 2018), https://www.national geographic.com/animals/.
38   Schultz, 100, 112.
39   *Moby-Dick*, 385.
40   Payne, "Melville's Disentangling of Whales," in *Moby-Dick*, ed. Parker, 703.
41   Delbanco, 177; Leyda, 427.

NOTES TO CHAPTER THIRTY

1   See Vincent, 389, and the list of flag symbols in Osborn, "Logbook of the *Charles W. Morgan*."
2   *Moby-Dick*, 572.
3   *Moby-Dick*, 539, 573; Bennett, vol. 2, 242; Vincent, 387-89. With thanks to Robert Madison.
4   Chris Elphick, John B. Dunning Jr., and David Allen Sibley, eds., *The Sibley Guide to Bird Life and Behavior* (New York: Knopf, 2001), 167; Peter Harrison, *Seabirds: An Identification Guide*, rev. ed. (Boston: Houghton Mifflin, 1983), 307-8, 310; Audubon, *Ornithological Biography*, vol. 3, 497. See also Walt Whitman, "To the Man-of-War-Bird" (1876) in *The Sea is a Continual Miracle: Sea Poems and Other Writings by Walt Whitman*, ed. Jeffrey Yang (Hanover: University Press of New England, 2017), 198.
5   Bennett, vol. 2, 243-44.
6   Walters, 18-19; Gregory S. Stone and David Obura, *Underwater Eden: Saving the Last Coral Wilderness on Earth* (Chicago: University of Chicago Press, 2013), 45-46; David Steadman, *Extinction and Biogeography of Tropical Pacific Birds* (Chicago: University of Chicago Press, 2006); Warren B. King, "Conservation Status of Birds of Central Pacific Islands," *Wilson Bulletin* 85, no. 1 (March 1973): 89, 101; Ian Fraser and Jeannie Gray, *Australian Bird Names: A Complete Guide* (Collingwood, Victoria: CSIRO Publishing, 2013), 57; Melville, "The Encantadas," *The Piazza Tales*, 134-35.

7 "Pelecanidæ," *The Penny Cyclopædia*, vol. 17 (London: Charles Knight and Co., 1840), 386.

8 Melville, *Typee*, 10. See also his poem "The Man-of-War Hawk" (1888), *Poems*, 230.

NOTES TO CHAPTER THIRTY-ONE

1 Tuake Teema, 22 July 2017, pers. comm.; Randi Rotjan, et al. "Establishment, Management, and Maintenance of the Phoenix Islands Protected Area," in *Advances in Marine Biology*, vol. 69, ed. Magnus L. Johnson and Jane Sandell (Oxford: Academic Press, 2014), 305.

2 Sea Education Association, "Whales and Tall Ships," ed. Chris Nolan, film by Jan Witting, 4 November 2015, www.youtube.com/watch?v=Egb9ZV6E3d4; Amber Kinter and Abby Cazeault, 5, 8 August 2017, pers. comm.

3 Haley, 111-22; Smith, Reeves, Josephson, and Lund, "Spatial and Seasonal Distribution of American Whaling and Whales in the Age of Sail," 10; Erin Taylor, "A Whale's Tale of the Phoenix Islands," New England Aquarium Phoenix Islands Blog (March-December, 2013), pipa.neaq.org; "3 May 1852," Logbook of the *Commodore Morris*, 1849-53.

4 Colnett, 28-29; Bennett, vol. 2, 172; McCauley, et al., "Marine Defaunation," 2; Walters, 232.

5 Deborah S. Goodwin, "Final Report for S.E.A. Cruise S274," (Woods Hole, MA: Sea Education Association, 2017), 5, 48.

6 Sandra Altherr, Kate O'Connell, Sue Fisher, and Sigrid Lüber, "Frozen in Time: How Modern Norway Clings to Its Whaling Past," Animal Welfare Institute, OceanCare, and Pro Wildlife (2016): 1-23; "Which Countries Are Still Whaling?" International Fund for Animal Welfare (accessed 12 March 2018), https://www.ifaw.org/united-states/our-work/whales/which-countries-are-still-whaling; Rachel Bale, "Norway's Whaling Program Just Got Even More Controversial," *National Geographic* (31 March 2016), www.news.nationalgeographic.com; "Special Permit Catches since 1985," International Whaling Commission, accessed 31 January 2019, https://iwc.int/table_permit; Rupert Wingfield-Hayes, "Japan and the Whale," *BBC News*, Tokyo (8 February 2016), http://www.bbc.com/news/world-asia-353 97749; Randall R. Reeves, "Hunting," *EMM*, 3rd ed., 492-96; J. G. Cook and P. J. Clapham, "*Eubalaena japonica*," The IUCN Red List of Threatened Species, 2018, http://www.iucnredlist.org/details/41711/0; J. G. Cook, "*Eubalaena glacialis*," The IUCN Red List of Threatened Species, 2018, www.iucnredlist.org/details/41712/0.

7 Shane Gero and Hal Whitehead, "Critical Decline of the Eastern Caribbean Sperm Whale Population," *PLOS One* 11, no. 10 (5 October 2016): 1; Brandon L. Southall, "Noise," *EMM*, 3rd ed., 642; Kathleen M. Moore, Claire A. Simeone, and Robert L. Brownell Jr., "Strandings," 945-47, *EMM*, 3rd ed.

8 Whitehead, "Sperm Whales in Ocean Ecosystems," 324-33.

9 *Moby-Dick*, 571, 572.

10 Leavitt, 29; see also Bennett, vol. 2, 221.

11 For more on this vortex, see Matthew Mancini, "Melville's 'Descartian Vortices,'" *ESQ* 36, no. 4 (1990): 315-27; and David Charles Leonard, "Descartes, Melville, and the Mardian Vortex," *South Atlantic Bulletin* 45, no. 2 (May 1980): 13-25.

12   Chris Nolan, 5 August 2017, pers. comm.

13   *Moby-Dick*, 457.

14   *Moby-Dick*, 568.

15   *Moby-Dick*, 573.

16   Kennedy Wolfe, Abigail M. Smith, Patrick Trimby, and Maria Byrne, "Vulnerability of the Paper Nautilus (*Argonauta nodosa*) Shell to a Climate-Change Ocean: Potential for Extinction by Dissolution," *Biological Bulletin* 223 (October 2012): 236–244; Goodwin, 47; Deborah Goodwin, 1 September 2018, pers. comm.; Ocean Portal Team with Jennifer Bennett, "Ocean Acidification," Ocean Portal, Smithsonian, 2017, http://ocean.si.edu/ocean-acidification; Kevin Krajick, "Ocean Acidification Rate May Be Unprecedented, Study Says," Lamont-Doherty Earth Observatory, 1 March 2012, www.ldeo.columbia.edu/news-events; O. Hoegh-Guldberg, et al. "Coral Reefs Under Rapid Climate Change and Ocean Acidification," *Science* 318 (14 December 2007), 1737.

17   Conservation International, "Establishing the Phoenix Islands Protected Area," (n.d.), 1; Stone and Obura, title page, 15–17; Max Quanchi and John Robson, *Historical Dictionary of the Discovery and Exploration of the Pacific Islands* (Lanham, MD: Scarecrow Press, 2005), xix; see also Dyer, *Tractless Sea*, 110 (the original ship name was *Canton*); *Moby-Dick*, 443; Colnett, ix; Elizabeth Dougherty, "Rise of the Reef Doctors," *BU Experts*, Boston University, 2017, medium.com/boston-university-pr; Randi Rotjan, Lecture, Sea Education Association, Woods Hole, 20 June 2017.

18   Stone and Obura, 9–12.

19   Anthony L. Andrady, "Persistence of Plastic Litter in the Oceans," 57–72, and Amy Lusher, "Microplastics in the Marine Environment: Distribution, Interactions and Effects," 260, in *Marine Anthropogenic Litter*, ed. Melanie Bergmann, Lars Gutow, and Michael Klages (New York: Springer, 2015).

20   Edith Regalado, "BFAR: Plastic, Steel Wires Killed Whale in Samal," *Philippine Star*, 20 December 2016, www.philstar.com. See also, for example, J. K. Jacobsen, L. Massey, F. Gulland, "Fatal Ingestion of Floating Net Debris by Two Sperm Whales (*Physeter macrocephalus*)," *Marine Pollution Bulletin* 60, no. 5 (May 2010): 765–67; Kristine Phillips, "A Dead Sperm Whale Was Found with 64 Pounds of Trash in Its Digestive System," *Washington Post* (11 April 2018), www.washingtonpost.com.

21   For the Sea Education Association's research on microplastics, see, for example, Kara Lavender Law, et al., "Distribution of Surface Plastic Debris in the Eastern Pacific Ocean from an 11-Year Data Set," *Environmental Science and Technology* 48, no. 9 (2014): 4732–38. For more on our modern relationship with plastic and connections to *Moby-Dick* and sea literature see Donovan Hohn, *Moby-Duck: The True Story of 28,800 Bath Toys Lost at Sea ...* (New York: Viking, 2011); and Patricia Yaeger, "Sea Trash, Dark Pools, and the Tragedy of the Commons," *PMLA* 125, no. 3 (2010): 523–45.

22   Steven L. Chown, "Tsunami Debris Spells Trouble," *Science* 357, no. 6358 (29 September 2017): 1356; Becky Oskin, "Japan Earthquake & Tsunami of 2011: Facts and Information," *LiveScience* (13 September 2017), www.livescience.com; James T. Carlton, et al., "Tsunami-Driven Rafting: Transoceanic Species Dispersal and Implication for Marine Biogeography," *Science* 357, nos. 1402–1406 (2017): 1–4.

23   Schultz, 110. See also Harvey, "Science and the Earth," 80; Randy Kennedy, "The Ahab Parallax: 'Moby Dick' and the Spill," *New York Times* (12 June 2010), www.nytimes.com.

24 Lewis Mumford, *Herman Melville* (New York: Literary Guild of America, 1929), 194.

25 Sanford E. Marovitz, "The Melville Revival," in *A Companion to Herman Melville*, 515–31; David Dempsey, "In and Out of Books," *New York Times*, 3 September 1950, 108; Paul Lauter, "Melville Climbs the Canon," *American Literature* 66, no. 1 (March 1994): 20.

26 Margaret Atwood, "The Afterlife of Ishmael," in *Whales: A Celebration*, ed. Greg Gatenby (Boston: Little Brown, 1983), 210.

27 On marine mammals and culture, see Whitehead and Rendell, 269–70. Catherine Robinson Hall, 3 July 2018, pers. comm.

28 John Vidal, "Pacific Atlantis: First Climate Change Refugees," *Guardian*, 25 November 2005, www.theguardian.com; The World Bank, "Water, Water, Everywhere, but Not a Drop to Drink: Adapting to Life in Climate Change-Hit Kiribati," 21 March 2017, www.worldbank.org.

29 John Walsh et al., "Ch. 2: Our Changing Climate," *Climate Change Impacts in the United States: The Third National Climate Assessment*, ed. J. M. Melillo, Terese (T. C.) Richmond, and G. W. Yohe (US Global Change Research Program, 2014): 44–45, doi:10.7930/J0KW5CXT.

30 The World Bank, "Water, Water, Everywhere."

31 Roberts, 349–62; Enric Sala and Sylvaine Giakoumi, "No-Take Marine Reserves Are the Most Effective Protected Areas in the Ocean," *ICES Journal of Marine Science* 75, no. 3 (2018): 1166–68; Anote Tong, "Foreword," in Stone and Obura, ix.

32 Kareati Waysang, c. 4 August 2017, pers. comm. See also Kayla Walsh, "Kiribati Confronts Climate Upheaval by Preparing for 'Migration with Dignity,'" Monga-Bay, 11 July 2017, https://news.mongabay.com/2017/.

33 Vidal, "Pacific Atlantis: First Climate Change Refugees"; Coral Davenport and Campbell Robertson, "Resettling the First American 'Climate Refugees,'" *New York Times* (3 May 2016), www.nytimes.com.

34 Batiri T. Bataua, et al., *Kiribati: A Changing Atoll Culture* (Suva, Fiji: Institute of Pacific Studies of the University of the South Pacific, 1985), 14.

35 "COP15 Kiribati Side Event—Song of the Friage Te Itei," posted by Marc Honore, presented at the United Nations Framework Convention on Climate Change, COP 15, 9 December 2009, https://www.youtube.com/watch?v=G5wEgGZhXrw&t=5s.

36 Spencer R. Weart, *The Discovery of Global Warming*, rev. ed. (Cambridge: Harvard University Press, 2008), 5.

37 *Moby-Dick*, 572.

38 Rachel Carson, *Under the Sea-Wind* (New York: Penguin, 2007), 162. On Carson, see, e.g., Susan Power Bratton, "Thinking like a Mackerel: Rachel Carson's *Under the Sea-Wind* as a Source for a Trans-ecotonal Sea Ethic," in *Rachel Carson: Legacy and Challenge*, ed. Lisa H. Sideris and Kathleen Dean Moore (Albany: SUNY Press, 2008), 79–93.

39 "NYC Flood Hazard Mapper," New York City Department of Planning, accessed 31 January 2019, http://www1.nyc.gov/site/planning/data-maps/flood-hazard-mapper.page; Matthew Bloch, Ford Fessenden, Alan McLean, Archie Tse, and Derek Watkins, "Surveying the Destruction Caused by Hurricane Sandy," *New York Times*, accessed 31 January 2019, www.nytimes.com.

SELECTED BIBLIOGRAPHY

BY HERMAN MELVILLE

Melville, Herman. *Correspondence*. Edited by Lynn Horth. Evanston: Northwestern University Press and The Newberry Library, 1993.

———. *Journals*. Edited by Howard C. Horsford and Lynn Horth. Evanston: Northwestern University Press and The Newberry Library, 1989.

———. *Mardi, and A Voyage Thither*. Edited by Harrison Hayford, Hershel Parker, and G. Thomas Tanselle. Evanston: Northwestern University Press and The Newberry Library, 1970.

———. *Moby-Dick or The Whale*. Edited by Harrison Hayford, Hershel Parker, and G. Thomas Tanselle. Evanston: Northwestern University Press and The Newberry Library, 1988.

———. *Omoo: A Narrative of Adventures in the South Seas*. Edited by Harrison Hayford, Hershel Parker, and G. Thomas Tanselle. Evanston: Northwestern University Press and The Newberry Library, 1968.

———. *The Piazza Tales and Other Prose Pieces, 1839–1860*. Edited by Harrison Hayford, Alma A. MacDougall, G. Thomas Tanselle, et al. Evanston: Northwestern University Press and The Newberry Library, 1987.

———. *Published Poems*. Edited by Robert C. Ryan, Harrison Hayford, Alma MacDougall Reising, and G. Thomas Tanselle. Evanston: Northwestern University Press and The Newberry Library, 2009.

———. *Redburn: His First Voyage*. Edited by Harrison Hayford, Hershel Parker, and G. Thomas Tanselle. Evanston: Northwestern University Press and The Newberry Library, 1969.

———. *Typee: A Peep at Polynesian Life*. Edited by Harrison Hayford, Hershel Parker, and G. Thomas Tanselle. Evanston: Northwestern University Press and The Newberry Library, 1968.

———. *White-Jacket, or The World in a Man-of-War*. Edited by Harrison Hayford, Hershel Parker, and G. Thomas Tanselle, et al. Evanston: Northwestern University Press and The Newberry Library, 1970.

MELVILLE'S MAJOR NATURAL HISTORY SOURCES
AND OTHER CONTEMPORARY WORKS

(For known editions that Melville used, see Bercaw [Edwards] and Sealts.)

Agassiz, Louis, and A. A. Gould, *Principles of Zoology*. Rev. ed. Boston: Gould and Lincoln, 1851.

Beale, Thomas. *The Natural History of the Sperm Whale*. London: John Van Voorst, 1839.

Bennett, Frederick D. *Narrative of a Whaling Voyage Round the Globe*. 2 vols. London: Richard Bentley, 1840.

Bowles, M. E. "Some Account of the Whale-Fishery of the N. West Coast and Kamschatka." *Polynesian* (4 October 1845): 82–83.

Brewster, Mary. *"She Was a Sister Sailor": The Whaling Journals of Mary Brewster, 1845–1851*. Edited by Joan Druett. Mystic: Mystic Seaport Museum, 1992.

Browne, J. Ross. *Etchings of a Whaling Cruise* [1846]. Edited by John Seelye. Cambridge, MA: Belknap Press, 1968.

Chase, Owen, et al. *Narratives of the Wreck of the Whale-ship Essex* [1821]. New York: Dover, 1989.

Cheever, Henry T. *The Whale and His Captors; or, The Whaleman's Adventures* [1850]. Edited by Robert D. Madison. Hanover, NH: University Press of New England, 2018.

Colnett, James. *A Voyage to the South Atlantic and Round Cape Horn Into the Pacific Ocean, for the Purpose of Extending the Spermaceti Whale Fisheries . . .* London: W. Bennett, 1798.

Cuvier, Baron Georges. *The Class Pisces*, with supplementary editions by Edward Griffith and Charles Hamilton Smith, vol. 10 of *The Animal Kingdom*. London: Whittaker and Co., 1834.

Dana, Richard Henry, Jr. *Two Years before the Mast: A Personal Narrative of Life at Sea*. New York: Harper and Bros., 1840.

Darwin, Charles. *Journal of Researches into the Natural History and Geology of the Countries Visited During the Voyage of H.M.S. Beagle*. 2 vols. New York: Harper & Brothers, 1846.

———. *On the Origin of the Species by Means of Natural Selection* [1859]. Edited by William Bynum. New York: Penguin, 2009.

Dudley, Paul. "An Essay upon the Natural History of Whales, with a particular Account of the Ambergris found in the *Sperma Ceti* Whale." *Philosophical Transactions* 33 (1724–25): 256–69.

Emerson, Ralph Waldo. "The Uses of Natural History (1833–35)." In *The Selected Lectures of Ralph Waldo Emerson*, edited by Ronald A. Bosco and Joel Myerson, 1–17. Athens: University of Georgia Press, 2005.

Good, John Mason. *The Book of Nature*. Hartford: Belknap and Hamersley, 1837.

Hamilton, Robert. *The Naturalist's Library: Mammalia. Whales, &c.*, vol. 7. Edited by William Jardine. Edinburgh: W. H Lizards, 1843.

Jackson, J. B. S. "Dissection of a Spermaceti Whale and Three Other Cetaceans." *Boston Journal of Natural History* 5, no. 2 (October 1845): 10–171.

Lawrence, Lewis H. Logbook of the *Commodore Morris*, 1849–1853. Falmouth Historical Society No. 2006.044.002.

Logkeeper. Logbook of the *Commodore Morris*, 1845–1849 (Capt. Silas Jones). Falmouth Historical Society No. 2013.076.09.

Martin, John F. *Around the World in Search of Whales: A Journal of the Lucy Ann Voyage, 1841–44*. Edited by Kenneth R. Martin. New Bedford: The Old Dartmouth Historical Society/New Bedford Whaling Museum, 2016.

Maury, Matthew Fontaine. *Explanations and Sailing Directions to Accompany The Wind and Current Charts*. 3rd ed. Washington: C. Alexander Printer, 1851.

———. *The Physical Geography of the Sea*. New York: Harper and Brothers, 1855.

———. "The Whale Fisheries . . . ," *New York Herald*, April 29, 1851, p. 3.

Morgan, Charles W. "Address before the New Bedford Lyceum." Charles Waln Morgan Papers, 1796–1861, MS 41, Subgroup 1, Series Y, Folder 1, New Bedford Whaling Museum (1830/37).

Olmsted, Francis Allyn. *Incidents of a Whaling Voyage* [1841]. Edited by W. Storrs Lee. Rutland, VT: Charles E. Tuttle Co., 1970.

———. *Relics from the Wreck of a Former World; or Splinters Gathered on the Shores of a Turbulent Planet*. New York: Henry Long and Brother, 1947.

Reynolds, Jeremiah N. "The Knickerbocker: Mocha Dick or the White Whale: A Leaf from a Manuscript Journal of the Pacific," vol. 13. New York: The Knickerbocker, 1839.

Scammon, Charles. *The Marine Mammals of the Northwestern Coast of North America* [1874]. New York: Dover Publications, 1968.

Scoresby Jr., William. *An Account of the Arctic Regions, with a History and Description of the Northern Whale-Fishery*. 2 vols. Edinburgh: Archibald Constable and Co., 1820.

———. *Journal of a Voyage to the Northern Whale-Fishery*. Edinburgh: Archibald Constable and Co., 1823.

The Society for the Diffusion of Useful Knowledge. "Whales." In *The Penny Cyclopædia of the Society for the Diffusion of Useful Knowledge*, vol. 27, "Wales–Zygophyllaceæ," edited by George Long, 271–98. London: Charles Knight and Co., 1843.

Weir, Robert. Journal aboard the *Clara Bell*, 1855–1858. Mystic Seaport Log 164.

Wilkes, Charles. *Narrative of the United States Exploring Expedition*. 5 vols., atlas. London: Wiley and Putnam, 1845.

SELECTED OTHER SOURCES

Baker, Jennifer J. "Dead Bones and Honest Wonders: The Aesthetics of Natural Science in *Moby-Dick*." In *Melville and Aesthetics*, edited by Samuel Otter and Geoffrey Sanborn, 85–101. New York: Palgrave Macmillan, 2011.

Bercaw [Edwards], Mary K. *Melville's Sources*. Evanston: Northwestern University Press, 1987.

Bender, Bert. *Sea-Brothers: The Tradition of American Sea-Fiction from "Moby-Dick" to the Present*. Philadelphia: University of Pennsylvania Press, 1988.

Berta, Annalisa, James L. Sumich, and Kit M. Kovacs. *Marine Mammals: Evolutionary Biology*. 2nd ed. Boston: Elsevier, 2006.

Berzin, A. A. *The Sperm Whale (Kashalot)*. Edited by A. V. Yablokov. Translated by E. Hoz and Z. Blake. Jerusalem: Israel Program for Scientific Translation, 1972.

Blum, Hester. *The View from the Masthead: Maritime Imagination and Antebellum American Sea Narratives*. Chapel Hill: University of North Carolina Press, 2008.

Bode, Rita. "'Suckled by the sea': The Materrnal in *Moby-Dick*." In *Melville and Women*, edited by Elizabeth Schultz and Haskell Springer, 181–98. Kent: Kent State University Press, 2006.

Bouk, Dan, and D. Graham Burnett. "Knowledge of Leviathan: Charles W. Morgan Anatomizes His Whale." *Journal of the Early Republic* 27 (Fall 2008): 433–66.

Burnett, D. Graham. "Matthew Fontaine Maury's 'Sea of Fire': Hydrography, Biogeography, and Providence in the Tropics." In *Tropical Visions in the Age of Empire*, edited by Felix Driver and Luciana Martins, 113–34. Chicago: University of Chicago Press, 2014.

———. *Trying Leviathan: The Nineteenth-Century New York Court Case That Put the Whale on Trial and Challenged the Order of Nature*. Princeton: Princeton University Press, 2007.

Callaway, David R. *Melville in the Age of Darwin and Paley: Science in* Typee, Mardi, Moby-Dick, *and* Clarel. Binghamton: State University of New York, 1999.

Clark, A. Howard. "The Whale-Fishery." In *The Fisheries and Fisheries and Fishery Industries of the United States*, edited by George Brown Goode, 3–293. Vol. 5, no. 2. Washington: Government Printing Office, 1887.

Creighton, Margaret S. *Rites and Passages: The Experience of American Whaling, 1830–1870*. Cambridge: Cambridge University Press, 1995.

Dyer, Michael P. *"O'er the Wide and Tractless Sea": Original Art of the Yankee Whale Hunt*. New Bedford: Old Dartmouth Historical Society/New Bedford Whaling Museum, 2017.

———. "Whalemen's natural history observations and the Grand Panorama of a Whaling Voyage Round the World." *New Bedford Whaling Museum Blog*, March 29, 2016. whalingmuseumblog.org.

Ellis, Richard. *The Great Sperm Whale: A Natural History of the Ocean's Most Magnificent and Mysterious Creature*. Lawrence: University Press of Kansas, 2011.

———. *The Search for the Giant Squid*. New York: Penguin, 1999.

Estes, James A., et al., eds. *Whales, Whaling, and Ocean Ecosystems*. Berkeley: University of California Press, 2006.

Flower, Dean. "Vengeance on a Dumb Brute, Ahab? An Environmentalist Reading of *Moby-Dick*." *Hudson Review* 66, no. 1 (Spring 2013): 135–52.

Foster, Elizabeth S. "Melville and Geology." *American Literature* 17, no. 1 (March 1945): 50–65.

Frank, Stuart M. *Herman Melville's Picture Gallery: Sources and Types of the "Pictorial" Chapters of* Moby-Dick. Fairhaven, MA: Edward J. Lefkowicz, 1986.

———, ed. *Meditations from Steerage: Two Whaling Journal Fragments* (The Commonplace Book of Dean C. Wright, Boatsteerer, Ship *Benjamin Rush* of Warren, Rhode Island, 1841–45, and Six Months Outward Bound: John Jones, Steward, Ship *Eliza Adams* of New Bedford, 1852). Sharon, MA: The Kendall Whaling Museum, 1991.

German, Andrew W., and Daniel V. McFadden. *The Charles W. Morgan: A Picture History of an American Icon*. Mystic: Mystic Seaport Museum, 2014.

Greenlaw, Linda. *The Hungry Ocean: A Swordboat Captain's Journey*. New York: Hyperion, 1999.

———. *Seaworthy: A Swordboat Captain Returns to the Sea*. New York, Penguin, 2011.

Harvey, Bruce A. "Science and the Earth." In *A Companion to Herman Melville*, edited by Wyn Kelley, 71–82. West Sussex, UK: Wiley-Blackwell, 2015.

Heflin, Wilson. *Herman Melville's Whaling Years*. Edited by Mary K. Bercaw Edwards and Thomas Farel Heffernan. Nashville: Vanderbilt University Press, 2004.

Hillway, Tyrus. "Melville and Nineteenth-Century Science." PhD diss., Yale University, 1944.

———. "Melville as Critic of Science." *Modern Language Notes* 65, no. 6 (June 1950): 411–14.

———. "Melville's Education in Science." *Texas Studies in Literature and Language* 16, no. 3 (Fall 1974): 411–25.

Hoare, Philip. *The Whale: In Search of the Giants of the Sea*. New York: Ecco, 2010.

Huggenberger, Stefan, Michel André, and Helmut H. A. Oelschläger. "The Nose of the Sperm Whale: Overviews of Functional Design, Structural Homologies, and Evolution." *Journal of the Marine Biological Association of the United Kingdom* 96, no. 4 (2016): 783–806.

Irmscher, Christoph. *Louis Agassiz: Creator of American Science*. Boston: Houghton Mifflin Harcourt, 2013.

Jackson, J. A., N. J. Patenaude, E. L. Carroll, and C. Scott Baker. "How Few Whales Were There After Whaling? Inference from Contemporary mtDNA Diversity." *Molecular Ecology* 17 (2008): 236–51.

Kelley, Wyn. "Rozoko in the Pacific: Melville's Natural History of Creation." In *"Whole Oceans Away": Melville and the Pacific*, edited by Jill Barnum, Wyn Kelley, and Christopher Sten, 139–52. Kent: Kent State University Press, 2007.

Laist, David W. *North Atlantic Right Whales: From Hunted Leviathan to Conservation Icon*. Baltimore: Johns Hopkins University Press, 2017.

Leyda, Jay. *The Melville Log: A Documentary Life of Herman Melville, 1819–1891*, vol. 1. New York: Harcourt, Brace, 1951.

Madison, R. D., ed. *The Essex and the Whale: Melville's Leviathan Library and the Birth of Moby-Dick*. Santa Barbara, CA: Praeger, 2016.

Marr, Timothy. "Melville's Planetary Compass." In *The New Cambridge Companion to Herman Melville*, edited by Robert S. Levine, 187–201. Cambridge: Cambridge University Press, 2014.

McCauley, Douglas J., et al. "Marine Defaunation: Animal Loss in the Global Ocean." *Science* 347, no. 6219 (2015): 1–7.

Morowitz, Harold J. "Herman Melville, Marine Biologist." *Biological Bulletin* 220 (2011): 83–85.

Onley, Derek, and Paul Scofield. *Albatrosses, Petrels, and Shearwaters of the World*. Princeton: Princeton University Press, 2007.

Olsen-Smith, Steven, "Melville's Copy of Thomas Beale's *The Natural History of the Sperm Whale* and the Composition of *Moby-Dick*." *Harvard Library Bulletin* 21, no. 3 (Fall 2010), 1–77.

Osborn, James C. Logbook of the *Charles W. Morgan*, 1841–1845. Mystic Seaport Log 143.

Otter, Samuel. *Melville's Anatomies*. Berkeley: University of California Press, 1999.

Parker, Hershel. *Herman Melville: A Biography, Vol. 1, 1819–1851*. Baltimore: Johns Hopkins University Press, 1996.

Philbrick, Nathaniel. *In the Heart of the Sea: The Tragedy of the Whaleship* Essex. New York: Viking, 2000.

———. *Why Read Moby-Dick?* New York: Viking, 2011.

Rediker, Marcus. "History from below the Water Line: Sharks and the Atlantic Slave Trade." *Atlantic Studies* 5, no. 2 (2008): 285–97.

Reeves, Randall R., Brent S. Stewart, Phillip J. Clapham, James A. Powell, and Pieter A. Folkens. *Guide to Marine Mammals of the World*. New York: Alfred A. Knopf, 2002.

Roberts, Callum. *The Unnatural History of the Sea*. Washington: Shearwater Books, 2007.

Roman, Joe. *Whale*. London: Reaktion, 2006.

Roman, Joe, and Stephen R. Palumbi. "Whales before Whaling in the North Atlantic." *Science* 301 (2003): 508–10.

Roper, Clyde F. E., and Elizabeth K. Shea. "Unanswered Questions about the Giant Squid *Architeuthis* (Architeuthidae) Illustrate Our Incomplete Knowledge of Coleoid Cephalopods." *American Malacological Bulletin* 31, no. 1 (2013): 109–22.

Rozwadowski, Helen M. *Fathoming the Ocean: The Discovery and Exploration of the Deep Sea*. Cambridge, MA: Belknap Press, 2005.

Sanborn, Geoffrey. "Melville and the Nonhuman World." In *The New Cambridge Companion to Herman Melville*, edited by Robert S. Levine, 10–21. Cambridge: Cambridge University Press, 2014.

Schultz, Elizabeth. "Melville's Environmental Vision in *Moby-Dick*." *Interdisciplinary Studies in Literature and Environment* 7, no. 1 (2000): 97–113.

Scott, Sumner W. D. "The Whale in *Moby Dick*." PhD diss., University of Chicago, 1950.

Sealts, Jr., Merton M. *Melville's Reading*. Rev. ed. Columbia: University of South Carolina Press, 1988.

Severin, Tim. *In Search of Moby Dick: The Quest for the White Whale*. New York: Da Capo, 2000.

Shoemaker, Nancy. "Whale Meat in American History." *Environmental History* 10, no. 2 (April 2005): 269–94.

Smith, Richard Dean. *Melville's Science: "Devilish Tantalization of the Gods!"* New York: Garland, 1993.

Smith, Tim D., Randall R. Reeves, Elizabeth A. Josephson, and Judith N. Lund. "Spatial and Seasonal Distribution of American Whaling and Whales in the Age of Sail." *PLOS One* 7, no. 4 (April 2012): 1–25.

Starbuck, Alexander. *History of the American Whale Fishery from Its Earliest Inception to the Year 1876*. Waltham, MA: Self-published, 1878.

Vincent, Howard P. *The Trying Out of Moby-Dick*. Carbondale: Southern Illinois University Press, 1965.

Ward, J. A. "The Function of the Cetological Chapters in *Moby-Dick*." *American Literature* 28, no. 2 (1956): 164–83.

Wallace, Robert K. "Melville, Turner, and J. E. Gray's Cetology." *Nineteenth-Century Contexts* 13, no. 2 (Fall 1989): 151–75.

Whitehead, Hal. *Sperm Whales: Social Evolution in the Ocean*. Chicago: University of Chicago Press, 2003.

Whitehead, Hal, and Luke Rendell. *The Cultural Lives of Whales and Dolphins*. Chicago: University of Chicago Press, 2015.

Wilson, Eric. "Melville, Darwin, and the Great Chain of Being." *Studies in American Fiction* 28, no. 2 (Autumn 2000): 131–50.

Würsig, Bernd, J. G. M. Thewissen, and Kit M. Kovacs, eds. *Encyclopedia of Marine Mammals*. 3rd ed. London: Academic Press, 2018.

Yaeger, Patricia. "Editor's Column: Sea Trash, Dark Pools, and the Tragedy of the Commons." *Proceedings of the Modern Language Association* 125, no. 3 (2010): 523–45.

Zoellner, Robert. *The Salt-Sea Mastodon: A Reading of Moby-Dick*. Berkeley: University of California Press, 1973.

WEBSITES

The Complete Works of Charles Darwin Online (Darwin Online). Edited by John van Wyhe. www.darwin-online.org.uk.
The IUCN Red List of Threatened Species. The International Union for the Conservation of Nature and and Natural Resources. www.iucnredlist.org.
The Melville Electronic Library. Hofstra University. Edited by John Bryant. www.hofstra drc.org/projects/mel.html.
Melville's Marginalia Online. Boise State University. Edited by Steven Olsen-Smith, Peter Norberg, and Dennis C. Marnon. melvillesmarginalia.org.
Moby Dick Big Read. Peninsula Arts with Plymouth University. www.mobydickbigread .com.
Mystic Seaport Museum. www.mysticseaport.org.
The New Bedford Whaling Museum. www.whalingmuseum.org.
Searchable Sea Literature. Williams College-Mystic Seaport. Edited by Richard J. King.
Whaling History. www.whalinghistory.org.

FIGURE CREDITS AND NOTES

If unspecified, public domain images are from Wikimedia Commons, Google Books, Hathi Trust, Internet Archive, or my own collection.

FIGURES

FIG. 1. This world map by Erin Greb was made from John B. Putnam, 1967, in *Moby-Dick*, 3rd ed., ed. Hershel Parker (New York: Norton, 2018), xix; revised with information from Heflin, *Herman Melville's Whaling Years* (2004), and Charles Robert Anderson, ed., *Journal of a Cruise to the Pacific Ocean, 1842–1844, in the Frigate* United States, *with Notes on Herman Melville* (New York: AMS Press, 1966).

FIG. 2. Although Browne illustrated some of the images in his narrative, this one of a whaleman aloft was by a previous artist.

FIG. 3. Courtesy New Bedford Whaling Museum. Dean C. Wright, "Commonplace Book," aboard the whaleship *Benjamin Rush* 1841–45 (KWM A-145), which can be read in Frank, *Meditations from Steerage: Two Whaling Journal Fragments*.

FIG. 4. Houghton Library, Harvard University, AC85.M4977.839b, courtesy of Melville's Marginalia Online.

FIG. 5. Seth King (Melville cartoon, design) and Skye Moret (design).

FIG. 6. Emese Kazár (2013).

FIG. 7. Courtesy Williams College Special Collections.

FIG. 8. Whale illustrations by Uko Gorter. The sources for this table of Ishmael's cetology are David W. Sisk, "A Note on Moby-Dick's "Cetology" Chapter," *ANQ: A Quarterly Journal of Short Articles, Notes, and Reviews* 7, no. 2 (April 1994): 80–82; Dyer, "Whalemen's Natural History Observations," New Bedford Whaling Museum Blog (29 March 2016); and Reeves, et al., *Guide to Marine Mammals of the World*. I chose Frederick Bennett's *Narrative of a Whaling Voyage Round the Globe* (1840) as a baseline nineteenth-century example because he is one of the most careful, comprehensive, and trusted of Melville's scientific sources for *Moby-Dick*. There was, however, a huge range in scientific and common names at the time, especially with the large rorquals and between the names for grampus, killer, and thrasher/thresher. For the narwhal, I used the scientific name in Melville's entry "Whales" in his *The Penny Cyclopædia*, 292. For "thrasher" as connected to "killer," I consulted Hamilton, *The Naturalist's Library*, 228, and Cheever, *The Whale and His Captors*, 55, 56, 173. For the twentieth-century mention of the "Algerine," see Murphy, *Logbook for Grace*, 146. For the modern scientific names, I've followed the convention in which parentheses indicate that the species has since been shifted to within another genus.

FIG 9. Erin Greb Cartography, after maps in the *Encyclopedia of Marine Mammals*, 3rd ed.

FIG. 11. Courtesy of The British Museum and Williams College.

FIG. 12. Courtesy of Mystic Seaport Museum.

FIG. 13. Sperm whale skin collected c. 2015 in the waters of Kaikōura Canyon, New Zealand by Marta Guerra Bobo. Author's photo, courtesy Guerra Bobo.

FIG. 14. Courtesy Mystic Seaport Museum.

FIG. 15. Courtesy Mystic Seaport Museum.

FIG. 16. Center for Research Libraries, National Archives, with thanks to McCaffery and Associates.

FIG. 17. American Geographical Society Library Digital Map Collection at the University of Wisconsin, Milwaukee.

FIG. 18. Courtesy David Rumsey Map Collection.

FIG. 23. Courtesy Williams College.

FIG. 24. Courtesy of the Falmouth Historical Society.

FIG. 25. Courtesy of the Nantucket Historical Association.

FIG. 26. Scoresby's invertebrates identified by James T. Carlton, pers. comm.

FIG. 28. The lengthy, detailed caption to this illustration of cutting in a cub sperm whale by Colnett is fascinating reading, found easily in the plates at the back of his narrative or in Madison, *The Essex and the Whale*, 224.

FIG 29. Courtesy of the New Bedford Whaling Museum, Log KWM 436.

FIG. 30. Courtesy of the New Bedford Whaling Museum.

FIG. 33. Courtesy of Mystic Seaport Museum.

FIG. 34. Courtesy Stefan Huggenberger. These figures first created and published in color in Huggenberger, et al., "The Nose of the Sperm Whale," 787, 788.

FIG. 35. Courtesy of Williams College and the Rubenstein Library, Duke University.

FIG. 36. Courtesy of Williams College.

FIG. 39. Courtesy of University of Otago Interloans.

FIG. 40. Courtesy of Duke University.

FIG. 41. Courtesy of Williams College.

FIG. 43. Courtesy of Hal Whitehead, adapted with permission from figures originally published in *Sperm Whales: Social Evolution in the Ocean*, 20, 130. The global population trajectory is from Whitehead's "best estimates of the population and model parameters." Information is limited between 1712 and 1800. See his p. 20 for his source material for the annual sperm whale catch and for leads as to the possible underestimation of both open-boat and modern numbers.

FIG. 44. This figure was created for this book in August 2018 by Jennifer Jackson, with data courtesy of David Laist and with help from Scott Baker. See Laist, *North Atlantic Right Whales*, 262. For the full discussion and statistics behind estimates of southern right whale abundance, see IWC, "Report of the Workshop on the Comprehensive Assessment of Right Whales: A Worldwide Comparison," *Journal of Cetacean Research and Management* (special issue) 2 (2001): 1–60, and J. A. Jackson, et al., "How Few Whales Were There After Whaling?," 236–51. As with the sperm whale figures, note the difference between "capture" and "kill," because thousands more whales died later from wounds after they were harpooned and escaped or they sunk before the whalemen were able to row them back to the ship to try out the oil.

FIG. 46. Courtesy of Williams College.

FIG. 50. Courtesy of Williams College and the Watkinson Library, Trinity College.

FIG. 51. Courtesy of Mystic Seaport Museum. For more on this image, see Dyer, *Tractless Sea*, 250–56.

FIG. 52. Specimen No. MCZ BOM 7914. Courtesy of the Harvard Museum of Comparative Zoology, photo and measurements by Mark Omura.

FIG. 53. Courtesy New Bedford Whaling Museum.

FIG. 54. Courtesy New Bedford Whaling Museum, NBWM 1938.79.3. On sperm whales holding calves in their mouths, see e.g., Whitehead, *Sperm Whales: Social Evolution in the Ocean*, 275–77, and Kurt Amsler, "Just Born," *AlertDiverOnline* (2015), www.alertdiver.com/sperm-whales.

FIG. 56. Dan Kitwood, Getty Images.

FIG. 57. Still from *Blue Planet II*, BBC Studios, 2017.

PLATES

PLATE 1. Courtesy Mystic Seaport Museum, Dennis Murphy, D2014-07-0210.

PLATE 2. Captain Ken Bracewell and crew of the *Rena*, 2014.

PLATE 3. Tony Wu.

PLATE 4. Amy Knowlton, Anderson Cabot Center for Ocean Life at the New England Aquarium. Taken under NOAA/NMFS Permit #15415.

PLATE 5. Flip Nicklin, Minden photography.

PLATE 6. Helmut Corneli, Alamy photography.

PLATE 7. Tim Smith, et al., first published in Smith, et al., "Spatial and Seasonal Distribution of American Whaling and Whales in the Age of Sail," 2.

PLATE 8. N. R. Fuller and Sayo-Art, first published in McCauley, et al., "Marine Defaunation," 1.

PLATE 9. Courtesy Daniel Aplin, 2018.

PLATE 10. Chris Fallows, first published in Chris Fallows, et al., "White Sharks (*Carcharodon carcharias*) Scavenging on Whales and Its Potential Role in Further Shaping the Ecology of an Apex Predator," *PLOS One* 8, no. 4 (2013): 7.

PLATE 11. Ocean Agency/ XL Catlin Seaview Survey, 2014.

PLATE 12. Andrea Westmoreland, Florida Keys, 2011, via Wikimedia Commons.

INDEX